ETHICS IN ACTION FOR SUSTAINABLE DEVELOPMENT

Ethics in Action Meeting on Education, Casina Pio IV, Vatican City, October 2017

ETHICS IN ACTION FOR SUSTAINABLE DEVELOPMENT

EDITED BY
JEFFREY D. SACHS,
MARCELO SÁNCHEZ SORONDO,
OWEN FLANAGAN,
WILLIAM VENDLEY,
ANTHONY ANNETT, AND
JESSE THORSON

FOREWORD BY
POPE FRANCIS AND
ECUMENICAL PATRIARCH BARTHOLOMEW

Columbia University Press
New York

This book was made possible by the remarkable leadership of Monsignor Marcelo Sánchez Sorondo, who hosted and guided the Ethics in Action initiative during his tenure as Chancellor of the Pontifical Academies. Ms. Gabriella Marino (Pontifical Academies) brilliantly coordinated the meetings that provided the foundation of this volume, in collaboration with Prof. Anthony Annett and Ms. Sharon Paculor (Center for Sustainable Development at Columbia University). Ethics in Action was supported by the UN Sustainable Development Solutions Network, Religions for Peace, the Center for Sustainable Development at Columbia University, the University of Notre Dame, the Blue Chip Foundation, the Fetzer Institute, Ms. Christina Lee Brown, and Ms. Jacqueline Corbelli.

Columbia University Press
Publishers Since 1893
New York   Chichester, West Sussex
cup.columbia.edu

Library of Congress Cataloging-in-Publication Data
Names: Sachs, Jeffrey, editor.
Title: Ethics in action for sustainable development / edited by Jeffrey D. Sachs, Marcelo Sánchez Sorondo, Owen Flanagan, William Vendley, Anthony Annett, and Jesse Thorson.
Description: New York : Columbia University Press, [2022] | Includes bibliographical references and index.
Identifiers: LCCN 2021052015 (print) | LCCN 2021052016 (ebook) | ISBN 9780231202862 (hardback) | ISBN 9780231202879 (trade paperback) | ISBN 9780231554756 (ebook)
Subjects: LCSH: Sustainable development—Moral and ethical aspects. | Sustainable development—Philosophy. | Social justice.
Classification: LCC HC79.E5 E755 2022 (print) | LCC HC79.E5 (ebook) | DDC 338.9/27—dc23/eng/20220113
LC record available at https://lccn.loc.gov/2021052015
LC ebook record available at https://lccn.loc.gov/2021052016

Columbia University Press books are printed on permanent and durable acid-free paper.
Printed and bound by CPI Group (UK) Ltd, Croydon, CR0 4YY

Cover design: Julia Kushnirsky
Cover painting: Eugène Delacroix, *Untergehende Sonne*

CONTENTS

FOREWORD

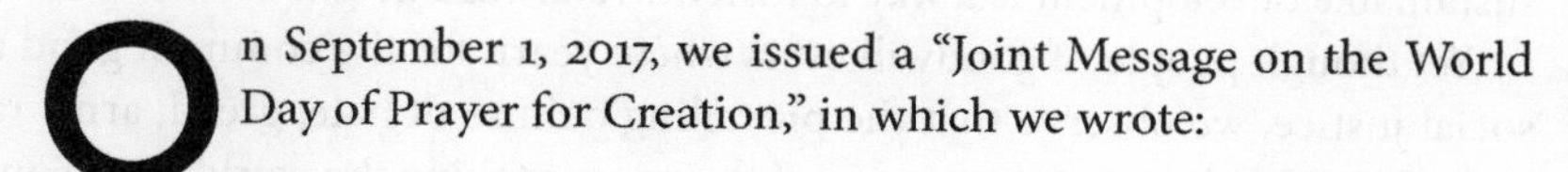

On September 1, 2017, we issued a "Joint Message on the World Day of Prayer for Creation," in which we wrote:

> The urgent call and challenge to care for creation are an invitation for all of humanity to work toward sustainable and integral development. . . . We urgently appeal to those in positions of social and economic, as well as political and cultural, responsibility to hear the cry of the earth and to attend to the needs of the marginalized, but above all to respond to the plea of millions and support the consensus of the world for the healing of our wounded creation. We are convinced that there can be no sincere and enduring resolution to the challenge of the ecological crisis and climate change unless the response is concerted and collective, unless the responsibility is shared and accountable, unless we give priority to solidarity and service.[1]

In this spirit, we hope that people in all parts of the world—especially students and young leaders—will contemplate and draw on the lessons and wisdom offered by the many contributors to *Ethics in Action for Sustainable Development*. In the writings in this volume, scholars and practitioners from a wide variety of religious traditions and academic disciplines find common ground in defending the Creation—our common home (*oikos*)—and in calling for human dignity and economic justice for all peoples across faiths, cultures, traditions, nations, and geographies. In these contributions, we see how the diverse religious traditions regard human beings as made body and soul in God's image, as part of the Creation.

It is our heartfelt conviction that there are profound connections among all living things and all human beings.[2] Therefore, we concur with the contributors to this volume that the world's Sustainable Development Goals (SDGs), adopted by all 193 member states of the United Nations, can help us to find a global path forward for the common good, and a method of worldwide collaboration to address the pressing social, economic, and ecological challenges of our age. The SDGs help to nurture a strong sense of the common good. They draw attention especially to the suffering of the poor and the suffering of the planet. From a religious perspective, "The way we relate to nature as creation directly reflects the way we relate to God as Creator. There can be no distinction between concern for human welfare, protection of the environment, and care for our salvation."[3]

Jesus taught, "You shall love the Lord your God with all your heart and with all your mind. This is the great and first commandment. And a second is like it: You shall love your neighbor as yourself. On these two commandments depend all the Law and the Prophets" (Mt 22.37–40). In our estimation, the pursuit of the global common good and social justice can be understood as a faithful response to this teaching. We may say that the world's shared commitment to sustainable development is a way to foster love toward neighbor and Creation.

Yet though people of goodwill everywhere yearn for the common good and social justice, we observe that the prevailing culture of hate, greed, arms race, and power that is not at the service of the poor is taking the world in the opposite direction: an upward spiral of war, violence, poverty, forced migration, and environmental destruction. We must, with all our hearts, speak out against this hate and violence, which is of course a dire negation of love, the common good, and social justice. We must, above all, recognize humanity's current peril:

> Step by step, we are moving towards catastrophe. Piece by piece the world risks becoming the theater of a unique Third World War. We are moving towards it as if it were unavoidable. But we must forcefully repeat: No, it is not inevitable! No, war is not inescapable! When we allow ourselves to be devoured by this monster represented by war, when we allow this monster to raise its head and guide our actions, everyone loses, we destroy God's creatures, we commit sacrilege and prepare a future of death for our children and grandchildren.[4]

We see these rising dangers in a pandemic that has claimed the lives of millions while leaving the poor without access to life-saving vaccines and medicines. We see the rising dangers in the multiple ecological crises—climate change, habitat destruction, and wanton pollution—that have all intensified in recent years. We see the rising dangers most starkly in the war against Ukraine, which escalates with frenzied calls for even more destructive weapons, and with no urgent call for peace, negotiations, and conciliation on behalf of the main actors.

These crises of poverty, disease, nature, and war are inseparable from the shifts in our global culture. Indeed, "A certain way of understanding human life and activity has gone awry, to the serious detriment of the world around us."[5] Consumerism, throwaway culture, and the technocratic paradigm together make us blind to our interdependence with one another as human persons and the environment. Those in positions of power and wealth have forgotten that there can be "no genuine progress that is founded on the destruction of the natural environment. . . . We are certain that there is an alternative way of economic structure and development besides the economism and the orientation of economic activity toward the maximization of profiteering."[6]

Above all, we must intensify the search and striving for the common good and social justice, and think of "one world with a common plan."[7] We must, with all urgency, strive to encounter the other in our midst, so that all together we may find our way back from the abyss. This special volume that brings together people of many faiths, cultures, and traditions thereby helps us to see how humanity together can put ethics to action in the service of the common good and social justice.

POPE FRANCIS   ECUMENICAL PATRIARCH BARTHOLOMEW

Notes

1. Pope Francis and Ecumenical Patriarch Bartholomew, "Joint Message on the World Day of Prayer for Creation," September 1, 2017, https://www.vatican.va/content/francesco/en/messages/pont-messages/2017/documents/papa-francesco_20170901_messaggio-giornata-cura-creato.html.
2. Ecumenical Patriarch Bartholomew, "Intergenerational and Interdisciplinary Solidarity," in *On Earth as in Heaven: Ecological Vision and Initiatives of Ecumenical Patriarch Bartholomew*, ed. John Chryssavgis (New York: Fordham University Press, 2012), 269–96.
3. Ecumenical Patriarch Bartholomew, "Message to the United Nations Conference of the Parties (COP24)," December 10, 2018, https://fore.yale.edu/files/message_to_cop24.pdf.
4. Pope Francis, "Introduction," *Against War: The Courage to Build Peace*, working translation published by *Vatican News*, April 13, 2022, https://www.vaticannews.va/en/pope/news/2022-04/pope-francis-war-introduction-book-courage-peace.html.
5. Pope Francis, *Laudato si'*, 2015, sec. 101, https://www.vatican.va/content/francesco/en/encyclicals/documents/papa-francesco_20150524_enciclica-laudato-si.html.
6. Ecumenical Patriarch Bartholomew, "Message for the World Day of Prayer for the Care of Creation," September 1, 2020, https://www.vaticannews.va/en/church/news/2020-09/bartholomew-i-message-for-world-day-of-creation-full-text.html.
7. Pope Francis, *Laudato Si'*, 2015, sec. 164.

ETHICS IN ACTION FOR SUSTAINABLE DEVELOPMENT

INTRODUCTION

JEFFREY D. SACHS AND OWEN FLANAGAN

The Ethics in Action initiative set out to find a moral consensus on the concept of sustainable development and a practical common agenda regarding the Sustainable Development Goals (SDGs). Religious leaders, philosophers, theologians, economists, and practitioners from a variety of disciplines were summoned to discern and discuss whether their respective religious and intellectual traditions—divergent in many ways—nonetheless shared a common concern for the social, economic, and environmental aims expressed in the framework of the SDGs. It is said that in polite company one doesn't talk about politics or religion. In this volume, we do pretty much both, yet find the company to be very polite indeed.

From Stockholm to Rio: A Brief History of Sustainable Development

The concept of sustainable development is a half-century in the making. In June 1972, the United Nations convened an international conference on the environment—the first of its kind—in Stockholm. The United Nations Conference on the Human Environment (also known as the Stockholm Conference) was a wake-up call for the world. In recognition of an impending collision between the environment and the economy, world leaders came together to discuss the rapid growth of the world economy and population in the context of a finite Earth and limited resources. They recognized, in the first global diplomatic gathering of its kind, that ever-increasing economic activity, including the increasing use of primary resources and burning of fossil fuels, would put the earth's resources and environment under greater and greater stress.

Not coincidentally, a pivotal book was published the same year: *The Limits to Growth*. This book presented a computer model commissioned by the Club of Rome to investigate the same challenge: global economic growth combined with the reality of limited physical resources. The authors argued that, if the geometric growth of the world economy were to continue with the technologies then in place, the resource burden on the planet would eventually become so great that the world economy would overshoot the carrying capacity of the planet, leading to a deep ecological, social, and economic crisis in the twenty-first century.

After 1972, many scholars went to work to understand the environmental threats ahead, and especially ways to combine economic progress with a safe environment. In 1987, a global commission led by Norway's prime minister and leading stateswoman, Dr. Gro Harlem Brundtland, called on the world to adopt the concept of sustainable development. The way Dr. Brundtland's commission put it in 1987 was that sustainable development means "meeting the needs of today's generation in a way that will enable future generations to meet their needs." In short, "Don't wreck the planet and leave a disaster for future generations."

Twenty years after the Stockholm Conference, the United Nations convened a follow-up conference, this time in Rio de Janeiro. The 1992 UN Conference on Environment and Development (also known as the Rio Earth Summit) was called with the aim of adopting and putting into motion a global action plan to solve the challenges that had been identified in 1972.

The Rio Earth Summit adopted the Brundtland Commission framework of sustainable development and arrived at three major environmental agreements: (1) The UN Framework Convention on Climate Change (UNFCCC), to fight human-induced climate change; (2) the Convention on Biological Diversity, to prevent an impending disaster of ecosystem destruction and species extinction; and (3) the UN Convention to Combat Desertification, to stop the spread of land degradation and deserts in the world's dryland regions.

In the ensuing years, vested-interest politics, geopolitical discord, unchecked greed, and shortsightedness thwarted the implementation of the three treaties, and the overall framework of sustainable development. Especially notable was the failure of the Kyoto Protocol, an agreement adopted in 1997 to realize the goals of the UNFCCC. The United States failed to ratify the Kyoto Protocol, and the protocol's aims were little heeded by many other countries. As a result, there was little action between 1992 and 2015 in implementing the climate change treaty. The United States, once a world leader on the environment, became a laggard, but hardly the only one. The three major environmental agreements of the Rio Earth Summit were most notable for the failure to promote real action.

New glimmers of diplomatic hope were raised in the year 2000 on the eve of the new millennium. UN Secretary-General Kofi Annan put before the world a set of objectives to address global poverty and its various manifestations in poor health, epidemic disease, hunger, illiteracy, and lack of access to safe water and sanitation. The world governments adopted the Millennium Development Goals (MDGs) as a framework of shared action for the period from 2001 to 2015. The MDGs offered some fresh, if fragile, hope that the world's governments could work together on the sustainable development agenda.

In 2012, forty years after the Stockholm Conference and twenty years after the Rio Earth Summit, the world's governments assembled once again in Rio at the UN Conference on Sustainable Development. Their purpose was to review progress on the sustainable development agenda set in Rio in 1992. The Rio+20 Conference was in many ways a downcast conference. All of the evidence presented at the conference showed the dramatic shortfall of results after 1992. While the MDGs offered a source of some hope and satisfaction that progress against poverty was feasible and taking place, the ledger on environmental sustainability looked especially grim. Not one of the three treaties was being properly implemented. Human-induced climate change was accelerating; the collapse of biodiversity was ongoing and dramatic; and the world's dryland regions continued to suffer from ongoing land degradation. The premonitions of overshoot first put forward in *The Limits to Growth* seemed to be ever closer to reality.

The Sustainable Development Goals and *Laudato si'*

Yet once again, the governments refused to give up. Drawing hope from the MDGs, and guidance from the sustainable development agenda adopted in 1992, the governments decided to adopt the SDGs as a post-2015 successor to the MDGs. By adopting the concept of SDGs in 2012, for implementation after 2015, the UN members states gave themselves three years to negotiate a new set of global goals based on the sustainable development concept. During this time, the UN member states deliberated, discussed, negotiated, and debated the new goals. They recognized that the world faced three interconnected challenges: promoting economic development, reducing the vast inequalities of income and wealth within their societies as well as across nations, and finally putting into place more decisive actions to head off major environmental disasters, including human-induced climate change, the collapse of ecosystems and biodiversity, and the mega-pollution choking the air, land, soils, waterways, and oceans.

Following three years of negotiation, on September 25, 2015, the UN member states unanimously adopted Transforming Our World: The 2030 Agenda for Sustainable Development, including the seventeen SDGs for 2030 (https://

sdgs.un.org/goals). The goals for the year 2030 can be summarized as follows: no poverty, zero hunger, good health and well-being, quality education, gender equality, clean water and sanitation, affordable and clean energy, decent work and economic growth, industry innovation and infrastructure, reduced inequalities, sustainable cities and communities, responsible production and consumption, climate action, healthy oceans, sustainable ecologies, peace and justice, and international cooperation to meet these goals. The 2030 Agenda thereby constitutes a holistic agenda that combines economic, social, and environmental objectives, with sustainable development as the core organizing framework for global cooperation.

Achieving the policy aims expressed in the SDGs requires money, scientific and technological savvy, and international cooperation, and especially the political efforts of communities ranging from the local level to the UN General Assembly. Yet these political efforts require strong cooperation and shared visions among highly diverse cultures and populations, both within and across nations. Moreover, the diverse peoples of the earth are members of many different religious traditions, with distinctive histories, rituals, theologies, and ethics. And there is also a large and growing group of people who are unaffiliated with any religion but who are also steeped in distinctive philosophical traditions; for example, post-enlightenment humanists in the North Atlantic and various stripes of morally serious socialists and communists in former Soviet countries and in China. To what extent do the world's religious and ethical traditions support the ethos of the 2030 Agenda? Might there be an overlapping moral consensus, sometimes called an unforced consensus, on the SDGs that spans these diverse sacred and secular traditions?

Through its tradition of social teachings, the Roman Catholic Church, for example, has called the world to higher visions of social equity and human flourishing. Notably, in 1891, Pope Leo XIII wrote a pathbreaking encyclical titled *Rerum novarum*, meaning "Of the New Things." Leo XIII highlighted the historical novelties of the industrial age and raised profound concerns about the challenges it would pose to human societies, thus beginning the modern era of Catholic social teaching: What ought we to do if millions of workers in hard labor and factories are put out of their jobs? How ought companies, workers, and employers behave?

In 2015, Pope Francis carried on the legacy of the Church's social teachings with the issuance of his remarkable encyclical *Laudato si'*. In this magisterial work, Pope Francis takes up the environmental crisis in its ethical dimensions and focuses on the duties of stewardship and responsibility that we have toward the planet and toward each other. He links the Church's conception of human dignity (*imago dei*) with our collective responsibility to take care of both each other and the physical earth. *Laudato si'* calls us to an integral ecology that recognizes the essential relationality or interconnectedness of all things.

Importantly, the powerful moral and spiritual statement of *Laudato si'* constitutes an institutional, religious expression of the challenges and ethos of the SDGs. In fact, the encyclical was published three months before the adoption of the SDGs, and Pope Francis opened the September 25, 2015, UN General Assembly session at which the SDGs were adopted, speaking to world leaders about the urgent need for a common plan for our common home, thereby deeply linking the remarkable religious and moral vision of *Laudato si'* with the powerful political and diplomatic statement of the 193 member states of the United Nations.

On December 12, 2015, just a few weeks after the adoption of the SDGs, the same 193 member states of the UN came together once again, in Paris, to adopt the Paris Agreement on climate change. The Paris Agreement is the world's operational agreement on how to address the dire crisis of human-induced climate change. The agreement put the climate agenda squarely within the broader context of sustainable development, meaning that climate change should be fought in a manner that also fosters economic progress and social inclusion. Climate change, though dire, is not a stand-alone objective but part of a larger ethical framework.

In *Laudato si'*, Pope Francis writes that our interdependence—with each other and with the planet—"obliges us to think of one world with a common plan." That is precisely what the 2030 Agenda and the Paris Agreement aim to achieve. And the objective of one world with a common plan is the driving force of our search for an overlapping consensus among the world's diverse religious and philosophical traditions.

Ethics in Action and the Possibility of Moral Leadership Based on a Shared Consensus

In the wake of the momentous declarations and agreements of 2015, a number of us around the world from major faiths, philosophical traditions, and the development practitioner community, decided that we should explore the possibility of a shared global ethics of achieving these objectives. We return to the question posed earlier: Might there be an unforced or overlapping moral consensus on the SDGs among the diverse sacred and secular traditions around the world?

Ethics in Action is an initiative designed to answer this question by assessing the degree of consensus among several great world religions and secular traditions on the SDGs, and to mobilize this consensus into moral capital that can be deployed to help each nation-state and the collection of nation-states to achieve the SDGs. Ethics in Action was hosted by the Pontifical Academy of Sciences and supported by the UN Sustainable Development Solutions Network,

Religions for Peace, the Center for Sustainable Development at Columbia University, and the University of Notre Dame. It counted on the generous philanthropic support of several founding benefactors, including the Blue Chip Foundation, the Fetzer Institute, Christina Lee Brown, and Jacqueline Corbelli.

Between October 2016 and December 2018, Ethics in Action convened religious leaders and theologians from major faiths, philosophers, academic experts, and practitioners from around the world to enter into a shared dialogue on supporting the common good and asked questions including the following: Which are the religious or ethical values that are common across the world's major faiths and ethical traditions that can be brought to bear to help motivate and expedite the worldwide process to achieve the goals set by the global community, including the crucial targets of the 2030 Agenda, the seventeen SDGs, and the Paris Agreement?

Ten thematic meetings were held, usually in Vatican City, to explore the following challenges for sustainable development: poverty, environmental justice, business, peace and conflict, migration and the refugee crisis, the challenges of Indigenous peoples, the future of work, corruption and governance, human trafficking and justice for the marginalized, and education. Each meeting was attended by the members of our core team—the Ethics in Action Working Group—and scholars and practitioners with expertise on the meeting's topic. At the end of each meeting, the group produced a consensus statement to capture the general moral congruency of our respective religious traditions or philosophical communities on these issues. This core group included representatives of seven world religions—Roman Catholic, Eastern Orthodox, Judaism, Islam, Hinduism, Buddhism, and Confucianism and other traditional Chinese religions—and a handful of secular thinkers.

One might compare the aim of Ethics in Action to an earlier UN initiative that also sought and located an unforced and overlapping moral consensus: the 1948 UN Universal Declaration of Human Rights (UDHR). Despite the fact that the concept of human rights has its origins in Western liberal thinking, thirty universal rights were enumerated with the help of representatives of many cultural, religious, and ethical traditions. As with the SDGs, all 193 members of the UN are now signatories to one or more treaties that advance the UDHR. Ethics in Action aimed to identify the same kind of overlapping consensus among the world's cultural and religious traditions to achieve "one world with a common plan."

Our meetings were ecumenical, but there are some caveats. First, the seven represented religions are not, of course, all the religions of the world—of which there are more than four thousand. By most measures, adherents of the seven included religions make up more than 50 percent of the world's population.

Second, all these religions have great variation internally, which was not always adequately represented. There are orthodox, conservative, and reform Jews; Sunni and Shia Muslims; Theravadan and Mahayanan Buddhists; Roman and Eastern Orthodox Catholics; as well as various admixtures, such as Jews for Jesus, Confucian Catholics, and secular humanist Buddhists.

Third, one large tradition not thoroughly represented in our deliberations was non-Catholic Christianity, which makes up about 40 percent of Christianity, including some vocal movements that, in America at least, are unlikely to be as friendly to the SDGs as the representatives we worked with.

Fourth, another under-represented group is the NONES, those who are unaffiliated with any organized religion ("none of the above"). In North America, about 21 percent of people are now religiously unaffiliated, and more than 30 percent of the populations of many countries are NONES, including China, the Czech Republic, France, Japan, the Netherlands, New Zealand, South Korea, and the United Kingdom.

Fifth, in the early twenty-first century, the leadership of most religions remains male. This is changing slowly, and we succeeded in including many women's voices in our deliberations and in this book. Nonetheless, there remain legitimate calls for long-lived religious traditions with patriarchal structures and traditions to be sensitive detectors of gender injustice and leaders in eliminating it in all forms.

This Book

Ethics in Action for Sustainable Development captures the intellectual depth and diversity of this initiative by collecting the religious and ethical thought of more than thirty of its leaders. You will walk through overviews of seven major religious traditions and their conceptions of the common good. You will read about the intersection of the SDGs, ethics, and philosophy. You will be able to reflect on the consensus statements, and you will hear from practitioners on specific challenges to sustainable development.

The situation is more urgent than it was in 2015 when the SDGs were adopted. The SDGs were to be achieved by 2030. We write in 2021, and at this time not one of the 193 nations that voted to adopt the SDGs is fully on track, as measured by benchmark data, to achieve the SDGs by 2030. And the world is currently beset by a major pandemic, intensifying climate disorders, and increasing inequalities.

But there is good news. There is in fact wide agreement among many world religions and secular ethical traditions that the SDGs are urgent moral imperatives, in addition to being wise instrumental goals. Achieving the SDGs is a way of acknowledging our common humanity, our interdependency, and the moral and religious aims of universal human flourishing. It is our hope that the story we tell about the political, moral, and religious consensus on the SDGs will help better mobilize the ethical will to achieve them.

PART I

ADVANCING THE COMMON GOOD

Shared Virtues and Visions of Well-Being

CHAPTER 1

THE VISION AND VALUES OF THE SUSTAINABLE DEVELOPMENT GOALS

JEFFREY D. SACHS

Introducing the Sustainable Development Goals

Transforming Our World: The 2030 Agenda for Sustainable Development, consisting of the seventeen Sustainable Development Goals, was adopted by all United Nations member states in September 2015 for the fifteen-year period from 2016 to 2030. The SDGs are the world's agreed goals to achieve sustainable development by 2030. In the context of the 2030 Agenda, "sustainable development" means the simultaneous fulfillment of economic rights, social inclusion, and environmental sustainability. The 2030 Agenda offers a detailed text that explains the purposes of sustainable development and the conceptual framework of the SDGs.

Economic Rights

Almost from the founding of the United Nations, and specifically with the adoption of the Universal Declaration of Human Rights (UDHR) in 1948, the world's governments have recognized the basic rights of all people, including political, civil, economic, social, and cultural rights. The Universal Declaration stands as the "moral charter" of the UN. It has given rise to a vast international law, including treaties, case law, UN General Assembly resolutions and initiatives such as the Millennium Development Goals (2001–2015), and the SDGS (2016–2030).

Economic rights are a subset of human rights, which also include political, civil, social, and cultural rights. Most of the economic rights in the UDHR are reflected in the SDGs. Key economic rights, and the associated SDGs, are as follows:

- Right to an adequate standard of living (SDG 1)
- Right to social security (SDG 1)
- Right to food (SDG 2)
- Right to health (SDG 3)
- Right to education (SDG 4)
- Right to safe water and sanitation (SDG 6)
- Right to safe and modern energy (SDG 7)
- Right to decent work (SDG 8)
- Right to development (SDG 9)
- Right to decent housing (SDG 11)

Social Inclusion

"Social inclusion" refers not only to social security and social insurance but also to the rights to fairness and justice of vulnerable groups within society and to strong social support networks (sometimes called social capital). Social inclusion is part and parcel of the SDGs, which include the following:

- Rights of vulnerable groups to land and other protections (SDG 2)
- Global citizenship and appreciation of cultural diversity (SDG 4)
- Gender equality (SDG 5)
- Reduced inequality of income and wealth (SDG 10)
- Right to freedom from violence (SDG 16)
- Right to equal access to justice (SDG 16)

We should also note dimensions of social inclusion not explicitly part of the SDGs, including measures to foster social support networks, voluntarism, values of generosity and honesty, and measures to overcome loneliness and isolation in society.

The goal to reduce income inequality (SDG 10) is notable. While the UN and international law have not defined a fair distribution of income, the high and rising inequality of wealth and income in the world today led to the adoption of SDG 10 calling for a reduction of inequality among and within nations. The 2030 Agenda notes that governments should assess inequalities according to "gender, age, race, ethnicity, migratory status, disability, geographic location and other characteristics relevant in national contexts."

The SDGs give special attention to vulnerable groups, including women and children, migrants, racial and ethnic minorities, the disabled, the geographically isolated, and Indigenous groups. Vulnerable groups are often poor and often lack access to justice. Their property and their rights as citizens and physical persons are often subjected to violence and abuse. Protecting the health, well-being, and legal rights of such individuals suffuses the SDGs and is particularly noted in SDG 16 in the call for "equal access to justice for all."

Environmental Sustainability

When the Universal Declaration of Human Rights was agreed in 1948 and key covenants on civil, political, economic, social, and cultural rights were agreed in the 1960s, there was still very little diplomatic focus on the growing costs and risks of human-induced environmental degradation. The first global conference on the environment was the UN Conference on the Human Environment, held in Stockholm in 1972. This was followed by the Brundtland Commission in 1987, which first described to the world's governments the concept of sustainable development, and then by the Rio de Janeiro Earth Summit in 1992 (known formally as the UN Conference on Environment and Development).

The Rio Earth Summit agreed on three major treaties: the UN Framework Convention on Climate Change (UNFCCC), the Convention on Biological Diversity (CBD), and the UN Convention to Combat Desertification (UNCCD). The UNFCCC aimed to stop human-induced global warming in order to avoid "dangerous anthropogenic [human-caused] interference in the climate system." The CBD aimed to stop the massive loss of Earth's biodiversity. The UNCCD aimed to stop the degradation of dry lands, known as desertification.

When the UN member states enthusiastically adopted the three multilateral environmental agreements at the Rio Earth Summit, it seemed that sustainable development had arrived as a guiding principle for global diplomacy. Yet it was not to be at that time. None of the three treaties was implemented in a serious way in the ensuing twenty years.

When the treaties came up for a twenty-year review in 2012, at a UN conference known as Rio+20, the considered technical view was that all three treaties had failed utterly to change the world's reckless headlong course toward global warming, species extinction, desertification (or degradation) of dry lands, massive pollution, and other environmental harms. It was the failure of the three treaties to be put into practice that motivated the Government of Colombia to recommend to the UN member states the adoption of new Sustainable Development Goals to help get the world on track to head off environmental disaster, and to do so in a context of economic rights and social inclusion.

As of 2015, when the SDGs were adopted, the UN member states had identified three critical categories of human-induced environmental disaster: climate change (global warming), the destruction of biodiversity and habitats (including freshwater depletion), and the pollution of the air, sea, and water from various industrial activities. With the onset of the COVID-19 pandemic in 2020, a fourth category—emerging zoonotic diseases—was added to the list of global priorities.

The SDGs therefore address the environmental threats facing the planet mainly in the following areas:

- Sustainable farm practices (SDG 2)
- Preventing and suppressing emerging zoonotic diseases (SDG 3)
- Sustainable water management (SDG 6)
- Sustainable cities and other human settlements (SDG 11)
- Sustainable consumption and production (SDG 12)
- Stopping human-induced climate change (SDG 13)
- Protecting Earth's marine ecosystems and biodiversity (SDG 14)
- Protecting Earth's terrestrial ecosystems and biodiversity (SDG 15)

In general, these goals are to be met through a combination of introducing new technologies (such as the shift from fossil fuels to renewable energy), improved global surveillance and public health systems (to contain emerging diseases), changing behavior (such as healthier diets that are also better for the environment), and improving regulations and enforcement (such as the improved design and enforcement of antipollution laws, overfishing regulations, limits on logging rights, the protection of endangered species, and other environmental protections).

Implementing the SDGs

The SDGs are a set of goals, not a plan of action. They are part of a UN General Assembly resolution rather than a legally enforceable treaty. In this sense, they are meant to be a tool for global guidance, advocacy, and accountability rather than a set of enforceable actions. And it is worth noting that even legally enforceable UN treaties are often not rigorously enforced in practice.

One can say that the SDGs are part of the long UN agenda dating back to the 1940s to implement the Universal Declaration of Human Rights, partly through treaties but mostly through global goals. A key innovation was the UN Development Decade (1960–1970), which put the UN member states on record as calling for the successful development of the poorer nations, including the newly decolonized nations. This was followed by the Second Development Decade (1971–1980), the Third Development Decade (1981–1990), and the Fourth Development Decade (1991–2000). By the late 1990s, the UN member states agreed that economic growth per se was not enough to achieve the development objectives and that a more holistic framework was needed. This in turn gave birth to the Millennium Development Goals (2001–2015) and now the Sustainable Development Goals (2016–2030).

Four Steps to Implementing the SDGs

The first step is garnering political commitment: Governments need to place the SDGs within the purview of their cabinets and the respective government departments. Governments should integrate the SDG targets and indicators within government visions, plans, and budgets.

The second step is planning: The SDGs require detailed, long-term plans of action. Some actions, such as the decarbonization of the energy system, will require consistent and steadfast actions until 2050 and accompanying policy "pathways" that extend to midcentury. Challenges such as providing universal health coverage, as called for by SDG 3, or ecosystem conservation and transformation similarly require detailed multiyear plans. Governments are often not adept at, or even interested in, plans of action with time horizons beyond the term of the sitting government, yet long-term plans lasting at least a decade or even far longer (e.g., in the case of energy policy) are essential for SDG success.

The third step is multistakeholder mobilization: SDG implementation will require the partnership and cooperation of business, community groups, religious organizations, academia, and other leading stakeholders in society, including governments at all levels, from local to national, transnational (such as in the European Union or the Association of Southeast Asian Nations), and global (through treaty law and UN declarations). The nature of the deep transformations required to achieve the SDGs makes a multistakeholder and intergovernmental approach essential.

The fourth step is financing: The SDGs are inevitably about the reorientation of economic resources, and hence financial flows, by both governments and businesses. Governments must direct additional revenues (from taxes, borrowing, and development assistance) toward the social protection of the poor, the provision of essential public services, investment in infrastructure, and the protection of the environment. Businesses must redirect current outlays away from socially and environmentally destructive practices, such as fossil fuel use, toward socially and environmentally sustainable ends, such as renewable energy.

Financing the SDGs

During the entire period of UN development efforts, dating back to the Development Decade in the 1960s, providing adequate financial support for the poorest countries has been an ongoing and largely unsolved struggle. From the start, the UN has tried to mobilize additional financing for the poorest countries in order for them to invest adequately in health, education, infrastructure, and other development priorities.

As early as 1960, the UN General Assembly established the UN Capital Development Fund to address the capital needs of the least developed countries. In 1970, the UN General Assembly called on the developed countries to provide at least 0.7 percent of their gross national income (GNI) as official development assistance (ODA). It is a goal that has been achieved only by a handful of donor countries, as of 2020 including Denmark, Germany, Luxembourg, Norway, Sweden, and the United Kingdom, among the member countries of the Organisation for Economic Co-operation and Development (OECD), and the United Arab Emirates, a major non-OECD donor country.

In the mid-1970s the developing countries led the call for a New International Economic Order (NIEO), which included higher prices for primary commodities and sufficient development assistance to end poverty. The NIEO was aggressively resisted by the United States, and the initiative was completely abandoned by the early 1990s.

In 2002, following the adoption of the UN Millennium Development Goals, the UN member states agreed on the Monterrey Consensus, which called on developed countries that had not done so to make concrete efforts toward the target of providing 0.7 percent of gross national product (GNP) as ODA to developing countries and 0.15 percent to 0.20 percent of GNP to the least developed countries, as reconfirmed at the Third UN Conference on the Least Developed Countries. In retrospect, the Monterrey Consensus had a very limited impact on the actual ODA flows.

In 2005, the G8 countries, meeting at the Gleneagles Hotel in Auchterarder, Scotland, pledged to double aid to Africa by 2010 and to increase overall ODA by at least US$50 billion by 2010. These Gleneagles commitments similarly were not met.

In 2015, in the lead-up to the SDGs, the UN member states adopted the Addis Ababa Action Agenda on financing for development, similarly aimed to structure the financing of the SDGs. The Addis Ababa Action Agenda recognizes the need for increased financing in several priority areas, including social protection and essential public services for all, scaled-up efforts to end hunger and malnutrition, bridging the infrastructure gap, sustainable industrialization, decent work for all, protecting ecosystems, and promoting peaceful and inclusive societies. It identifies action areas for financing, including domestic public resources (mainly budget revenues), private business financing, and international development cooperation.

Poignantly, while the Monterrey Consensus had called on *developed countries that have not done so* to reach the target of providing 0.7 percent of GNI in official development assistance, the Addis Ababa plan merely reaffirmed the commitment by "many developed countries" to achieve the 0.7 percent target. The main reason for the change in language between 2002 and 2015 is that by 2015, the U.S. government had explicitly repudiated the intention to ever reach 0.7 percent of GNI in ODA. As of 2019, U.S. official development assistance was languishing at around 0.17 percent of GNI, a shortfall of roughly $100 billion per year from the 0.7 percent target.

The SDG Financing Shortfall

The stark reality is that the world's low-income developing countries (LIDCs) cannot afford to finance the SDGs with their own resources. A recent project in 2018 by the International Monetary Fund (IMF) and the UN Sustainable Development Solutions Network aimed to clarify the spending gap for the fifty-nine LIDC countries eligible for IMF concessional assistance.[1] These countries include 1.7 billion people with per capita incomes generally below $1,700 per year (while average incomes in the developed countries are $40,000 or more). The world's poorest countries are included in the LIDC group and are highly concentrated in sub-Saharan Africa.

The IMF study demonstrated that on average, the LIDCs would have to spend an additional 14 percent of their gross domestic product (GDP) in public outlays (that is, through budgetary expenditures) in order to meet SDG targets for health, education, water and sanitation, and electricity. On the other hand, these countries could be expected to mobilize only around 5 percent of GDP in additional budgetary revenues. The implication is a shortfall in budget financing on the order of 10 percent of GDP for all LIDC countries, which amounts to $300 billion to $400 billion per year for this group of nations. The shortfall would be even greater for the poorer countries within the group, because the increment in needed budget outlays is even larger than 14 percent of GDP.

The IMF helpfully put the shortfall in perspective by noting that $300 billion to $400 billion per year is on the order of a mere 0.9 percent of the combined GDP of the advanced economies and 0.5 percent of world GDP. That is, the SDG financing shortfall for 1.7 billion people comes to less than one-half of 1 percent of world output! Yet raising that incremental sum, such a modest proportion of global output, has proven impossible. The rich countries are instead turning their backs on the poorest countries. The United States is worst in this regard, willfully cutting development aid despite being by far the largest rich nation in the world.

These IMF findings have continued to be supported by further excellent studies by the IMF's Fiscal Affairs Department into 2021. The main point is that the low-income developing countries will be able to achieve the SDGs only if they receive considerably more financial support than is currently extended, but still a modest amount relative to the income and wealth of the rich countries, perhaps another $300 billion to $500 billion per year in grants and loans compared with an annual income of around $50 trillion.

A Case Study: The Fight Against AIDS, Tuberculosis, and Malaria

[While this section was written before the COVID-19 pandemic, the principles continue to apply and could be applied in a suitably modified form to the control of COVID-19.]

Three epidemic diseases—AIDS, tuberculosis, and malaria—currently claim the lives of around 1.7 million people per year and cause suffering for many more. Yet each could be substantially controlled, bringing deaths to near zero. Scientific action plans have been identified for each disease to demonstrate how the epidemics could be decisively ended by 2030, in line with the aspirations of SDG 3: healthy lives for all.

The good news is that the cost of comprehensively controlling these diseases is not very high, roughly $101 billion for the three-year period from 2020 to 2022, or about $34 billion per year.[2] The domestic budget revenues of the affected developing countries could provide around $46 billion of the needed $100 billion, leaving a three-year financing gap of about $55 billion. Of that amount, $37 billion is currently expected from existing donors between 2020 and 2022, leaving a three-year funding gap of some $18 billion, or just $6 billion per year. Yet finding that $6 billion is proving extremely difficult, even though funding this effort would mark an enormous difference in lives saved, new cases of the diseases, and prospects for ending the epidemics.

The Wisdom of Saint Ambrose, the Major Faith Traditions, and Global Wealth Today

In his magisterial encyclical *Populorum progressio*, Pope Paul VI wrote movingly and persuasively about the obligations of the rich to help the poor. The 1967 encyclical emerged during the first UN Development Decade, when dozens of former colonies in Africa and Asia were winning their political independence for the first time since the nineteenth century. Pope Paul VI called on the world to help these countries with justice, compassion, and mercy, given their extreme poverty and high disease burden.

In the encyclical, Pope Paul VI elucidates the Church's doctrine of the universal destination of goods, which holds that the world was created for everybody, not just for the rich. Human rights and dignity must take precedence over private property rights. Private ownership is never inviolate but instead is subject to the moral law of universal dignity and human needs. In this context, Pope Paul VI quotes Saint Ambrose (339–397), one of the most renowned doctors of the Church: "You are not making a gift of what is yours to the poor man, but you are giving him back what is his. You have been appropriating things that are meant to be for the common use of everyone. The earth belongs to everyone, not to the rich."[3]

Pope Paul VI goes on to explain:

> Private property does not constitute for anyone an absolute and unconditioned right. No one is justified in keeping for his exclusive use what he does not need, when others lack necessities. In a word, "according to the traditional doctrine as found in the Fathers of the Church and the great theologians, the right to property must never be exercised to the detriment of the common good." If there should arise a conflict "between acquired private rights and primary community exigencies," it is the responsibility of public authorities "to look for a solution, with the active participation of individuals and social groups."[4]

The doctrine of the universal destination of goods is more vital than ever today because the global economy has created an astounding "winner take all" dynamic that is leading to unprecedented wealth accumulation among the world's richest individuals. As Oxfam has demonstrated, the world's richest twenty-six individuals have a combined net worth of $1.6 trillion, equal to the combined wealth of the bottom half of the planet (3.8 billion people). *Forbes* magazine calculated that there were 2,208 billionaires as of March 2018, with a combined net worth of $9.1 trillion. As of March 2021, the number had risen to 2,755, with a combined wealth of $13.1 trillion. Also as of March 2021, Jeffrey Bezos, the founder and part owner of Amazon, has an estimated net worth of $182 billion. The number of billionaires in the world has risen almost sixfold since 2000, and their estimated combined wealth, adjusted for inflation, has increased roughly ninefold in just twenty-one years.

The billionaires might reflect upon the Hindu tradition's doctrine of renunciation. The first two verses of the *Isha Upanishad* read, "This entire universe, moving and unmoving, is enfolded in God. Renounce and enjoy."[5] In other words, because the universe is a sacred reality, humanity is called to enjoy the planet, its resources, and any wealth derived from it with open and generous hands. Exploitation and hoarding reflect a greed incompatible with charity, gratitude, and enjoyment of the planet's bounty.

Distributive justice and compassion are similarly at the heart of Judaism, Islam, Buddhism, and the other great world faiths. Jewish biblical law enjoins the landowner to set aside part of the harvest for the poor: "When you reap the harvest of your Land, you shall not completely remove the corner of your field during your harvesting, and you shall not gather up the gleanings of your harvest. [Rather,] you shall leave these for the poor person and for the stranger. I am the Lord, your God."[6]

With its key concept of interrelatedness, Buddhism also emphasizes the importance of compassion and donation. Acknowledging our interdependence with others promotes the virtues of compassion, self-giving, and generosity. The Buddhist tradition has also generated its own mode of economic reflection, which highlights simplicity, minimizing suffering and violence, and the simplification of desires, against the typical consumerism of modern economies.

In Islam, the practice of *zakat* calls on those with incomes above a minimum to give a portion (often one-fortieth) to the poor. *Zakat* is one of the five pillars of Islam, regarded as second in importance only to prayer. The word *zakat* itself derives from the word for "purification": *Zakat* constitutes the purification of one's wealth through giving. Alms in this sense are given not only for the sake of the poor but also for the moral and spiritual health of the one who gives.

Indeed, Saint Ambrose's vision of justice, applied to merely 2,755 of the world's 7.6 billion people, could readily solve the extreme deprivation of the world's poorest 1.7 billion people. If the combined net wealth of the world's billionaires could be treated as an endowment put to work to fight extreme poverty, the $13.1 trillion would generate an annual flow of income of some $650 billion per year, assuming an annual payout rate of 5 percent. This sum exceeds the financing gap identified by the IMF for the fifty-nine low-income developing countries! This $650 billion, if well directed, could ensure universal basic health coverage; universal education through secondary school; the end of AIDS, tuberculosis, and malaria; and access to clean water, sanitation, and modern energy services.

The SDGs and Six Deep Societal Transformations

According to the UN Sustainable Development Solutions Network, the seventeen SDGs can most usefully be understood as requiring six transformations in society.[7] These transformations, if successfully achieved, would enable every part of the planet to achieve sustainable development, including the targets of the seventeen SDGs.

Education, Gender, and Inequality

SDG 4 calls on all countries to achieve universal secondary school completion by 2030. This will require increased investments in early childhood development and prekindergarten programs, which have been shown to boost the cognitive and emotional development of children, with persistent effects into adulthood. Preschool education also helps reduce inequalities in opportunity among children. In parallel, countries will need to ensure primary and secondary school completion for all, which, among other things, will require enhanced teacher training and curriculum development.

Another pillar of this transformation is promoting gender equality (SDG 5) and social inclusion (SDG 10). Among other endeavors, achieving this goal will require measures to end discrimination in the workplace, as well as other anti-discrimination policies and standards. Equal access to high-quality education, health care, and other services is critical for reducing inequalities, as are social safety nets.

In addition to improved education, social safety nets, and antidiscrimination measures, improved labor standards form another pillar for reducing inequalities (SDG 10). The International Labour Organization has developed detailed standards that every country and employer should meet. Of particular importance should be efforts to end all forms of modern slavery, human and organ trafficking, and child labor, which continue to be prevalent in poor and rich countries alike (SDG 8, Target 8.7).

Health, Well-Being, and Demography

The SDGs also focus on universal health coverage (SDG 3) and the social and environmental determinants of health and well-being. They frame health as a basic need and human right. A central pillar of the health, well-being, and demography transformation is universal health coverage, which will also contribute directly to SDG 5 (gender equality) and SDG 1 (no poverty). Universal health coverage requires a core, publicly financed health system that integrates preventive, therapeutic, and palliative services. Health systems also require integrated information systems and real-time epidemic and disease surveillance and control. In many countries, community health programs have been shown to improve health outcomes significantly.

The health system must offer a range of services. Of critical priority are interventions for maternal, newborn, and child health. To control the spread of infectious diseases, health systems must offer effective prevention and treatment. And they need to integrate noncommunicable disease control, including mental health treatment and basic surgery.

Clean Energy and Industry

The world has a growing need for energy but must drastically reduce carbon dioxide emissions. The climatology is clear. To have at least a 50 percent chance of remaining below an increase of 1.5 degrees Celsius in global warming requires that net greenhouse gas emissions reach zero by 2050. Yet as of 2021, the greenhouse emissions trajectory puts the planet on a course to reach a disastrous warming of 2.7 degrees Celsius.

Available national and global pathways for decarbonizing the energy system suggest three major pillars for action. First, countries need to ensure universal access to a fully decarbonized (that is, zero-carbon) electricity grid and other clean fuels (SDG 7). This will require a rapid and decisive shift from fossil fuels (coal, oil, and gas) to zero-carbon sources, including wind, solar, hydro, geothermal, ocean, and others. Second, countries need to improve energy efficiency in final energy use in areas including transport (e.g., lighter and more efficient vehicles, car sharing, autonomous vehicles), buildings (heating and cooling, thermal insulation), industry, and household appliances. Third, countries need to electrify current uses of fossil fuel energy outside of power generation, such as in internal combustion engines (through electric or hydrogen vehicles), boilers and heaters in buildings (through heat pumps), and various industrial processes, and countries need to use green fuels (such as hydrogen) produced with zero-carbon electricity for uses that cannot be electrified.

In addition to energy decarbonization, all nations must sharply cut their use of industrial pollutants of the air, water, and land (SDG 12). Key industrial pollutants include methane, nitrous oxide, and sulfur dioxide, as well as organic and other inorganic pollutants. Water management, life-cycle approaches, and other tools of circular economy (such as reuse, recycling, and eliminating waste) can increase resource efficiency and decrease pollution. The circular economy also provides a framework for twenty-first-century industrialization strategies in sub-Saharan Africa and elsewhere.

Sustainable Food, Land, Water, and Oceans

The fourth transformation needed is in land use and food systems. The current patterns of land use, mainly related to food production, are unsustainable in three ways. First, today's agricultural systems are major drivers of environmental change. They account for about a quarter of greenhouse gas emissions (SDG 13) and over 90 percent of scarcity-weighted water use (SDG 6), and they are the major drivers of biodiversity loss (SDGs 14 and 15), eutrophication through nutrient overload, and water and air pollution. At the same time, the food system is vulnerable to environmental changes now underway, through the increasing severity of droughts, floods, disease, and land degradation caused, in part, by climate change. Similarly, most ocean and freshwater fisheries are overexploited.

Finally, today's food system is not aligned with healthy diets and thus contributes to persistent hunger, widespread malnutrition, and a growing obesity pandemic.

The implications are clear. The world will need a major transformation of food systems and land use to mitigate human-caused environmental degradation, build resilience into food production, and achieve better health outcomes. This in turn will require efficient and resilient agricultural systems, the conservation and restoration of biodiversity, a shift to healthier and more plant-based diets, and improved land-use regulation and management.

Smart Cities and Transport

Cities today are home to around 55 percent of humanity and 70 percent of global economic output. By 2050, cities will be home to around 70 percent of humanity and perhaps 85 percent of global economic output. What happens in cities, therefore, will determine the well-being of most of humanity and the prospects for sustainable development. It is therefore no accident that the world's national governments assigned SDG 11 to sustainable cities, meaning cities that are economically productive, socially inclusive, and environmentally sustainable.

As a first priority, cities need to develop sustainable urban infrastructure. This includes an efficient transport system; universal access to reliable and low-cost electricity, safe water, and sewerage; recycling and other sustainable waste management; and high-speed, low-cost broadband connectivity for all to support businesses and public service delivery. These elements should be deployed according to a plan that takes account of likely population growth. Safe and open green spaces, infrastructure for cycling and walking, and higher-density settlements increase resource efficiency and quality of life. Smart urban networks can provide real-time monitoring and management of safety, traffic, energy use, and other services.

Digital Technology and E-governance

The greatest single technological enabler of sustainable development in the coming years will be the digital revolution, constituted by the ongoing advances in computing, connectivity, information digitization, machine learning, robotics, and artificial intelligence (AI). The impact of the digital revolution rivals those of the steam engine, internal combustion engine, and electrification in its pervasive effects on all parts of the economy and society. The rapid pace of advance continues, with imminent breakthrough prospects for AI, quantum computing, virtual reality, and 5G broadband, as well as other technological advances, including the following:

- Universal access to high-quality, low-cost mobile broadband
- Digital transition and connectivity of all government facilities
- Online national systems for health care and education
- Online e-finance and e-payments to facilitate trade and business services
- Universal online identification for official purposes (e.g., banking, taxation, registration)
- Regulatory security for online identity and privacy
- Income redistribution to address income inequalities arising from digital scale-up
- Tax and regulatory systems to avoid the monopolization of internet services
- Online data governance and interoperability provisions
- Democratic oversight of cutting-edge digital technologies

The Decisive Role of Multireligious Action to Achieve the SDGs

Multireligious action is essential to the achievement of the SDGs. Every major religion is committed to the core values espoused by the SDGs: human dignity and flourishing, the rights of the poor, social justice, and peace. Every major religion has unique and significant assets to bring to the fulfillment of the SDGs, including a profound code of ethics; a daily dialogue with all parts of society; face-to-face engagement with billions of people around the world, including the world's poorest people; vital institutions of education, health, charity, and environmental protection; and the ability to teach and disseminate the vital information needed for global success in sustainable development. Here is a brief summary of some of the key practical pathways for religious engagement with the SDGs.

Ethics in Action

The world's major religions are the repositories of humanity's core moral codes and ethical guideposts. Fortunately, there is a deep congruence among the religions in the core ethical precepts regarding human dignity, the rights of all people to have their basic needs met, the essential value of compassion and service to others, and respect for and stewardship of the natural environment. The Ethics in Action initiative demonstrated the ability of a multifaith leadership group to formulate agreed ethical principles and guidelines for action to achieve the SDGs, which the various contributors to this present volume explore.

This multifaith consensus should now be further strengthened and broadened so that religious leaders across all major faiths and regions will become leaders of the SDGs in line with the deep teachings of their respective faiths.

Religions and Multistakeholder Engagement

Religious communities can convene the key stakeholder groups in society whose participation is needed to achieve the SDGs. The Catholic Church, for example, has convened scientists, mayors, judges, ethicists, and faith leaders of many religions to support integral and human development, guided by Pope Francis's call for a plan for our common home in his encyclical *Laudato si'*. The Pontifical Academy of Sciences played a decisive role in gathering experts to support the SDGs and the Paris Agreement on climate change, and Pope Francis was key in helping to build a global consensus around both the SDGs and the climate agreement. Other religions are now convening religious leaders and scientists to work hand in hand on sustainable development initiatives, with powerful benefits for society.

Work with Local Congregations

Achieving the SDGs will require that people around the world know of the goals and understand their relevance and potential benefits for their own families, communities, and nations, as well as for the world. The SDGs should be explained from the pulpits of all faiths, to help all of humanity to understand their human rights and the global quest to end poverty, promote social justice, and protect the natural environment.

Direct Service Provision

All major faiths play a vital role in the direct service provision of education, health care, and social support services, often with special attention paid to the most vulnerable, including the young, the elderly, the disabled, migrants, minority groups, and the marginalized. In so doing, religious communities strive to live their commitment to "leave no one behind," even when societies are misguided enough to cast the vulnerable aside.

The SDGs are first and foremost about ensuring vital social services for all, including social protection for the extreme poor (SDG 1) and the hungry (SDG 2); health care for all (SDG 3); education for all (SDG 4); water and sanitation for all (SDG 6); and modern and safe energy services for all (SDG 7).

Religious providers of these social services should team up with governments and international donor agencies in a systematic way to ensure that the universal aspirations of the SDGs can be successfully fulfilled, truly leaving no one behind.

Public Education and Awareness

As Pope Francis has frequently reminded us, our greatest vulnerability today is the "globalization of indifference," meaning humanity's neglect of even its own survival. We are lost in a world of online imagery, substance and behavioral addictions, political demagoguery, commercial distractions, and rampant consumerism to the point that we neglect the essential needs of our communities, to say nothing of the needs of the poor and vulnerable. We are manipulated by fear rather than inspired by compassion.

The world's religions have a unique role to play in overcoming the globalization of indifference by joining together in the clarion call for human survival and well-being; by demonstrating common bonds across races, religions, classes, and ethnicities; and by proving through their good works the ability of our societies to leave no one behind.

In *Laudato si'*, Pope Francis invited a dialogue among all people, believers of all faiths and nonbelievers alike, in search of a path to authentic human and sustainable development. He noted that our interdependence obliges us to search for a common plan for humanity and the planet. By raising the voices of all of the world's great faiths in unison, the multifaith community will be able to overcome the indifference that holds us hostage, drown out the haters and the fearmongers, and open the way to a new generation of collective action for the common good that will inspire people around the world.

In this regard, religious leaders and congregants can work with the United Nations, governments, and civil society to realized SDG Target 4.7, which calls for education for sustainable development for all. Specifically, Target 4.7 calls for a global educational effort that embraces the core values of sociality, trust, respect, and a culture of peace, which are championed by the world's faiths:

> *SDG Target 4.7* By 2030 ensure all learners acquire knowledge and skills needed to promote sustainable development, including among others through education for sustainable development and sustainable lifestyles, human rights, gender equality, promotion of a culture of peace and non-violence, global citizenship, and appreciation of cultural diversity and of culture's contribution to sustainable development.

Notes

This chapter was adapted from a paper written by the author for the seventh World Assembly of Religions for Peace held in August 2019 in Lindau, Germany.

1. Vitor Gaspar, David Amaglobeli, Mercedes Garcia-Escribano, Delphine Prady, and Mauricio Soto, "Fiscal Policy and Development: Human, Social, and Physical Investments for the SDGs," IMF Staff Discussion Note No. 19/03, January 23, 2019, https://www.imf.org/en/Publications/Staff-Discussion-Notes/Issues/2019/01/18/Fiscal-Policy-and-Development-Human-Social-and-Physical-Investments-for-the-SDGs-46444.
2. The Global Fund to Fight AIDS, Tuberculosis and Malaria, "Investment Case Update" (PowerPoint presentation, Geneva, Switzerland, January 18, 2019).
3. Paul VI, *Populorum progressio* (1967), sec. 23.
4. Paul VI, *Populorum progressio* (1967), sec. 23.
5. Translation from scholar and Ethics in Action member Anantanand Rambachan.
6. Leviticus 23:22.
7. Jeffrey D. Sachs, Guido Schmidt-Traub, Mariana Mazzucato, Dirk Messner, Nebojsa Nakicenovic, and Johan Rockström, "Six Transformations to Achieve the Sustainable Development Goals." *Nature Sustainability* 2 (2019): 805–14.

CHAPTER 2

A SOCIAL MOVEMENT TO MAKE THE LAST FIRST

MARCELO SÁNCHEZ SORONDO

The Ethics in Action initiative seeks to identify the major challenges presented by contemporary global society and the possible paths we might take to achieve a more equitable, fair, and peaceful society that respects the human dignity of each person. Happily, we now have a road map to follow. Pope Francis has published the encyclical *Laudato si'*, which can serve as our guide, and the United Nations has approved the Sustainable Development Goals, which seek to end poverty and hunger; achieve food security; ensure healthy lives and promote well-being for all; guarantee inclusive, equitable, and quality education; guarantee access to clean water and sustainable energy; promote sustainable economic growth and decent work for all; achieve cities and human settlements that are inclusive, safe, resilient, and sustainable; guarantee sustainable consumption and production patterns; and take urgent measures to combat climate change and its effects. We can rightly say that these are the new imperatives that arise from faith and reason, from religion and civic values. Ethics in Action intends to study how to implement these new imperatives arising from the pope's magisterium and the work of the representatives of all the nations that unanimously approved the SDGs.

As I am a Catholic bishop having served for nearly twenty years at the Pontifical Academies of Sciences and of Social Sciences, it is only natural that I will frame my reflection on the motivation to achieve these goals from the point of view of a Christian philosopher. However, much of what I am going to say is simply based on the human person and therefore will hopefully speak to all people, irrespective of personal religious identification (or lack thereof).

Heidegger coined the celebrated idea that human beings had been "thrown" into the world. While it is certainly true that no one asks, or even plans, for their own birth, I would sooner say that we are called individually into the world to live. But we are not called to live individually. Rather, human beings are called into existence to become truly themselves only when they are in relation to the other (and of course, in its highest expression, in relation to that Other who is God).

I therefore suggest that one great starting point in working toward making the world a better place is to be found in drawing attention to this fact—and in the building of what might be called an interpersonal, inter-relational capital throughout society.

How might this be done? Let's be a little radical.

The Greek for "to suffer" is *paskhein*, and the cognate noun is *pathos*. The Greek prefix for "together" is *syn*. To suffer together, therefore, is *syn-pathos*. In fact, to feel misery out of *sympathy* with the suffering of another, from one's heart (*cor*) is *miseria-cordia*—*misericordia*—the Latin word from which we get "mercy."

The idea is not that we should be filled with sympathy for others, with warm fuzzy sentiments, and then carry on doing exactly what we were doing before. Rather, we should be moved by these emotional responses; we should not stand by watching as a stranger is robbed and abandoned, left to die by the side of the Road to Jericho. Compassion should motivate us to *do something*. To borrow from Aristotelian terminology, we should move from *potentia* to *acta*.

When we are driven in our compassion by the suffering of another, *how* should we govern our response? This is the central question to which I will return.

The three texts I have chosen to act as a foundation for my proposal for building a more compassionate society all come from the Gospel: the Golden Rule, the Beatitudes, and the Judgment Protocol.

The most common model of how to respond, I think, is summarized by what is popularly known as the Golden Rule: "Do to others as you would have them do to you."[1] Most religions have their own variant of this principle. However, it is clear that a society in which individuals genuinely wanted to receive nothing from others would also—coherently—offer nothing in return! Or, at the other end of the spectrum, a society in which people genuinely wanted to give everything to others would also—again entirely coherently—expect to receive everything in return. Both are equally consistent applications of the Golden Rule.

Therefore, in order to complete the Golden Rule, the Beatitudes (and in fact, also the famous Judgment Protocol of Matthew 25) round off the evangelical exposition on compassion. In the same way that Paul Ricœur spoke of the other as another self, or oneself as another, Christianity has something unique to offer regarding the fullest interpretation of the Golden Rule. The Danish Lutheran philosopher Søren Kierkegaard developed his concept of the "moment"; that is, the precise moment that God gives a person a certain grace (e.g., for knowledge of God, conversion, repentance) that precedes an interior change of disposition. Whether this grace is accepted or rejected depends on the person, as it is governed by their free will.

Let's apply Kierkegaard's insight to one of the most well-known and well-loved parables in the Gospel:

> Jesus replied, "A man fell victim to robbers as he went down from Jerusalem to Jericho. They stripped and beat him and went off leaving him half-dead. A priest happened to be going down that road, but when he saw him, he passed by on the opposite side. Likewise, a Levite came to the place, and when he saw him, he passed by on the opposite side. But a Samaritan traveler who came upon him was moved with compassion at the sight. He approached the victim, poured oil and wine over his wounds and bandaged them. Then he lifted him up on his own animal, took him to an inn and cared for him. The next day he took out two silver coins and gave them to the innkeeper with the instruction, 'Take care of him. If you spend more than what I have given you, I shall repay you on my way back.' Which of these three, in your opinion, was neighbor to the robbers' victim?" He answered, "The one who treated him with mercy." Jesus said to him, "Go and do likewise."[2]

Conventional thinking would suggest that the Good Samaritan was the benefactor and the stranger the recipient, and of course this is certainly true. But I do not think this is the whole picture. In fact, a profoundly Christian reading of the parable reveals that on another, higher, meta-level, something else is taking place.

This is the presence of another transfer going in the opposite direction—and I believe an appreciation of this dynamic is crucially important but very rarely given sufficient attention. This is the insight that the victim is himself a donor of something essential! He gives to the Samaritan the opportunity to be merciful, to be compassionate. He is the means by which the Samaritan can be the recipient of the Kierkegaardian *moment*.

Who, then, actually gives the gift of greater value? This idea is counterintuitive to the values of our contemporary, mainly selfish, culture and to the implementation of *Laudato si'* and the SDGs. And it is this insight into motivations that lies at the heart of Pope Francis's mission. It is this that he wants the Church to understand more profoundly.

When we respond with compassion to a neighbor who has fallen on hard times, we may become the willing recipients of a grace. The material recipient becomes the donor of the grace of our having been able to see the face of Jesus Christ in the least of one of His brothers or sisters and having been able to respond accordingly. A radically Christian interpretation of compassion therefore helps us to see that when giving of ourselves, through the action of Divine Grace, we become the recipients of a gift far more valuable.

The Golden Rule finds its echo in the formulation "love one's neighbor as oneself," and, as Our Lord taught, it is at the heart of the Old Testament:

> "Teacher, which commandment in the law is the greatest?" He said to him, "You shall love the Lord, your God, with all your heart, with all your soul, and with all your mind. This is the greatest and the first commandment. The second is like it: You shall love your neighbor as yourself. The whole law and the prophets depend on these two commandments."[3]

But what about the New Testament? What does the Gospel have to add? Can it perfect, in a Christian way, an already perfect formulation? The most influential manifesto in all politico-religious language, the most revolutionary discourse, the most relevant, the most human and the most divine, the shortest and the most profound, that any man has ever pronounced during the course of human history, is that of the Sermon on the Mount, of our Lord Jesus Christ:

> Blessed are the poor in spirit, for theirs is the kingdom of heaven.
> Blessed are they who mourn, for they will be comforted.
> Blessed are the meek, for they will inherit the land.
> Blessed are they who hunger and thirst for righteousness, for they will be satisfied.
> Blessed are the merciful, for they will be shown mercy.
> Blessed are the clean of heart, for they will see God.
> Blessed are the peacemakers, for they will be called children of God.
> Blessed are they who are persecuted for the sake of righteousness, for theirs is the kingdom of heaven.
> Blessed are you when they insult you and persecute you and utter every kind of evil against you [falsely] because of me.[4]

I add to the Beatitudes the Judgment Protocol because, in consideration of compassion, the two texts complement each other very well:

> When the Son of Man comes in his glory, and all the angels with him, he will sit upon his glorious throne, and all the nations will be assembled before him. And he will separate them one from another, as a shepherd separates the sheep from the goats. He will place the sheep on his right and the goats on his left. Then the king will say to those on his right, "Come, you who are blessed by my Father. Inherit the kingdom prepared for you from the foundation of the world. For I was hungry and you gave me food, I was thirsty and you gave me drink, a stranger and you welcomed me, naked and you clothed me, ill and you cared for me, in prison and you visited me." Then the righteous will answer him and say, "Lord, when did we see you hungry and feed you, or thirsty and give you

> drink? When did we see you a stranger and welcome you, or naked and clothe you? When did we see you ill or in prison, and visit you?" And the king will say to them in reply, "Amen, I say to you, whatever you did for one of these least brothers of mine, you did for me." Then he will say to those on his left, "Depart from me, you accursed, into the eternal fire prepared for the devil and his angels. For I was hungry and you gave me no food, I was thirsty and you gave me no drink, a stranger and you gave me no welcome, naked and you gave me no clothing, ill and in prison, and you did not care for me." Then they will answer and say, "Lord, when did we see you hungry or thirsty or a stranger or naked or ill or in prison, and not minister to your needs?" He will answer them, "Amen, I say to you, what you did not do for one of these least ones, you did not do for me." And these will go off to eternal punishment, but the righteous to eternal life.[5]

One may wonder why—when there are so many potential meritorious actions available—the Lord proposes works of mercy toward our neighbors as the action plan and criterion for salvation.

By performing acts of mercy, one works toward one's own salvation: the offer being that if one abstains from sin and does penance, one is released from sin and may be saved through almsgiving, on Judgment Day, when Christ declares that inasmuch as we did these things to the least of his brethren, we did them to him. Saint Augustine claims that we all sin in this world, but not all of us condemn ourselves, for those who do penance and perform acts of mercy are saved.

In essential terms, the Beatitudes and the Judgment Protocol are more concrete existentially than the Golden Rule, which is always maintained through an abstract view of the other (or, as Paul Ricœur would say, of oneself as another). These two texts, on the other hand, which speak about the other in their existential situation of suffering—the poor, those who weep, those who suffer, the pure of heart, the merciful, those who look for justice and who suffer for justice, and those who feed the hungry, give drink to the thirsty, give shelter to the homeless, clothe the naked, tend to the sick, and visit the imprisoned—demonstrate concrete suffering and a correspondent compassion that is not present in the Golden Rule.

In definitive terms, in the Beatitudes and the Judgment Protocol, unlike in the Golden Rule, the other is that suffering being that the Gospel never ceases to place at its center. Suffering is defined not only as physical suffering, as mental or moral pain, but also by the diminution or destruction of the capacity to be and to act, to be and to be able to do, which are felt as an attack on the integrity of the dignity of the person.

One can extract two corollaries from these two texts, Matthew 5 (the Beatitudes) and Matthew 25 (the Judgment Protocol). First, one could say that from a philosophical point of view, Pope Francis, as regards the great subject of evangelization, being able to start from truth or human good—which is justice—prefers to follow the Beatitudes, which speak about the poor, the afflicted, the righteous, and the peacemakers. In other words, if we reduce the subject of the beginning to transcendentals and their mutual membership and conversion, "*quodlibet ens est unum, verum, bonum*,"[6] without neglecting the transcendental of truth, which was emphasized by Popes Saint John Paul II and Benedict XVI, perhaps Pope Francis begins with that of good, which today is that of justice and the Beatitudes, as did Christ in the Sermon on the Mount. We can say with Saint Thomas, "The object of the intellect is the first and the most important in the genus of formal causes: indeed, its object is being and truth. But the object of the will is the first and the most important in the genus of final causes: indeed, its object is good, within which are included all ends."[7] We can conclude that attraction to good, to happiness, and to perfection has priority as regards all attitudes of the conscience. Therefore, to begin with human good, which is justice, not only seems suited to human anthropology but also demonstrates the intensity of the social destination and obligations of the Gospel.

The second corollary is that in essential terms, the Beatitudes and the last are more concrete existentially than the Golden Rule. This last, both in its positive and negative meanings, quoted in the Gospel as "do not do to others" or "do to others," is always maintained in the abstract view of the other or of oneself as another (Ricœur). The Beatitudes, on the other hand, which speak of the other in their existential situation of suffering—the poor, the weeping, the suffering, the pure in heart, the merciful, the one who looks for justice and suffers for justice—in definitive terms, the last, demonstrate a human and social concretion of suffering that is not present in the Golden Rule. Today, therefore, we are called, following Pope Francis, to see how these recommendations of the Lord can be thought about and structured in the social order. Blessed are those who know how to think of and organize a global society in which the last are the first!

Notes

1. Luke 6:31.
2. Luke 10:30–37.
3. Matthew 22:36–40.
4. Mathew 5:3–11.
5. Matthew 25:31–46.
6. Kant, *Critique of Pure Reason*, § 12.
7. Aquinas, *De Malo*, q. 6, a. un., cor.

CHAPTER 3

VIRTUES ACROSS TRADITIONS

Common Ground?

OWEN FLANAGAN

A key assumption of Ethics in Action is that achieving the goals expressed in the Sustainable Development Goals requires economic resources, creative public policy, smart technical solutions, and moral capital. Indeed, all seventeen SDGs express ethical values. Morally attuned individuals with a vision of a common good and the will to achieve, sustain, and maintain that common good are as important as the economic, scientific, and political resources required to achieve the SDGs. The unanimous agreement in 2015 on the SDGs by the 193 member states of the UN is a good sign. But assenting to noble values is one thing; achieving and sustaining them is another.

Thus, it would be good to know the following: What is the state of the world's moral capital or moral capacity? Are the governments and citizens of those 193 nation-states committed to taking the actions required to achieve the SDGs? In particular, are the virtues, ethical principles, and moral and political values necessary for a worldwide effort to support sustainability and flourishing in place or available to the peoples of the earth who are called upon to enact collectively the SDGs?[1]

In this chapter, I provide an overview of one aspect of moral capital that will be needed for any such overall assessment by asking, What values and virtues are provided by some of the world's great spiritual and religious traditions, as well as their secular kin, that might in the present moment advance the SDGs? The traditions discussed are the three Abrahamic traditions, Judaism, Christianity, and Islam, as well as Hinduism, Buddhism, Native American Anicinapek philosophy, Confucianism, and a few varieties of secular ethics (virtue ethics, deontology, and utilitarianism/consequentialism). Each of these traditions contains multitudes, and thus I take the views discussed to be samplings of the moral capital these traditions provide. Sometimes, I will begin by marking the orientation or sect represented by my informants, so the reader has a sense of which variant of each tradition is being discussed. I sometimes quote from the sources but do not reveal their identities. All are true believers, representatives, and/or scholars of the traditions they represent.

My analysis consists of four parts:

(1) An inventory of the tradition-specific sources or grounding (God, prudence, suffering, human nature, universal rights, culture) of morality in general and the virtues in particular
(2) The main virtues advocated by several religious and secular traditions
(3) The conception of the common good and its scope (tribal, nation-state, universal) across this sample of traditions, as well as the ethical resources provided by the traditions surveyed for achieving the SDGs
(4) An analysis of the resources and obstacles to the development of a global virtue theory and a global set of values that would provide the requisite moral capital to achieve the SDGs

This chapter contains information gathered from questionnaire answers from leading representatives of various faith and secular traditions and from conversations at a two-day meeting on the virtues across traditions held at the Garrison Institute in New York in autumn 2019.

Sacred Ground

I asked my colleagues to answer this question: What is the experience of the "Sacred" that grounds virtue ethics in your tradition?

Grounding is an epistemic concept and has to do with justification, in particular with where justification bottoms out. It also expresses ontological commitments. If one ought not kill because prudence or a social contract condemns it, then (as far as the justification of the commandment that "thou shalt not kill" goes) prudence and a social contract are the ontological ground that explain and warrant the commandment. If one ought not kill because God commanded it, then God is, as philosophers say, ontologically fundamental; God grounds, rationalizes (in the good sense), or justifies the prohibition on killing. Same commandment, different ontological ground.

There is agreement that scientific theories are grounded in some mixture of observation, experiment, rigorous testing, and consistency (often with the rest of science). There is disagreement about what ethical theories (comprising claims about the correctness of certain values, principles, norms, virtues, and conceptions of well-being) are grounded in or justified by. Some candidates are God, reason, prudence, human nature, happiness, suffering, culture, power, and admixtures.

Most of my informants are religious leaders, religious scholars, philosophers, or all of these. Here are some ways they think about grounding:

Grounding in Islam

(Two informants: A. is a Muslim university Chaplain; B. is a scholar of Sufi mysticism)

A. begins his answer with two caveats. First, no particular view can represent the wide range of Muslim voices. Second, there is no sharp distinction between "religious/sacred" and "secular" in Islam, since hardly anything "secular" is free from the connection to the "religious/sacred." For this reason, Islam advises Muslims to "feel the presence of the divine in their lives and make an attempt to keep and sustain this divine presence as successfully as they can."

B. says that the sacred ground of Islam is Allah, the divine, who grounds the practices of spiritual chivalry (*futuwwah* in Arabic; *javanmardi* in Persian and Urdu ethics) followed by Ṣūfis in order to attain spiritual perfection. The aim is to orient oneself and one's life toward "the manifestation of the Light of the original human nature (*fitrat Allah*)," wherein one experiences one's true or best self as desirous of the Good, which is perfectly exemplified by Allah.

Like A., B. insists on the integration of the sacred and the secular, perhaps the absorption of the secular by the sacred, by emphasizing that belief in Allah should be woven into Muslims' existence so that service to others, which is service to Allah, comes before any consideration is given to self.

Grounding in Judaism

(Two informants: C. is a scholar and Orthodox Jew; D. is a rabbi and leader of international interreligious dialogue)

C. begins with a caveat that we should not essentialize Judaism: "Jews do not have dogma—Jews have debates." Thus, according to her, while the Hebrew scriptures are the core source of Judaism, Judaism has a discursive and interpretive tradition as revealed in the text of the Talmud(s). C. emphasizes the integration of the sacred and the secular in the Jewish tradition: "Every act—eating, waking, seeing the ocean—has a blessing of praise, which sacralizes every act."

According to D., the Hebrew Bible describes God as "absolute sanctity (from which all sanctity is derived) that we are called to emulate (Leviticus 19:1)" and "just, righteous, compassionate, and loving." D. says that since human beings are created in the image of God and our world is the Divine Creation, it is our supreme responsibility to ensure the flourishing of humanity (especially regarding the vulnerable) and the ecosystem. Any failure in this responsibility would be to offend the very source of the sanctity of all creation. C. and D. agree that Judaism teaches that care of vulnerable people and the environment is a matter of serving God.

Grounding in Christianity

(Two informants: E. is a Catholic cardinal; F. is an evangelical Christian leader)

According to E., (Roman Catholic) Christianity understands God as the "Creator of heaven and earth, of all things visible and invisible." E. emphasizes that the present world is "only preparation for the world to come." This world involves a "life eternal" and is "worth striving to attain, at whatever costs to us in the present world."

F. describes the (evangelical) Christian God as "wholly other" owing to the radical incomprehensibility of God to our finite mind. Because this world is God's creation, it is holy. This world inherits its holiness from its Creator, who both in scripture and in natural law directs us to what is morally correct. F. writes, "Because all of created reality is under the sovereign control of God, sacredness infuses every aspect of reality that is consecrated by those who recognize its divine origins. Sacredness implies being set apart or consecrated toward a holy end. This entails that we are obligated to care for all of creation."

Grounding in Hinduism

(G. is a Hindu and a scholar of Hinduism)

G. emphasizes that Hinduism is not a monolithic religion but a family of religions. In all its forms, Hinduism is not monotheistic. So, the ground of morality is not the wisdom or will of one God. Every being is the locus of the sacred. Scholars disagree about whether Hinduism should be classified as polytheistic, panentheistic, or pantheistic.

There is no distinction between the sacred and the secular in Hinduism. G. writes, "Hindus see the sacred in trees, rivers, mountains, Earth, mother, father, teacher, spouse, and so on. Millions of deities in Hinduism are linked to various qualities, virtues, and powers." As the whole world is sacred and for the benefit of all, Hinduism teaches that whoever takes more than their share is a thief.

Divine Reality is described as comprising three main attributes, *Satchidananda*, that orient us towards what is good. Such an orientation involves recognition of the sanctity of all life, the power of knowledge, especially knowledge of the divine, the joy of peace, and the good of common flourishing.

Thus, ethics in Hinduism flows from a variety of sources. Some codes of social order (such as ideas of caste) claim to be derived from a sacred cosmic order and a belief in the infusion of all reality with the divine.

Another important ground of morality emphasized in Hinduism is found in the theology or metaphysics of the unity of all beings and the tradition of honoring unifying relations. The concept of debt (*rna*) to those who are related to us (e.g., parents, Earth, the natural elements) is the underlying normative ground.

One distinctive aspect of Hindu ethics, according to G., is that it "does not rely on external fear or any rewards in heaven to motivate or enforce morality. Morality is directly connected to the spiritual goal of realizing the unity of all creation, an ethic of care for the other who is in their essence one's very own self." One reply is that most Hindus do in fact believe in a series of afterlives whose quality depends on the moral quality of one's present life. That is not the same idea as heaven (and hell), but it shares similar features: promising liberation and release (*moksha* and *nirvana*) or reincarnations that involve descents into worse lives.

Grounding in Buddhism

(H. is a Buddhist scholar)

Whereas Hinduism has a creator God, *Brahma*, Buddhism does not. Or, as the Buddha puts it in one of his earliest sermons, questions that we would think of as theological or metaphysical are not answerable by us and thus not worth asking or opining on. Buddhism can be understood as atheistic or agnostic if we use Western categories (although in many Buddhist sects all sorts of supernatural beings abound, and rebirth is a common belief in many sects). But there is certainly no creator God, nor any person-like being who is omnipresent, omniscient, all good, all loving, and so on. What there is, and all there is, is impermanence and flux.

The Buddha insists that what grounds morality is the most obvious fact: that suffering abounds. This is the truth of *dukkha*, the first noble truth. This truth, this obvious existential fact, invites the question as to how, if it is possible, *dukkha* can be overcome or alleviated. The Buddha's answer is that much of *dukkha* is caused by the grasping human ego (the second and third noble truths) and that this grasping ego can be overcome or tamed by practices of wisdom (seeing what matters and what doesn't), practices of meditation and mindfulness (to still the grasping ego), and ethics, culminating in compassion and lovingkindness for all sentient beings (the fourth noble truth). Buddhist scriptures are not the word of any God; they are compilations of "recipes" for overcoming suffering, primarily by overcoming the grasping ego, the source of most suffering. These facts may explain why Buddhism is so attractive in "spiritual, but not religious," precincts in the West.

Grounding in Native American Philosophy

(I. is the grand chief of the Anicinapek Tribe in Northern Quebec, Canada)

Grand Chief I. describes the sacred ground as understood by the nomadic *Anicinapek* peoples in terms of human relations with the natural world. The root metaphor of Anicinapek philosophy is that "in nature everything is circular." The Four Cardinal Points are taught by the elders of the Anicinapek First Nation to the youth as a way of orienting oneself both geographically and morally: (i) East: the sunrise teaches us that "everything can always start anew" and thus the ever-present possibility that one can "reinvent oneself"; (ii) South: the abundance of the sun at its zenith teaches that "we must first learn to respect oneself before we can respect others"; (iii) West: the sunset teaches us to let things go and accept departures; (iv) North: The purifying cold wind of winter evokes experiences and possibilities of "healing, peace, and inner freedom." Anicinapek philosophy sees nature and its work as sacred ground. Nature teaches, and humans are called to accommodate its teachings. The sun and moon are the divine grandmother and grandfather. Anicinapek ethics is grounded in the wisdom of nature.

Grounding in Confucianism

(J. is a philosopher and Confucian scholar)

J. writes,

> The most salient characteristic of the Sacred in Confucianism is its concern for the welfare of humanity. Early Confucians believed that Heaven was conscious, designed the world to benefit human beings (primarily by instilling in them a moral heart-mind), pays attention to the lives of human beings (so that those who have been wronged can appeal to Heaven), and on occasion acts in the world to nudge forward the cause of the good. The value Confucian thinkers put on recognizing and embodying oneness and establishing harmony among "all under Heaven" points toward a form of life consistent with the SDGs. For example, the early Confucian Xunzi (fourth century BCE) argued for the need to establish a happy symmetry among human beings and between humans and the natural world.

He adds,

> While Confucians regard Heaven (*Tian*) as the ultimate source of ethical warrants, these are not conceived of in terms of commandments handed down by a

lawgiver who stands apart from and injects order into a wholly separate world. In other words, there is no hint of God's aseity in Confucianism. So, there is no inclination to separate piety to Heaven from the good of humanity or the world. Rather, Heaven is conceived more in terms of being the ultimate ancestor of humanity, whose interest is inseparable from his family and the world in which they live. This is why, from the very earliest periods, there was a tendency to see the different senses of the word 天, which can mean Heaven, the heavens, or Nature, as overlapping or interpenetrating but never wholly distinct.

Grounding in Secular Ethics

(K. is a professional philosopher)

Proponents of secular ethics do not ground public morals in supernatural or sacred sources because they believe that there are no supernatural sources, that humans cannot communicate with or know such sources, or that common ethics or political life should not be based on parochial, non-shared, non-naturalistic/supernaturalistic metaphysics or theology.

A true believer in some religion might advocate a secular grounding of morals and politics because they see that only if there is an "overlapping consensus" at the level of basic values, rather than at the level of shared religious belief, which is not in the cards at present, can there be social order and harmony. Indeed, many secularists are true believers, and many are agnostics or atheists. (Note: Most theists are a-theists about all religions but their own.)

K. introduces four modern secular ethics: Neo-Aristotelian character-based virtue theory, Kantianism (deontology), utilitarianism/consequentialism, and socialism. The first grounds morality in human nature, in the universal psychological fact that we all seek to be happy, to flourish, to achieve *eudaimonia*. The second grounds morality in the faculty of reason, which calls upon us to be consistent and universal, treating everyone as an end worthy of dignity and respect. The third grounds morality in the recognition that one's concern for one's own well-being warrants concern for the well-being of all (including future generations, and possibly all beings capable of experiencing pleasure and pain). The fourth grounds morality in the rightness of equal ownership of the earth's bounty and equal responsibility for the earth's well-being. Equality of well-being is warranted because unequal distributions are the result of amoral (sometimes immoral) forces such as good and bad luck (in matters of location of birth, social class, and geography) and unequal exercises of power and domination.

Other forms of secularism, such as egoism (every ego for itself), relativism (individual and cultural) and nihilism (no ethical values can be rationally justified or nothing matters ethically), suggest that secular grounding is not ontologically deep enough to sustain a concern for the common good. One response is that some varieties of religious ethics (possibly neo-Aristotelian virtues ethics as well) are, in practice, tribal and parochial and not genuinely universal in scope, whereas enlightenment philosophies such as Kantian-Rawlsian deontology, Bentham-Mill-Singer-style utilitarianism, and Marx-Engels socialism are universal in scope, and, in addition, offer public reasons why we should care for all humanity.

Secular moral theories vary in terms of whether they ontologically ground morality in reason, human nature, prudence, social contracts, culture, or admixtures. The advantage of each, it is said, is that it is accessible to all people; it is not esoteric or parochial, as is grounding in a theology or non-naturalistic metaphysics.

The Vision of the Good Person and Their Virtues

I asked my informants the following: What is your religious tradition's understanding of the "good person"? What spiritual practices has your tradition developed to support individuals in becoming more virtuous? Within your tradition, what role do spiritual communities play in supporting this process? What are the highest religious virtues (e.g., faith, hope, love, mercy, compassion), and how do they strengthen the development of the whole person?

In this section, I provide an overview of the main virtues endorsed by our sampling of traditions. We can think of the virtues, both spiritual and ethical (often admixtures), described in this section as warranted or entailed by whatever it is that grounds morality in each tradition as discussed in the previous section. The motivational force of a morality in some large measure results from, first, the attachment to or conviction (creedal and/or emotional) of what it claims is the ground or source of morality, and to the moral community's effectiveness in inculcating its moral beliefs and favored virtues, principles, values, norms, and rules, and then sustaining that morality generationally. Remember: My aim in this chapter is to provide an inventory of the moral sources available to us diverse modern people. I am not offering an assessment of how well actualized the available sources are in the hearts and minds of the people who claim to be adherents of these traditions.[2]

I concentrate specifically on the picture of a good person, an individual of good character, where character is conceived as comprising an interactive set of dispositions to perceive, feel, think, judge, and act as a good person would. The specific dispositions are the virtues. Readers will see that many informants use the term *virtue* in a wider sense than is standard among virtue ethicists to refer to goals and states of mind (such as a sense of duty), as well as consequentialist and deontic principles that are also part of the fabric of their moral conceptions.

Islamic Virtue

In discussing the sacred ground of morality in Islam, both informants insisted on the complete interpenetration of the sacred with the secular, such that there is no autonomous zone of the secular, or better, everything is sacred. For this reason, what we might want to think of as a distinction between theological and moral virtues is not clearly demarcated in Islam.

According to A., a good Muslim abides spiritual practices that mostly involve "living a prayerful life." The Five Pillars of Islam are the bundle of practices that make up prayerfulness: the *Shahadah* (declaration of faith), *Salat* (prayer), *Zakat* (charity), *Sawm* (fasting during Ramadan), and *Hajj* (pilgrimage to Mecca).

A. says goodness involves being "compassionate, merciful, forgiving, resilient, nonjudgmental, God-centered, resilient, and hospitable."

B. lists seven religious virtues: belief in one God (*tawhid*); God-consciousness (*taqwa*); integrity through moral discipline (*qanata*); empowering courage (*batasha*); persevering in goodness (*sabara*); empathic altruism (love, mercy, compassion: *ihsan*); being fair and just (*adl*; also, right-minded: *rashad*).

B. writes that the character traits at which one aims include generosity, humility, loyalty, courage, forgiveness, and gaining mastery over the negative aspects of the self, such as greed, jealousy, anger, lust, and envy.

According to B., moral goodness (*murawwah*) involves *Wisdom* (faith in the One God), *Courage* (trust in the One God), *Temperance* (charity toward God's creatures), and *Justice* (putting reality in its proper perspective). A good person is fair and just and benefits others; such a person is at the stage of spiritual chivalry (*futuwwah*).

It is important to note that the virtues of wisdom, courage, temperance, and justice as described in the previous paragraph are not the same as those with the same names in Aristotelian ethics, other religious ethics, or in the varieties of secular ethics. The disposition to be, for example, courageous or just, is motivated by a proprietary conception of the divine (*Allah* is not the Trinitarian God) rather than, say, the aim of realizing human nature for its own sake. Also, the content of the virtues and actions that one ought to follow are different. For example, justice in Aristotelian ethics is not a matter of "putting reality in its proper perspective"; nor is temperance a matter of "charity toward God's creatures." This is a matter of considerable significance, as we think about whether and to what degree there is common ground in the virtues across traditions.

Jewish Virtue

According to C. (an Orthodox Jew and divinity school scholar), a good person in Judaism is "a human being acting in accordance with a divinely commanded

Law, and an awareness of a Lawgiver, within a community (for there is always a community) in which the Law is received and interpreted." Goodness in Judaism involves conformity to law and is not a matter of personal choice. C. says that Jewish ethics, despite its virtue ethical elements, is primarily a deontological system that follows the requirements of law or "commandments." The tradition also emphasizes virtues such as kindness, compassion, sharing, and care for the younger and the older. A complete understanding of the Law and the practice of the commandments (*mitzvot*) are the central goals of Judaism, for which virtues such as modesty, control of one's passion, and empathy are important means.

D. (an internationally famous rabbi who works on interreligious dialogue) defines the good person as "the one who contributes to human flourishing in keeping with the Divine Attributes." He emphasizes loving one's neighbor as the supreme principle of Judaism. A good person loves their fellow human beings just like God does.

C. points out that the holiday of Yom Kippur gives each individual a chance to reflect on one's practice of *midot*, or virtuous behaviors. The reflection includes first asking our fellow humans for forgiveness before asking for forgiveness from God.

Because the sacred–secular distinction is not at home in Judaism, neither C. nor D. makes a sharp distinction between religious and moral virtues. C. says that justice and judgment are key civic virtues, and D. lists love, mercy, righteousness, and justice as the highest virtues in Judaism. Interestingly, both emphasize the prohibition on idolatry in Judaism, although it is not a "virtue," strictly speaking.

Christian Virtue

E. (a Catholic cardinal from the Global South) says that the good person in the Christian tradition is righteous, which is expressed in a life of love of God and love of neighbor. This life involves seeking and doing the will of God and obeying his commandments, which are understood as reflecting the natural law.

According to E. and F., liturgy and the regular practice of prayer, self-awareness, and self-discipline are required for a morally good life. F. (an American evangelical leader) emphasizes that Christian moral development requires faithful obedience to Christ and his word as expressed in the sacred scriptures.

E. lists love, expressed in compassion, care, and forgiveness as the central personal virtues in Christianity and adds that "love means loving your neighbor as God loves him." Charity is the highest religious virtue: "Faith, hope, and charity; and the greatest is charity" (Saint Paul). Christian moral virtues also include honesty, truthfulness, justice, and fairness.

F.'s central Christian virtues are "love, joy, peace, patience, kindness, goodness, faithfulness, gentleness, [and] self-control." F., like E., sees love as the highest religious virtue, which serves as the foundation to all other virtue.

Hindu Virtue

Hinduism claims that there are four appropriate goals for human life, which, taken together, are necessary for flourishing: *dharma* (leading a virtuous life; meeting one's obligations), *artha* (having resources to enable the flourishing of humans and the sustenance of other living beings), *kama* (harmonious social relations, friendship, and love), and *moksha* (freedom or liberation from wants and self-centeredness).

G. lists the following as virtues in Hinduism: compassion, giving/charity, self-restraint, a sense of duty, steadfastness, rigorous self-inquiry, compassion, forgiveness, and a constant search for oneness with the reality in which we find ourselves. These five virtues are also highly prized: *ahimsa* (nonviolence), *satya* (truth), *brahmacarya* (control of the senses), *asteya* (non-stealing) and *aparigraha* (non-possessiveness). Among these virtues, *ahimsa* or nonviolence—non-exploitation in any form—is the highest moral virtue, and its spirit infuses much Hindu ethical thinking.

Various spiritual practices that help cement commitment to this complement of virtues include festivals for worshipping rivers, trees, and Mother Earth, as well as religious sacraments related to childbirth and ancestor worship. Finally, many Hindu texts speak of situations in which virtues conflict with each other, rather than denying the possibility of such conflicts.

Buddhism and Virtue

Buddhism conceives of morality as grounded in taming the ego in the service of alleviating suffering (*dukkha*) and possibly to bring happiness in its stead. There are two kinds of virtues in Buddhism: the conventional and the perfectionistic. The conventional virtues (*sila*) are explicitly mentioned in the fourth noble truth and include honesty, no gossiping or backstabbing, sexual propriety, no use of intoxicants, and no work that harms any sentient being. These virtues are supplemented by the four perfectionist virtues at which we should all aim:

- *Karuna:* the disposition to feel compassion for all suffering and the desire to help alleviate it
- *Metta:* the disposition to wish to bring well-being where there is ill-being
- *Mudita:* the disposition to feel sympathetic joy rather than envy for others when things go well for them, even in what are called "zero-sum games"
- *Upekkha:* the disposition to feel equally for all others (as one feels for oneself and those near and dear) and to accept with serenity how the world unfolds

Many Buddhists take vows to achieve these virtues. In Buddhism, meditation and mindfulness are not meant to be used as personal hygiene for high-strung people who wish to take the edge off their high-achieving lifestyle. The aim of meditation, like everything else in Buddhism, is to alleviate the suffering (*dukkha*) of all sentient beings. Meditation on the impermanence of the self and its desires is therapy that prepares and then sustains a life of work to overcome the greedy, rapacious ego and thus overcome suffering for oneself and others. A *bodhisattva* is an enlightened person who works for the elimination of suffering for all sentient beings.

Native American Anicinapek Philosophy and Virtue

According to I., the Anicinapek people's great understanding of their relationship with nature enabled them to survive in the forest for millennia. What is fundamental to their teaching is knowing how to observe nature, to respect it, to love it, and to feel loved by it. As shown in their language—for example, they call the moon "grandmother" and the sun "grandfather"—they feel connected to the gifts of creation in a filial way, full of love. Through their relational philosophy and experiences, they became "masters in the art of loving together," which is a matter of life and death to them.

Prayer that comes from "the heart, according to the inspiration of the moment," is emphasized as an example of an important spiritual practice in this tradition. Silent retreat time in nature is also encouraged because nature is our teacher. Some of the ceremonies give a chance to take a break and build common connections (e.g., sunrise ceremonies or sacred pipes ceremonies), and others celebrate the great passage of existence (e.g., the welcome ceremony for a new child and the ceremony of departure to the spirit world). Nature is a circle, with entries and exits, and ever-present prospects of renewal.

The main virtues inculcated and endorsed by the Anicinapek people are attunement to nature and love and respect for Mother Earth and fellow humans.

Confucianism and Virtue

J. writes,

> There is a variety of terms designating different levels of moral achievement (e.g., the worthy, the noble person, and the sage), but the latter two are most often used to designate the highest level of spiritual achievement. The noble person or sage is very much conceived of in terms of the consummate human being: someone who fulfills a variety of different, interlocking roles (e.g., father, son, husband, official, friend, etc.). The noble person or sage is characterized by a set of virtues, which include importantly abiding the rites/rituals/ceremonies,

respecting the elders, and filial piety. But care or benevolence (*ren* 仁, appropriately translated as "humanity") is presented as the key or dominant virtue. In order to be fully humane, one must care for all the world but in a graded fashion, beginning with one's own family members.

Secular Ethics and Virtue

Secular moral theories, both classical and modern, could reject the premises assumed in the questions posed because secular traditions are defined as not being religious traditions: What is your religious tradition's understanding of the "good person"? What spiritual practices has your tradition developed to support individuals in becoming more virtuous? Within your tradition, what role do spiritual communities play in supporting this process? What are the highest religious virtues (e.g., faith, hope, love, mercy, compassion), and how do they strengthen the development of the whole person?

The reason has to do with what I said about grounding in the first section of this chapter. Secular theories, of which there are many, attempt to avoid the grounding virtues, values, principles, and norms of religion as conventionally understood. They do so because they think there are no such credible sources or because they think that even if there are such sources, people will never agree on them—even within the Abrahamic triumvirate, *Yahweh* is not the trinitarian God, who is not *Allah*.

The good news is that the secularists can agree that faith, hope, love, mercy, and compassion are virtues. They are not religious ones, because they are not or need not be grounded in any supernatural source. The future is open, have hope for it, and do what must be done to make it better; suffering is real, attend to it, and so on.

Many classical Chinese and Greek philosophers—not religious in any modern sense—independently understood human nature as having a proper function (*ergon*) and a goal described in terms of some *telos*—for example, attunement to the good, the true, and the beautiful (Plato); *eudaimonia* (Aristotle); or becoming a *junzi* (a gentle person), who is benevolent, just, follows the rituals, is wise, and is a pious family member (Confucius, Mencius, Xunzi). According to K., while the moral virtues in these traditions are (almost) entirely relational and interpersonal, they were originally intended for relatively small, circumscribed communities, such as a *polis* or perhaps a nation-state, rather than the entire world.

A universalist secular ethics emerged forcefully in the European Enlightenment of the eighteenth century when ethical and political theorists self-consciously and explicitly began to speak of the scope of ethics as including equally all occupants of the planet (some parts still unknown), as well as future generations. This universalizing movement was no doubt catalyzed in the West by powerful resources in the Abrahamic traditions. But the project of the Enlightenment was to restate the rationale for the wisdom of the Golden Rule without invoking any divinity.

The dominant modern secular theories, deontology and utilitarianism, propose that virtues are important (both Kant and Mill have theories of virtue) but that universal principles such as the categorical imperative or the principle of maximizing utility are also required. Many other practices are also valuable, such as freedom of conscience and freedom of speech, which are not virtues (although we can embody these values virtuously or viciously). The virtues most prized in secular precincts in the West are likely justice as fairness and compassion/kindness. And if one were to provide a complete inventory of secular virtues and vices, one might reasonably expect to find almost that each is endorsed by the sacred and premodern theories discussed earlier; further, what was once and is now considered vicious would have considerable, but not perfect, overlap. Who would deny that love, compassion, benevolence, respect, temperance, moderation, and honesty are virtues? Thus, the secularist will endorse largely the familiar inventories of virtues and vice but for reasons different from those of the religious traditions. The secularist will claim that their reasons are public and shareable—as they depend on facts about human nature, prudence, rational consistency, social contracts, and discoveries about human flourishing—and thus can ground an overlapping and unforced consensus across traditions that are otherwise theologically and metaphysically at odds.

Contemporary political philosophy is dominated by discussions not of virtues but of the relative weight of such values as freedom, equality, justice, authority, benevolence, and justice. Everyone agrees that good character among citizenry matters. But the consensus is that this is not enough to make for a truly effective or robust universal or global ethics. In addition to teaching our children well, teaching them to be good, we will need, as political societies, to work out complex views about complex matters of value, and put these into effect as matters of public policy.

For example, everyone agrees that we want children to be just and to give others an equal chance. We may know what this means, more or less, in close interpersonal relations, for example, on the schoolyard, and thus how to teach our charges to behave. But at the social level, citizens must develop action-guiding theories about justice, fairness, and equality in at least four arenas: distributive or economic justice, procedural justice, restorative justice, and retributive justice. Does justice require equality in opportunity or in outcome? Liberals say equality of opportunity; democratic socialists say equality of outcome. Similar disagreements exist on the content of types of justice that are not solved by virtuous individuals. These require complex worked-out policy decisions, as well as action plans or procedures.

We know that people who are judged as reasonable, fair, and compassionate disagree about such matters at the political level. One cannot solve these problems inside any virtue theory. One can resolve them perhaps at the level of public philosophy, and then return to virtue theory with recommendations for how to build the requisite dispositions in the youth. A virtuous citizenry is a good thing, but one needs commitment to explicit, well-defined values and principles to ensure the desired outcomes outside of close interpersonal relations.

One weakness of secular moral theory as compared to the religious traditions is that secularists have no places to go—no synagogues, mosques, temples, sanghas, or churches—and no common rituals to affirm their core beliefs and principles or to collectively inculcate the young in a common morality. Schools have largely abandoned teaching ethics, in part because of the great success of secularist vigilance to keep out religious ethics out of schools. But this may have been too much vigilance. The problem to watch out for in societies that see the need to separate church and state sensibly involves resistance to any attempt to impose a sectarian view of the grounds of morality (requiring creedal agreement to a specific theology and its attendant rites and rituals), as well as wariness about imposing nonshared moral beliefs that require that theology. But it does not require similar vigilance about teaching a common morality endorsed across a multiplicity of sacred and secular traditions, if there is one.

The Common Good and the SDGs

In the first part of this chapter, I reported my colleagues' views of the grounds of morality. In this section, I link those views with the tradition-specific views of virtues, both spiritual and moral, insofar as that the distinction is made, and to the traditions' pictures of a good person. I then discuss how the picture of good or virtuous persons links to ideas of a good community, nation-state, and, ultimately, world. In particular, I am interested in whether our informants think their traditions have the moral resources to achieve the SDGs.

I asked my colleagues the following: What is your religious tradition's vision of the "good society"? What are the key characteristics of the good society? What is your religious tradition's understanding of the "common good" that is essential to building the good society? Does achieving the common good entail achieving the SDGs?

Islam, the Common Good, and the SDGs

B. sees fairness and justice as the core of the good society in Islam. The common good for society is the *maqasid al-shari'ah* (universal purposes), which include protecting and promoting religion and life; respect for the environment, the mind, and the family; the right to wealth; and *haqq al mal* (economic justice). According to B., the civic virtues in Islam will be conducive to achieving the SDGs if Islam's universal principles for governing society and spiritual chivalry for the individual are emphasized. According to B., solidarity in the Muslim tradition is directly related to SDG 16: "Promote peaceful and inclusive societies for sustainable development, provide access to justice for all and build effective, accountable and inclusive institutions at all levels." One wonders whether the goals of "promoting religion" and "promoting peaceful and inclusive societies" are compatible. This is a recurring problem for all religious traditions.

A. says that a good society is achieved only when good individuals are committed to the greater good of the collective, and when the individual and the collective are intertwined. The common good is often explained in terms of *Hasanah* or *Khayr*, which express the thought that every good thing achieved is ultimately done so for the sake of achieving the communal, not individual, good. A. emphasizes the relationship between a good person and the common good by citing Muhammad: "You are not a believer if you go to bed full when your neighbor is starving." A. adds that Muslims' vertical connection with the sacred—achieved through various spiritual disciplines such as revelation, belief, prayer, worship, and creedal rectitude—is not complete without the horizontal connection with fellow human beings, especially those who are less privileged. As the concept of neighbor has expanded to include the entirety of the global human community, A. says, Islam commits the believers to solving global and social issues, including environmental problems. He mentions the concept of *Waqf*, which involves religious philanthropic initiatives, as a promising sphere of public civic virtue in Islam that might be expanded and tapped as a way of achieving the SDGs

Judaism, the Common Good, and the SDGs

C. (an Orthodox Jew and divinity school scholar) insists that justice is the key civic virtue in Judaism and says that a good society—in which wise Jews can live—should have just organizations and internal systems, including a court of justice, a communal charity fund, and a synagogue. In Judaism, every Jew is responsible for every other Jew. In this tradition, humans are responsible for taking care of the earth as proper stewards. The natural world is not sacred in Judaism (in contrast to Hinduism and Native American philosophy). Nonetheless, humans should take steps to make the world a safer and fairer place for human habitation and fecundity. C. states that it is notable that Judaism (in its founding revelations) is not universal in scope and is focused on the good of Jews, although it teaches Jews to treat non-Jews kindly as well.

To describe a good society in Judaism, D. (an internationally prominent rabbi involved in interreligious dialogue) contrasts it with the society of Sodom and Gomorrah, which was full of greed, selfishness, and exploitation and a lack of hospitality and sensitivity to others. The idea of the common good, according to D., is expressed in the Hebrew Bible in "the ordinances for the care and maintenance of the poor and vulnerable, who are the responsibility of society as a whole." This leads in turn to injunctions for the agricultural system as a whole. It is remarkable that Judaism emphasizes structures securing the economic and social good of the community or society as a whole. Like C., D. also highlights the inextricable relationship between an individual's flourishing and that of society as a whole. The remark that "when one loves God it is impossible not to love His creatures" shows how the close relationship among God, creation, and human society in Judaism can provide a source to address environmental and developmental challenges. As Jewish civic virtues that can help to achieve the SDGs, D. mentions extensions of existing practices such as *Tzedakah* (charitable tithing).

In relation to the SGDs, C. says that Judaism can see achieving the SGDs as a matter of paying debt, since "the naturally occurring world is unequal and unjust in its terrors and grace, [and] the task of the Law is to make it just." According to her, we should help the poor because any surplus is the result of divine order and belongs to the poor.

Christianity, the Common Good, and the SDGs

E. (a Roman Catholic cardinal from the Global South) characterizes a good society in Christianity as one in which "people live together in justice, peace, and harmony, under God," which involves teaching each generation a common morality of love and the Golden Rule. Because all humans are a family under God, Christians seek the well-being of one other and the achievement of the virtues that are necessary for a good life. E. sees God as the primary reference point for the common good. He mentions the virtues of solidarity and care for others as important for achieving the good society. According to E., Christianity is well suited as a resource to advance the goals of the SDGs, since "love thy enemy" expresses the idea of limitless solidarity.

For F. (an evangelical Christian leader in the United States), a good Christian society is one that is compassionate, just, and humane but not necessarily officially Christian. A good society has a social structure that helps individuals pursue their penultimate ends (not the ultimate end, which is eternal reward)—"to treat one another fairly, to ensure that families are allowed to thrive, to make room for private industry, to limit the scope of the state from enveloping all areas of society, to allow individuals to order their lives in accord with their conscience." Christians pursue the collective welfare of society based on the belief that all people are created in God's image. F. lists these civic virtues (or attitudes or values) as necessary for building the common good: seeing every person as made in God's image and deserving of love, respect, kindness, and forbearance; principled pluralism and religious liberty; civility; justice; equality; dignity; a limited role for the state's involvement in human affairs; an appreciation for a doctrine of human rights (emanating from Christian anthropology); and stewardship. These virtues will help Christians to assume responsibility for achieving the SDGs insofar as they understand how love for the world is entailed by the authority of the Bible.

Both E. and F. believe that Christian convictions will help us become good people who contribute to achieving the SDGs. F. believes that Christians will become aware of and attentive to the world around them and care about remedying injustice. They will want to advance the SDGs to the extent that they are responses to human brokenness.

F. writes that Christians believe that we are commanded to "seek the welfare of the city to which God has called us." He explains the Christian idea of subsidiarity as the responsibility we have to people who are near to us, as we are not omnicompetent: "Proximity determines responsibility." He finds a close link between the good person and good society in Christianity and says that a society is good "insofar as it encourages or nourishes the virtues and ethical conditions necessary for communal flourishing." Like E., he finds the Christian idea of solidarity well expressed in the lesson of loving our neighbor and being a Good Samaritan.

Hinduism, the Common Good, and the SDGs

G. depicts the ideal society in Hinduism through the epic of *Ramayana*. Ram Rajya (the kingdom of Rama) is founded on the principles of truth and compassion, and there, hunger, ill health, and environmental degradation do not exist. Gandhi derived from this ideal a vision of *Swaraj* (self-rule), which encompasses a vision of a just and inclusive society. According to G., the Hindu belief that the sacred is present in everything grounds concern for everything. The concept of *karma* (actions and the consequences of actions) also promotes right actions. According to G., Hinduism values each individual as a unique person who constitutes an important part of the whole community, based on the polycentric nature of the Hindu conceptions of the good person and the good society. The common good in Hindu tradition is encapsulated in the idea of *Vasudhaiva Kutumbakam*: the whole earth is your family. G. emphasizes that Hinduism sees humans as part of the environment and never sees nature merely as an object of human use.

As the civic virtues directly related to the *dharma* principle in Hinduism—such as *seva* (service) and *danam* (giving)—G. mentions giving, volunteering, philanthropy, and stewardship as examples. The Hindu perspective considers the guest as divine and a life as one of repaying sacred debts, which grounds prosocial actions. G. mentions Vinoba Bhave's *sarvodaya* (uplift of all) movement as an example of aims inspired by Hinduism aligning with the SDGs. The Hindu values of nonviolence, giving, sharing, truth, civility, and helping the destitute are at the core of a good society and can contribute to achieving the SDGs.

G. emphasizes the value of "moral heroes" in advancing noble ends. Moral heroism on the part of some visible leaders may be essential to raising more awareness of issues affecting some of the SDGs, such as caste, gender, and class inequities. For example, Gandhi, a moral hero who pursued *moksha* (spiritual freedom), lived for India's political freedom from a foreign regime and fought against social injustices and inequalities.

Buddhism, the Common Good, and the SDGs

It is noteworthy that the secretary-general of the United Nations at the time the SDGs were unanimously enacted was Ban Ki-moon. Ban has not shared his personal religious views, but his mother was Buddhist. So it was with some familiarity that, in 2016 on the Day of Vesak, the highest Buddhist holy day, a year after the SDGs had been enacted, he said, "The fundamental equality of all people, the imperative to seek justice, and the interdependence of life and the environment are more than abstract concepts for scholars to debate; they are living guidelines for Buddhists and others navigating the path to a better future."

Ban went on to cite the story of Srimala, a woman who pledged to help all those suffering from injustice, illness, poverty, or disaster, and suggested that Srimala's spirit could animate global efforts to realize the 2030 Agenda for Sustainable Development and the Paris Agreement on climate change and to promote universal human rights.

Buddhism, recall, calls on all people to enter the bodhisattva's path by vowing to self-cultivate the virtues of compassion (*karuna*) and lovingkindness (*metta*), which are entirely universal in scope. The vow is to alleviate suffering for all sentient beings and to work to bring well-being it its stead. Buddhism encourages vegetarianism for this reason, and in addition, perhaps uniquely among modern spiritual traditions, calls out acquisitiveness and greed at the point an individual has enough. After that, especially in the present state of the world, more is too much.

Native American Anicinapek Philosophy, the Common Good, and the SDGs

According to the notion of circles in Anicinapek philosophy, a good society is one in which everyone has a role with specific responsibilities and exists for the common good. I. compares this idea to concentric circles. The sacred objects are at the heart of these circles, and the next around them are the guardians of the sacred objects: "male and female elders, medicine men and medicine women, designated to protect, share, and convey our philosophy and traditional medicine." Around the guardians is a circle of children and then a circle of mothers, supported by a circle of grandmothers. The men, who protect all those in their circle, form the last circle around all circles. This structure shows how the members of good Anicinapek society take care of each other with specific, proud responsibilities entrusted to each member. Grand Chief I. says proudly, "We feel that [the SDGs] are finally joining what we have always believed!"

Confucianism, the Common Good, and the SDGs

J. reports that the good Confucian society is peaceful, prosperous, and characterized by feelings of care and concern. The primary model of the ideal Confucian society often has been a loving and harmonious family writ large. In the later tradition, this was augmented by the idea of the universe as a vast single body with one's consciousness caring for and seeking to nurture each and every part. This idea is described in moving detail in the "Evolution of the Rites" chapter of the *Book of Rites* in terms of the Age of Great Harmony.

As for a source in Confucianism (it matters that Xi Jinping, the president of China, has been reclaiming Confucian roots for the People's Republic of China) that can support the SDGs, the classical Confucian Xunzi called for a happy symmetry and argued, based upon this ideal, for policies that strictly limited the seasons and amounts of natural resources that could be harvested each year. The purpose of such policies was to ensure that both human beings and the natural world could flourish in a sustainable way. The Ming dynasty Confucian Wang Yangming (1472–1529) argued, "The great person forms one body with Heaven, earth, and the myriad things." The message of this statement is that we should care for and nurture each and every thing in the world just as we care for and nurture each and every part of our own bodies.

With respect to framing the moral project as growing weak seeds of fellow feeling into robust concern for the well-being of all, J. says this of the Confucian view:

> Solidarity is a concept most at home in cultures that share a fairly robust conception of individualism. We don't normally argue for the need to establish "solidarity" among family members or among the various parts of one's own body. The very conception of a family implies a prior commitment to the various "subsidiary levels" within these larger wholes. . . . there is a strong holistic

or organic connection between the self and the rest of the people, creatures, and things in the world. A proper understand of the self entails caring in graded but still significant ways for all other people, creatures, and things. One's own good and happiness is not wholly separable and to a large degree *requires* working for the good and happiness of the rest of the world.

Secular Ethics, the Common Good, and the SDGs

What about secular ethical resources to advance the common good in general and the SDGs in particular? The good news is that the SDGs are, in fact, a perfect example of the idea that John Rawls referred to as an overlapping consensus and that Charles Taylor calls an unforced consensus. This idea favors finding common ground, first at the level of the bottom line; for example, everyone should receive a high school education, sexual slavery is wrong, and poverty should be eradicated. The grounds upon which an agreement is based do not matter; for example, God commanded it, it is smart prudentially, it will save us money, we promised each other, we will be happier.

Such an unforced but not unanimous consensus was achieved in the 1948 Universal Declaration of Human Rights, which is mentioned in the SDGs as playing a grounding role: "We reaffirm the importance of the Universal Declaration of Human Rights, as well as other international instruments relating to human rights and international law. We emphasize the responsibilities of all States, in conformity with the Charter of the United Nations, to respect, protect and promote human rights and fundamental freedoms for all, without distinction of any kind as to race, colour, sex, language, religion, political or other opinion, national or social origin, property, birth, disability or other status."

Unanimous agreement in 2015 on the SDGs by the UN's 193 member states shows that, beneath any creedal disputes or disputes about ritual rectitude, about who is chosen and not, whether there is an afterlife, or about what the afterlife is like or who reaps its rewards, there is an ethical and political consensus on the values embedded in the SDGs. Few countries in the world are theocracies, but as discussed earlier, the world's great spiritual traditions offer resources to extend concern for the well-being of one's own ethnic or religious tribe to the well-being of all, as the SDGs require.

Three of the great secular moral theories—deontology, utilitarianism/consequentialism, and socialism—straightforwardly entail the SDGs because they avow universal rights and principles of equality and justice that are universal in scope. That said, fundamental disagreements exist in secular quarters of nation-states such as the United States about what, for example, justice demands. Libertarians will say it demands maximal freedom, liberals often favour equality in opportunity, and socialists will call for equality of outcomes.

According to K., secular philosophers generally agree that having virtuous citizenry is one thing, and political justice and equality are another. For example, a sage ruler might be able to impose right principles of justice and universal sharing of the earth's bounty regardless of whether the citizens have a virtuous sense that these principles are right and good. However, it is likely that a just society will be best achieved and sustained, and the SDGs will be best abided, only if citizens are virtuous and rationally committed to certain explicit principles of global well-being, and if these principles are avowed and enacted at the levels of both the nation-state and its foreign policy.

Conclusion: Resources and Obstacles for the SDGs Across Traditions

In this chapter, I have provided an overview of some of the moral capital that is available across a sample of sacred and secular moral traditions as described by experts and that can be tapped to energize efforts to achieve the SDGs. There is a fair amount of good news in terms of the moral resources of various sacred and secular traditions that endorse love, compassion, justice, and the flourishing of humanity. William Vendley, the long-time director of Religions for Peace, sees an emerging consensus among religious leaders for what he calls shared well-being. This is good, if true. Better than the alternative.

It is worth noting, however, that this emerging consensus on the ethical goal of shared well-being, which is also expressed in secular political terms in the UN's unanimity on the SDGs, does not mean that the moral and political capital necessary to achieve the SDGs actually exists. The reason for this becomes apparent when the leaders of various traditions are asked to identify ideals and high-minded values available in their traditions. These ideals and lofty goals compete with other forces internal to almost every religious tradition to think of one's fate as shared first and foremost with co-religionists, as well as with political realities and ideologies that commonly conceive nation-states as the units of concern and nation-state borders as the places where moral obligation ends. Furthermore, many people, possibly most people on Earth, including members or advocates of every secular and sacred tradition discussed in this chapter, likely, and in some sense rightly, think it empirically false that the well-being of each depends on the well-being of all. One has to truly believe, to embrace as a normative conviction, that one is not well while others are ill, for this truth to be so.

Sometimes, the scope of moral concern is not a nation-state but a particular tribe, religion, ethnicity, or identity group within a nation-state, or one that exists across nation-states. Solidarity with and focus on the well-being of a nation-state, tribe, ethnicity, or identity group can make sense for a host of reasons, including practicality and rectifying past injustices. But this narrower focus is also a way of saying that it is each boat on its own bottom, that while each boat would wish all other boats to stay afloat, whether each one does is up to its passengers.

Another question about the moral capital available to achieve the SDGs pertains to whether love and generosity or justice are the moral resources we tap. Shared well-being could be conceived as valuable but not obligatory. It is good for individuals, groups, nation-states, and international organizations to be generous. Charity is good, but it is supererogatory.

The demands of justice are typically more onerous than those of maintaining a spirit of generosity. Justice requires well-being to be shared and ill-being to be mitigated across the earth. Secular sources generally do better at theorizing justice with this sort of universal scope than do their sacred brethren, in part because secular moral and political philosophy were created to address problems of global justice in a world of cultural and religious differences.

One virtue needed for achieving the SDGs that was not mentioned by my expert informants is sacrifice: sacrifice by individuals and sacrifice by nation-states. Almost all traditions discussed in this chapter have theories about the importance of sacrifice, but it is a virtue that needs emphasizing and rehabilitating. One reason for this is to help overcome the situation just mentioned: While sacred and secular moral theories reach consensus about universal care, nation-states theorize partial care for themselves, religions are often parochial, most ethicists live upper-middle-class lives, and very wealthy individuals offer private philanthropy as a matter of generosity, rather than as a matter of justice. Furthermore, the amount contributed to the well-being of others is almost never thought to require a nation-state to lower its standard of living. In the case of most preachers, professional ethicists, and private philanthropists, the well-being of others would certainly not be thought to require a sacrifice that would have a serious impact on one's lifestyle. In addition to following the collective wisdom of our sacred and secular traditions on love, compassion, justice, and shared well-being, we may need to reemphasize the moral obligation of the most fortunate, including the most fortunate nation-states, to make genuine sacrifices if we are to achieve the SDGs for all. The need to attend to moral resources across traditions for the virtue of sacrifice is vivid in the preamble to the declaration of the SDGs: "As we embark on this great collective journey, we pledge that no one will be left behind. Recognizing that the dignity of the human person is fundamental, we wish to see the goals and targets met for all nations and peoples and for all segments of society. And we will endeavor to reach the furthest behind first."

Notes

1. "Transforming Our World: The 2030 Agenda for Sustainable Development," UN Department of Economic and Social Affairs, September 25, 2015, https://sustainabledevelopment.un.org/post2015/transformingourworld.
2. There are such assessments. See Christian Miller's two volumes: *Moral Character: An Empirical Theory* (New York: Oxford, 2013) and *Character and Moral Psychology* (New York: Oxford, 2014). The actual state of moral character is not very good. The quality of character in terms of conventional virtues, such as honesty, takes a normal, bell-shaped distribution. A small percentage of people are very good, an equally small percentage of people are reliably bad, and most of us are OK but erratic.

CHAPTER 4

SECULAR ETHICS, MORAL CAPITAL, AND THE SUSTAINABLE DEVELOPMENT GOALS

OWEN FLANAGAN

Secular Countries and Well-Being

Once again, in the 2019 *World Happiness Report*, the North Atlantic, specifically Finland, Denmark, Norway, Iceland, and the Netherlands, leads the world. Happiness is measured by a mix of subjective data—how people feel about or judge the quality of their lives—and by objective quality-of-life factors such as health, quality of air and water, economic equality, the strength of the welfare system, education, social inclusion, and governmental integrity and efficiency. Together, these subjective and objective factors contribute significantly to human flourishing, what Aristotle called *eudaimonia*.[1] In the *World Happiness Report*, "happiness" is a convenient, if somewhat misleading, term for the combined set of subjective feelings and judgments—the happiness part (https://plato.stanford.edu/entries/happiness/)[2]—and objective conditions of sustainability and flourishing that individuals might not even be aware of as goods they possess—the well-being part (https://plato.stanford.edu/entries/well-being/). The happiest countries in the world—in this complex sense of happiness—are also the furthest along the way to achieving the SDGs, which the UN unanimously passed in 2015 as goals for 2030. Jeffrey Sachs has said, "It is literally the truth, that sustainable development is the path to happiness."[3]

Because the SDGs are not merely national goals but global ones, these happy countries cannot just pat themselves on their backs and wait for the rest of the world to catch up. Rather, the combination of their progress and achievements on measures of happiness, well-being, and sustainability, together with their legal and moral obligations as signatories to the 2015 SDGs agreement, demand that these high-well-being countries try to do for the world what they have (to some extent) done for themselves.

Moral Capital Without Religion

It is widely agreed that achieving the SDGs requires economic resources, creative public policy, and smart, but not always intuitive, technical solutions. The happiest countries are economically well off and have well-functioning governments. Thus, they are able to implement smart and successful public policies. It is also agreed that achieving the SDGs, and sustaining them, requires moral vision and moral commitment, what I'll sometimes call, following David Brooks, "moral capital." Brooks defines moral capital as "the set of shared habits, norms, institutions and values that make common life possible."[4]

The measures used in the *World Happiness Report* and the similar ones used to measure progress toward achieving the SDGs all embed and express moral and political values: that poverty is bad and should be eradicated; that all people deserve quality health care; that great disparities in wealth are bad; that education is a necessary condition for well-being, and so on. A large part of the rationale for the SDGs is moral, and thus it makes sense to think that moral capital must play a role in creating the energy and will to achieve them, and furthermore that moral capital will be required, human nature being as it is, to sustain the SDGs if and when they are achieved.

What forces in the world's happiest countries now produce the moral capital available to create and sustain the conditions for happiness and well-being and to produce forward momentum on the SDGs? Remember the five happiest countries are Finland, Denmark, Norway, Iceland, and the Netherlands. The next five are Switzerland, Sweden, New Zealand, Canada, and Austria. The answer is that whatever is producing the moral capital of the world's happiest countries, it is not a shared supernatural or transcendental vision. These countries, Austria aside, are among the most agnostic and atheist on Earth. (China is the most secular country and ranks ninety-third in happiness.)

This situation presents a challenge to those who think that moral capital requires religious sources and that religion provides the only, or the best, or at least good and sufficient, reason to be moral.[5] "Religion" comes from the Latin word *religio*, which means "to bind." In many parts of the world, religion is thought to bind a people by a shared theology, which normally involves belief in a deity or deities and a theory about the afterlife (or a series of afterlives). God or the gods (in their goodness and wisdom) create and/or warrant moral values; a good afterlife (or afterlives) is determined by the moral quality of one's life, not by wealth, fame, physical strength, or beauty.

It is a relatively common belief among religious people in the United States (America is one of the most religious countries on Earth and ranks only nineteenth in happiness) that atheists have no reason to be moral,[6] lack the motivation to be moral, and thus, given the way human nature is, cannot be trusted to be moral.[7]

We can sketch a logically valid argument that this is not so. A logically valid argument is one in which the conclusion of the argument must be true, it is necessarily true, if the premises are true:

(1) Achieving the ends measured in the *World Happiness Report* and those expressed as goals in the seventeen SDGs requires (in addition to money and technical solutions) commitment to the goals of ending poverty, environmental sustainability, health, ending human trafficking, education, and so on because they express important ethical values.
(2) The countries with the highest well-being (and those furthest ahead in achieving the SDGs) are secular, not religious.
(3) If the expressed moral values and political commitments of these high-happiness and high-SDG countries are involved causally in producing the good outcomes they have already achieved, then moral goals, moral courage, and moral commitment—moral capital—do not require theological or supernatural foundations.
(4) The expressed moral values and political commitments of these countries are involved causally in producing the good outcomes on measures of happiness and the SDGs.
(5) Therefore, high-minded moral goals, moral courage, and moral commitment—moral capital—do not require theological or supernatural foundations.

Some have suggested that the evidence warrants an even stronger conclusion, namely that secularism and secular ethics provide more reliable and efficient support for the progressive ends embedded in the SDGs than do sectarian religious ethics. In this chapter, I leave that possibility to the side,[8] and explain what secular ethics is and how its insights can ground and advance the aims of the SDGs.

Varieties of Secular Ethics: Four Types

"Moral capital" refers to all moral resources available to a people for use in motivating and cultivating a morally good life and for energizing moral perceptiveness, moral sensitivity, and moral progress. Moral capital includes whatever natural dispositions humans possess to live as gregarious and cooperative social animals, as well as regimens of moral education and cultivation (in families, schools, and the common social atmosphere), plus values, rules, and norms, invented or discovered, that are conducive to flourishing. Secular moral philosophers have located powerful sources of moral capital in three veins. One involves inculcating in each generation the moral habits or perceptual-cognitive-affective skills—called virtues—that previous generations have discovered are good and that suit us for a high-quality life in relations with other individuals. The other two veins involve locating and arguing for values and principles that can help us to be ethically excellent members of nation-states or, ultimately, of the world, and that are concerned with nonlocal well-being, the welfare of others with whom one shares a common fate but whom one likely will never meet.

Virtue Theories

Socrates's question is, "How shall I live? How shall we live together?" One answer familiar from classical Greece, Rome, and China is that a good human life requires good character, where character is conceived as comprising a set of intellectual and ethical dispositions to perceive, feel, think, and act in the appropriate way depending on the situation. On an uncrowded subway it is perfectly fine for a busy student to take a seat in order to continue reading Aristotle's *Nicomachean Ethics*. But once the train fills up and seats are sparse, and an elderly person with shopping bags boards, a virtuous student, despite being absorbed in reading, sees the person and gives them their seat. In a person of good character, this is almost automatic. Virtues are like the skills and virtuosity of a slalom skier, basketball player, cyclist, or rock climber who adjusts expertly to whatever subtle obstacles and novel situations come their way.[9]

A good person is benevolent when benevolence is called for, just and merciful when these are required, kind, honest, courageous, respectful of elders, and so on. A person is morally good if they possess the full complement of virtues. Furthermore, being virtuous is a necessary condition for living a good human life, for flourishing, for achieving *eudaimonia*. No one flourishes if they are not virtuous.[10]

The account of vice is the same. The moral well-being of the youth depends entirely on the moral health of the adults and the wider social ecology. If the adults are greedy, teach the youth that this is a dog-eat-dog world, and even that there are ways to insulate oneself (go to the finest schools, live in the safest communities, and make sure that governmental laws and tax rules are on one's side), the youth will get the message.[11]

When secular virtue theorists are asked what makes individual virtues, such as kindness, courage, fairness, and humility, good, and what makes a life of virtue the worthiest choice, they answer that virtues are excellences by which we achieve our human nature or, not incompatibly, they are ideal tools (or "good enough" tools) to achieve individual and communal flourishing.

Aristotle thought that the human end (*telos*) was to achieve happiness (*eudaimonia*) by developing one's intellectual capacities and one's own special talents, as well as becoming virtuous, where this meant possessing a unified set of dispositions to be generous, courageous, friendly, modest, and so on. Why? When you look around and survey the human landscape, you can see that the exemplary souls are the ones who develop their rational, creative, and moral capacities to the fullest. The development of our individual natural talents and our shared moral potential fulfills our nature. Just as good birds fly and sing, good wolves hunt in packs, and good bees pollenate plants and dance to tell compatriots where the nectar is, so too good humans develop and actualize their intellectual and moral potentials. A eulogy that honestly surveys a life and tells the story of a person who realized their potential and lived a life of exemplary virtue speaks of a *eudaimon* (*eu* = happy; *daimon* = spirit), an ideal specimen of humanity.

A different, but closely related idea, is that virtues are tools or instruments for successfully navigating social life given that the natural and social world present obstacles, as well as opportunities, to flourishing. Flourishing requires effort, ingenuity, compromise, and learning that putting off childish demands that the world satisfy our desires now brings greater happiness. Just as it is good to build a bridge or a boat to get across a river if that is where the food is, it is good to be compassionate and just in order to have maximally harmonious relations with others, and to make possible friendship and love (Aristotle says that no one would choose to live without friends).

On the first view, the warrant for acquiring and abiding virtues is that doing so realizes a potential that uniquely suits our kind of animal, *Homo sapiens*.[12] On the second, not incompatible, view, virtues are reliable tools, methods, and strategies for producing what we all sensibly want—to live without fear and insecurity and to be happy. On the second view, virtues suppress or rechannel dispositions to selfishness, while also amplifying good, but originally restricted, positive dispositions, for example, to love and share but only with close kin. Virtues are developed or inculcated in the youth because they are necessary conditions for peace and order, commerce, civic friendship, love, work, and *eudaimonia*. Virtues are means to these ends, and they are justified in terms of how reliably they produce these ends. The ends justify the virtues. But as often happens in such cases, virtues come to be valued for their own sake. And thus we speak of virtue being its own reward.

Virtue theory, as I have described it, is entirely secular and naturalistic.[13] Virtues are described in terms of psychological, sociological, and anthropological concepts, and the reasons offered for why being virtuous is good refer only to human nature, human relations, flourishing in this life and in this world, and things humans value for their own sake, such as happiness, friendship, and love.[14]

One might wonder, Would a world in which most people are virtuous also be a world in which the SDGs were endorsed by everyone, and in which achieving the SDGs had the moral capital needed to support them? My view is that such a world would not produce sufficient moral capital.

Virtues are dispositions in individuals. Achieving the SDGs requires social coordination, economic and technical wisdom, high-minded universal ethical values, commitment to universal action-guiding moral principles, and collective action. Social coordination, money, scientific and technological knowledge, and moral wisdom are not in individuals, or at best are only partly so. So virtues matter, but they are not enough.

One can imagine a nation of morally good individuals who are parochial and thus do not believe that their benevolence should extend beyond local precincts. And even if and when they have the thought that they should care or, what is different, share equally with all, they do not know how to be generous to geographically or temporally distant others.[15]

The widely recognized point is that a collection of good, that is, virtuous, people might not implicitly or explicitly endorse the SDGs, because the SDGs express values and goals that go far beyond the demands of ordinary virtue.[16] Many conventionally virtuous people do not believe in economic equality (most nowadays believe in equality of opportunity, not outcome) or in responsibility for the well-being of others beyond one's own nation-state borders. And even if a group of individuals held a global, rather than parochial, view of whose good concerns them, they might be lacking in the practical wisdom needed to solve complex problems of global collective action.[17]

These last few observations lie behind the idea that a nation governed by excellent laws and institutions can withstand, for a time, widespread vice and selfishness among its people. Once ethically progressive tax laws and medical and social security institutions are put in place, such laws and institutions can sustain themselves (but only for a while), even when popular sentiment turns greedy and vicious.

Deontology and Utilitarianism and Socialism

The natural thought is that virtues are not enough to create the conditions for happiness and well-being, or to create the moral capital to achieve the SDGs. In addition, principles are needed. The SDGs express and endorse values. Why did the 193 nations of the United Nations unanimously agree on the seventeen SDGs? Because the SDGs express universal values, worthy aims, excellent moral goals, and wise prudential ends.

Here, for example, are two values and principles embedded and expressed in the SDGs. The first principle posits human equality; the second posits an obligation to act so as to promote good lives for all:

- Every human has a right to live well; everyone is equally worthy and equally deserving of living a good human life.
- Each human has an obligation to help others live a good human life.

Versions of these two principles became commonplaces in the thinking of a certain secular elite during the European Enlightenment and post-Enlightenment from the seventeenth through nineteenth centuries. Versions of both are expressed in the thinking of the great utilitarians Bentham, Mill, and Sidgwick, who endorsed the following principle:

> One ought always to act to maximize the greatest amount of happiness for the greatest number of people in the long run.

Kant, the great deontologist, advocated the categorical imperative, which claims that reason requires the following:

> One ought never treat another person as a means or thing or instrument of one's will, but always as an end, a being worthy of dignity and respect for their own sake.
>
> One ought to do to another or for another only what one can will universally as the right thing to do.

Socialists like Friedrich Engels and Karl Marx put their universal maxim in these terms:

> There ought to be redistribution for the good of all from those with the natural and social luck to have plenty to the unlucky in need.[18]

My claim is not that these principles were discovered or invented by these secular thinkers. They were not. Versions of them appear in both secular and sacred sources. Mill, an atheist, says that utilitarianism endorses what Jesus of Nazareth endorsed—and one can hear the message of the "Sermon of the Mount" in the moral vision of socialism, whose founder, Karl Marx, thought religion was "the opiate of the masses." The point is that these universal principles, along with the language of natural rights and duties, emerged in modern times, and they are justified in natural—psychological, social, conventional, and prudential—terms, rather than in supernatural ones. Rights are natural; contracting or agreeing that there are rights minimizes pain and suffering and produces order and harmony; rights provide deontological (inviolable) constraints on the pursuit of the good so that no one becomes a means (e.g., a slave) to another's well-being; humans naturally want to be happy and to minimize pain; we ought to value consistency so that if I want to live well, I ought to acknowledge that desire in others; peace, order, and harmony are great goods, and so on.

The modern secular idea is not that these universal values or principles are warranted because God or the gods approve of them or made them good. Rather, it is that God or the gods approve what is good; they have good taste.[19] And even if there is no God, these universal laws or principles are true. Their truth depends not on any supernatural or transcendental sources but on features of human nature, human social relations, human ingenuity, and the earth and its resources.[20]

Political Virtues and Political Values

Values, norms, and principles are not virtues. Thus, the values expressed by the utilitarian (consequentialist), deontological (Kantian), and socialist principles are not virtues, although they could, as it were, be inculcated in the form of dispositions to think and act in ways that enact the relevant principles. Such virtues would be global in scope and thus something new under the sun. Until recent times, virtues have almost always been advocated as equipment suited to convivial life in a particular community or nation-state, not in a global community. Furthermore, as standardly conceived, virtues do not involve obligations as much as particularistic "know-how" and skills that embed (often implicitly) knowledge of local norms. The language of rights, duties, and obligations, especially global ones, is most at home and much more well theorized in the utilitarian/consequentialist and deontological traditions with their universalistic aspirations than it is in virtue theory.

A secular theorist who wants to see how far virtues alone can get you might work to create socialization schemes that try to teach the youth to have the disposition to perceive, feel, think, and judge that every situation that falls short of satisfying the SDGs ought to be remedied and to take action to do so. The virtue or virtues required would be heroic. An objection is that such a heroic conception would require an individual virtue or set of virtues to do too much work. Such a virtue would require superhuman cognitive abilities to track where the SDGs are or are not being met and practical capacities to make a difference that most ordinary individuals do not have.[21]

Another idea is that the virtue or virtues required to enact and sustain the SDGs are virtues of nation-states or international political institutions (and in the case of happiness and well-being measures, virtues of these international institutions plus local communities and nation-states), rather than of individuals. This suggestion raises questions of whether a nation-state or collection of states can be literally virtuous rather than metaphorically virtuous. Remember, a virtue is defined normally within virtue theory as a well-honed disposition in an individual—a habit or a skill—to perceive, feel, think, and act in the right way as a particular situations calls for.

Can a village, town, nation-state, or international body be virtuous? Can such entities be individuals and display what is called collective intentionality? It seems possible.[22] Plato, Aristotle, Confucius, and Mencius seem to have thought that both individuals and states can be virtuous. Not any old collection of individuals can be conceived as a collective individual. But such entities as fire and police departments, states with constitutions, universities with genuinely unified missions, and corporations with certain clear product lines and corporate aims and values statements seem like entities with aims, ends, and mechanisms of decision-making (not necessarily democratic) for which it makes sense to speak of the organization's values, goals, intentions, and principles. Such entities have the sort of structure for which it makes sense to ascribe common aims to them, which, for example, the entity composed of all the people walking on the sidewalk in New York City at this moment does not have.

Thus, it makes sense to say that America believes that all humans are created equal. It says this in founding documents. Do all Americans believe this? Maybe. But it doesn't matter as far as the claim goes, since the belief is ascribed to the higher-order entity—America—not to its parts, its citizens. If the United States reliably displayed, presumably through its institutions—for example, in laws passed by Congress and in Supreme Court decisions—the disposition to detect and correct unequal treatment whenever and wherever it occurs, then we might plausibly say that America is virtuous in this respect.

One might wonder whether America, the country, believes in the principle that all people on Earth are created equal and are all equally deserving of good lives. Let's suppose the answer is yes. Do Americans also believe that we individually and collectively have an obligation to do what is necessary, or what is different, all we can do to make those who are not well off better off? The answer we know is this: |First, people will want to know how much is required of them (how much overall and how much from each individual)? Once they are told, they will often deny that neither the individual nor the nation-state has an obligation to give that much. That much is too much. Once again, we confront a mismatch, disharmony, or disequilibrium between the values and principles embedded in and enacted by institutions and those endorsed by individuals or other units that compose the relevant collective entity.

Earlier I explained why we ought not to be overconfident that a society of virtuous individuals will be sufficient to create the moral vision or moral motivation necessary to achieve the SDGs. It will not unless the virtues are supplemented by convictions about universal values and obligations of the sort endorsed in secular traditions, such as consequentialism, deontology, socialism, as well as some versions of contract theory (in which we stipulate universal rights and corresponding duties and agree or contract to abide our creation). But now I have pointed to another worry: The universal values and obligations endorsed by these theories are not the actual values of most people. What this means is that in addition to whatever virtues we do in fact work to inculcate and maintain, we also need to work more directly on values, working to make the rationales for the universal ethical principles in the works of Mill, Kant, and Marx more visible and more appealing.

Moral Education and Building Moral Capital

The SDGs express global aspirations. They aim at the good of all humanity. Achieving the SDGs almost certainly requires, in the sense of "obliges," certain nation-states to give more than others so that people who are worse off (including future generations) can be and do better. This is not merely philanthropic work; it is not above and beyond the call of duty. According to the principles of Enlightenment and post-Enlightenment secular moral theory, it is a duty. The burden is on the well-off to give to those who are worse off for the sake of the greater good. Meeting obligations involves sacrifice.

"America first" was the motto of the previous ruling party in the United States, and the disposition to perceive international affairs in these terms was a disposition enacted by the administration of Donald Trump. But if nation-states can hold good values and display these virtues, they can also hold bad values and display vices. In a Hobbesian world, it is every ego for itself. In a world of "[insert name of country] first," it is every nation-state for itself. In such a world, universal values, principles, and the dispositions of individuals and nation-states to work for the common good are undermined. Concepts like the common good, shared fate, and obligations to all humanity and future generations recede from view.

What is to be done? Moral capital consists of currency in several denominations: There are virtues. There are role models for virtues. There are moral beliefs and principles such as that everyone equally deserves a good life And there is a distinct action-guiding principle that everyone has an obligation to do what is necessary so that others can live well. There is the social and political atmosphere, which encourages or discourages various virtues and high-minded values. How can secular ethics contribute to the creation of moral capital?

There is one opening that I think we can use to advantage. This opening is well suited to the American situation because of two distinct views in America about ethics. The first is that ethics needs a religious foundation, that only a theology provides a reason to be moral, and that nothing else can provide a reason to be moral. This view is widely held but is entirely false, as the very existence of secular ethics and morally serious atheistic and agnostic individuals and nation-states shows. The second is that the separation of church and state demands that ethical teaching be kept out of schools if it endorses the views of a particular religion.

These two beliefs taken together have led to suspicion about almost all regimens of moral education in schools and other public institutions. Why? Well, if ethics is grounded in or made sense of by religion, if moral capital is produced only by religions, then any recommendation for moral education in schools will (on purpose or not) be a tool for indoctrinating children into a value of some religion or other. Both the left and the right are absurdly watchful for attempts to smuggle in religion via ethics, and moral education has become an ideological hot potato.

The result: Only eight states in America currently require something called civics education, and by the time students are in high school, much of that is devoted to career and college counseling, nothing very normative.

The results of recent initiatives like Ethics in Action, which reveal an overlapping consensus about values, specifically ones that support the SDGs, among religious traditions suggest that liberal society can encourage explicit instruction in the overlapping consensus of values without being a partisan of any particular religion. But an agnostic or atheist might object that this still involves teaching that religion is necessary for morality, which is false.

However, this situation can be seen as a gift. The fact that the secular traditions of virtue ethics, deontology, consequentialism, and socialism vindicate the values expressed and endorsed in the SDGs without any religious foundation at all means that there are secular methods to produce vast amounts of moral capital. In addition, that resistance to teaching morality in schools has no rationale, since we have proof that ethics does not require any particular religion, indeed any religion at all. This ought to be consolation in an increasingly secular world.

One might say, "Well, still, isn't it kind of pushy to endorse certain values and virtues?" It isn't, if one thinks that the values are the right ones and can show that they are. This can be done using the tools of moral philosophy. Furthermore, the fact that 193 countries endorsed the values behind the SDGs means every nation in the UN has explicitly affirmed them. It would be in terribly bad faith if these very nation-states resisted inculcating the values they affirm and were shy about endorsing the values of global responsibility and sustainability in schools and in public political discourse. One thing we know from history is that the moral fiber of the youth depends primarily on the moral tone set by one's family, schools, community, and nation-state, and on the direct moral instruction on offer. This means that we adults have special responsibilities to educate and reeducate ourselves in ethics and to insist on and model the enlightened values that support global happiness, well-being, and sustainability.

Notes

1. The seventeen SDGs require meeting 169 targets that are themselves measured by 232 indicators, which allow measurements of progress toward meeting the goals. No country is currently on track to meet the goals, although the happiest countries are the furthest ahead.
2. In the 2019 *Gallup Global Emotions* report (https://www.gallup.com/analytics/248906/gallup-global-emotions-report-2019.aspx), which reports on global happiness in terms of positive versus negative emotions, Latin American countries score highest. So, whereas the North Atlantic countries, on average, do very well on the question of life satisfaction and on objective measures such as government corruption, quality of education, per capita income, and social support, Latin American countries, on average, excel on measures of mood, emotions, and ratio of positive to negative emotions.
3. "Sustainable Development: A Path to Happiness," *Environment News Service*, July 10, 2018, https://ens-newswire.com/sustainable-development-a-path-to-happiness/.
4. David Brooks, "How to Repair Moral Capital," *New York Times*, October 20, 2016, https://www.nytimes.com/2016/10/21/opinion/how-to-repair-moral-capital.html. Moral capital is called "moral capacity" by Annett et al. (2017). They write, "To succeed, sustainable development needs 'moral capacity' as much as financial or technical capacity" (2017, 3). See Anthony M. Annett, Jeffrey D. Sachs, Marcelo Sánchez Sorondo, and William F. Vendley, "A Multi-religious Consensus on the Ethics of Sustainable Development: Reflections of the Ethics in Action Initiative," Kiel Institute for World Economy Economics Discussion Papers, no. 2017–56 (2017).
5. A widely accepted epistemic rule of thumb is that one ought not justify an observation, norm, practice, or principle by observations, norms, practices, or principles that are more questionable than the norm, practice, or principle itself. Some secularists (agnostics and atheists) think that attempts to provide religious foundations for ethical principles always violate this principle.
6. Dostoevsky and Nietzsche (the first devout, the second an atheist) both worried that moral and existential nihilism might follow a European loss of faith. The statement that "if there is no God, then everything is allowed" could be an accurate sociological prediction without any philosophical necessity. That is, people who are used to a theology being used to make sense of their moral commitments might feel lost if the belief of that theology are taken away. But people raised in secular cultures might not feel the same angst about morals and meaning upon learning that their parents and teachers think supernatural beliefs lack warrant.
7. Jordan W. Moon, Jaimie Arona Krems, and Adam B. Cohen, "Religious People Are Trusted Because They Are Viewed as Slow Life-History Strategists," *Psychological Science* 29, no. 6 (2018): 947–60, https://doi.org/10.1177/0956797617753606.
8. Secularists, of course, include atheists and agnostics. Many religious believers are also secularists for pragmatic reasons. They think that widespread religious disagreement means that the state cannot use any particular religion, including their own, to justify its moral or legal demands. One idea is that in diverse cultures in which there are many religious and cultural traditions, we ought to look for points of agreement about what is right and wrong, good and bad—what John Rawls called "an overlapping consensus." Annett et al. (2017) explain that there is a "multi-religious consensus" among religious leaders that the SDGs express common values.

9. The two most accessible books on virtue ethics are Rosalind Hursthouse's *On Virtue Ethics* (Oxford: Oxford University Press, 1999) and Julia Annas's *Intelligent Virtue* (Oxford: Oxford University Press, 2011).
10. The claim that no one flourishes if one is not virtuous can be read as an empirical claim, a normative claim, or a hybrid claim to the effect that everyone who has genuinely flourished is virtuous and this is as it should be. It is also the reason that *eudaimonia* is normally translated as "flourishing," rather than as "happiness." Ordinary language allows that happy people can be bad, whereas a *eudaimon* cannot be bad. A *eudaimon* might not feel happy, even if they have reasons for self-esteem and self-respect.
11. Perverse incentives and private philanthropy: A story from recent times is used to provide moral camouflage for greed. The story takes as a premise the existence of a class of billionaire philanthropists. These billionaire philanthropists are role models of individuals who made a lot of money (in some cases unintentionally; they were just immensely creative) and now give a lot of that money away to good causes. This allows young ambitious types to tell themselves and others that they will be generous when they make it. In some cases this is sincere; in other cases, it provides moral cover for rapacious acquisitiveness and insatiable appetites. Aristotle actually endorsed a version of this cover story in his admiration for men who were "great-souled" and "magnanimous." Both, according to Aristotle, are virtues.
12. The idea that we are animals (fully embodied, not part soul and part body) is key for many secular ethicists. It challenges the idea that morality has a necessary, objective, transcendental, or supernatural source. Hursthouse writes, "The fact . . . that our nature is such that the virtues, as we know them, suit human beings . . . is a highly contingent one. It is a contingent fact that we can, individually, flourish or achieve eudaimonia, contingent that we can do so in the same way as each other, and contingent that we can do so all together, not at each other's expense. If things had been otherwise then, according to the version of ethics presented . . . morality would not exist, or would be unimaginably different" (*On Virtue Ethics*, 22).
13. Every ethical tradition ever known attends to the moral socialization of the youth, which is guaranteed to produces some reflection on virtues. Thus, numerous secular and sacred variants aim at virtue. In Thomas Aquinas's virtue ethics, the natural law arguments of Aristotle, according to which ethics in some respects is conforming to what nature ordains, are retold in a sacred idiom. What Aristotle saw as our natural proper function to develop reason and virtue (*ergon*) and of the human *telos* as achieving *eudaimonia* were recast as expressions of divine activity. The end of virtue was recast as more than *eudaimonia*: as eternal heavenly life. Many anthropologists and ethicists believe that being reliably ethical is very hard and thus that stories of fantastical rewards might motivate when more mundane ones don't or won't. See Ara Norenzayan, *Big Gods: How Religion Transformed Cooperation and Conflict* (Princeton, NJ: Princeton University Press, 2013).
14. Every philosophical and folk morality has a theory of which virtues make for excellent character. But which virtues are prioritized and which are trumped when conflicts arise varies across traditions. Being fair has pride of place in secular liberal precincts; compassion for all sentient beings is the highest Buddhist virtue; Confucians emphasize familial love, respect for elders, and abiding the rituals. See Owen Flanagan, *The Geography of Morals: Varieties of Moral Possibility* (New York: Oxford University Press, 2017).
15. A common criticism of virtue theories is that they are vague and not clearly action guiding. Most everyone thinks that generosity is a virtue and that it should abide the mean, in the sense that both excess and deficient generosity could be vicious. But in

common sense morality, how much a good person, conventionally conceived, should share is not a lot. One should be generous, and miserliness is a vice. But what exactly "generosity" and "miserliness" mean are unspecified. Is tithing 10 percent enough? Hursthouse (1999) suggests that one standard for what ought to be done is to do what the generous person would do. But suppose I think that Aristotle, Jesus, and Bill Gates are all generous; who should serve as my exemplar?

16. To see this, I recommend gathering friends whom you think are virtuous and asking them how much they are willing to pay (in taxes, through efficient philanthropies, etc.) to help the worst off.
17. The "effective altruism" movement is a utilitarian-inspired movement that provides empirical information about how to do good most effectively.
18. There is a dispute about whether, especially in Marx's writings, we should read the principle as a moral "ought" or as a prediction about what the needy will demand and eventually get from the rich, or both.
19. This "Euthyphro Dilemma" occurs at the beginning of Western philosophy in Plato's dialogue *Euthyphro* and suggests an abiding puzzle about all theological justifications of ethical practices.
20. One might worry that a virtue ethics grounded in a theology will not typically model universality or globality but rather tribalism and exclusivity. Why? First, insofar as creedal rectitude or orthodoxy grounds the right ethical view, and given that there are numerous creedal disputes between, and even within, most religious traditions, there will be visible divisions between those who get things right and those who do not (heretics and such). Second, religious rituals are overtly tribal; they mark members of the same religious and ethical community off from members of other communities. Together, the demand of creedal rectitude and membership restrictions on ritual performance model in-group/out-group distinctions in the very religious practices that might be claimed to model or encourage universality and global inclusion. Of course, one worry for secular ethics is that secularists have no common places to gather and affirm their universal values at all.
21. The principle of minimal realism says that the demands of ethics should be psychologically realizable. See Owen Flanagan, *Varieties of Moral Personality: Ethics and Psychological Realism* (Cambridge, MA: Harvard University Press, 1991). If the new global virtues require affective and cognitive capacities and practical knowledge that individuals do not or cannot normally possess, then the virtues being recommended are heroic and violate minimal psychological realism.
22. One could then adapt Hursthouse's idea that we can use virtuous individuals as exemplars in deciding what to do and ask at the level of collectives, What would a virtuous state or country do?

CHAPTER 5

THE CURRENT RESURGENCE OF INTEREST IN THE CIVIL ECONOMY PARADIGM

STEFANO ZAMAGNI

How can one explain the resurgence of public interest over the past quarter of a century in the civil economy paradigm, after its virtual disappearance from economic discourse for almost two centuries? How is it that the transition from national economies to a global economy has rekindled interest in the study of the civil economy? Since the first half of the nineteenth century, the civil vision of the economy had disappeared from both the field of scientific study and from political and cultural debate. Various reasons can be given for this fact, and in this chapter I am going to focus on what I consider to be the two most important such reasons. However, before moving to that, let me devote few lines to clarify what civil economy is.

Civil economy represents a tradition of economic and philosophical thought rooted proximately in civic humanism (in the fifteenth century) and more remotely in the works of Aristotle, Cicero, Aquinas, and the Franciscan school of theology. Its golden age, when it was at the height of its influence as a school of economic thought, came in the Italian Enlightenment, specifically the Neapolitan and Milanese one. (The University of Naples created the world's first chair in economics in 1753 with the denomination of "chair in civil economy.") While Adam Smith and David Hume in Scotland were developing the principles of political economy, Genovesi, Gaetano Filangieri, Ferdinando Galiani, and Giacinto Dragonetti in Naples and Pietro Verri, Cesare Beccaria, and Gian Domenico Romagnosi in Milan were developing civil economy. The Scottish and Milanese-Neapolitan schools had many features in common: the polemic against feudalism (the market seen above all as the way out of feudal society); the praise of luxury as a force for social change, with little concern for the "vices" of the consumers of luxury goods and much more for the benefits for all of society; a great ability to comprehend the cultural revolution that the growth of trade was bringing about; the recognition of the essential role of trust in a market economy and for cultural progress; and the modernity of their views of society and the world.

Yet there is also a crucial difference between the school of political economy and that of civil economy. Smith, even while acknowledging that people have a natural tendency toward sociability ("sympathy" and "correspondence of sentiments"), does not consider genuine, noninstrumental sociability or relationality to be relevant to how markets work. For Smith, and for the tradition that, following him, became the official thought in economics, the market is the means for building relations that are genuinely social (no civil society can exist without the market) because it is free of vertical bonds and unchosen status; but the market is not, per se, the locus of all-round relationality. That mercantile relations are impersonal, with mutual indifference, is not a negative but a positive, civilizing characteristic, in Smith's eyes. This is the only way the market can produce wealth and progress.

In other words, friendship and market relations belong to two distinct spheres. The existence of market relations in the public sphere (and there only) ensures that in the private sphere friendship is genuine, freely chosen, and unconnected with status. For the civil economy, school, the market, the enterprise, and the economy are *themselves* the place for friendship, reciprocity, gratuitousness, and fraternity. Civil economy rejects today's increasingly common notion, taken for granted, that the market and the economy are radically different from civil society and are ruled by different principles. Instead, the economy *is* civil, the market *is* life in common, and they share the same fundamental law: mutual aid. For Antonio Genovesi's "mutual aid" is something more than Smith's reciprocal advantage. For the latter, a contract will suffice; for the former, *philia*, perhaps *agape*, is needed. To civil economy, the actual "golden rule" of the market is reciprocity, since contracts, businesses, and exchanges are matters of cooperation and of common advantage, that is, forms, albeit different one from another, of reciprocity. In place of Smith's assumption of a peculiar human prosperity "to truck, barter and exchange one thing for another," Genovesi grounds his analysis of markets on an assumed human inclination toward mutual assistance, as evidenced by his famous phrase "*Homo homini natura amicus*," to be contrasted with Hobbes's "*Homo homini lupus*."[1]

The Disappearance of the Civil Economy Paradigm

The first reason for the disappearance of the civil economy paradigm was the rapid explosion in the popularity and influence of Jeremy Bentham's utilitarianism among European circles of high culture: A few decades after its publication in 1789, Bentham's key work had gained a hegemonic position within economic discourse. It was utilitarian philosophy, rather than protestant ethics (although some observers continue to argue to the contrary), that saw the emergence within economic science of the hyper-minimalist anthropology of *homo oeconomicus*, and with it the social atomist approach. The following significant passage from Bentham's work *An Introduction to the Principles of Morals and Legislation* is clear: "The community is a fictitious body, composed of the individual persons who are considered as constituting, as it were, its members. The interest of the community, then, is what? The sum of the interests of the several members who compose it."[2]

The second reason was the full establishment of an industrial society following the first Industrial Revolution. An industrial society is one that produces goods, with its attendant machinery playing a predominant role in setting the rhythm of people's lives. Energy largely replaced muscle power and accounted for the incredible leap forward in productivity, which in turn was often accompanied by mass production. Energy and machines transformed the nature of labor, and the capacities of individuals were broken down into their most elementary components. Hence the need for coordination and organization. Consequently, the world has developed as one in which people are seen as "things," since it is far easier to organize things than it is to organize individuals, and in such a world, the person and the role they plays are two distinct things. Organizations, in particular companies, rather than people, take up the key roles, and this happens not only within the factory but also in society as a whole. This is where the underlying meaning of Ford-Taylorism lies: that is, in the (successful) attempt to theorize this model of social order and putting it into practice. The advent and success of the "production line" was accompanied by the spread of consumerism; hence the schizophrenia characterizing modern times: on the one hand, the diminishing importance of labor (alienation owing to the depersonalization of the worker's role); on the other, in order to counterbalance this loss, consumption becomes increasingly opulent. Marxist thought and its political expressions during the course of the twentieth century attempted (with variable but limited success) to offer an alternative model of society.

The great challenge, both political and cultural, that the civil economy faces, is that of superseding the traditional capitalist market economy model without, however, forgoing those benefits that such a model has guaranteed to date. In fact, contrary to what some people would argue, it is not true that if the civil economy wishes to preserve and extend a social order based on market forces, it is necessarily obliged to accept (or suffer) the traditional capitalistic form of such. This was not the case at the beginning, as is widely acknowledged. Nowadays it is commonly believed that the so-called turbo model of financial capitalism has lost its momentum, and thus now is the right time for a fundamental rethinking of the concept of the market.

There can be no doubt that the Fourth Industrial Revolution today represents the most compelling reason for reexamining the civil economy paradigm. As we are all well aware, the success of convergent technologies—those technologies resulting from the synergetic combination of nanotechnology, biotechnology, information technology, and cognitive science (referred to collectively as NBIC)—is radically altering not only the mode of production but also, and above all, social relations and the cultural matrix of our society. We still do not know how digital technology and the culture governing such technology are going to change the nature of capitalism in the coming years. What we do know, however, is that a "great transformation," of the type envisaged by Karl Polanyi, is taking place, and this transformation is having a considerable impact on the meaning of human labor (as well as on the question of the destruction of jobs), on the relationship between the market and democracy, and on the ethical aspects of human action.

The promise of the strengthening, and thus of a transformation, of both humanity and society, as a result of the convergent NBIC technologies, accounts for the incredible interest in technological science within a variety of fields, ranging from the scientific to the cultural, and from the economic to the political. The objective pursued is not only that of strengthening the powers of the mind, or of increasing our diagnostic and therapeutic capacities in relation to a vast variety of diseases, or even of improving the ways in which information is controlled and manipulated. What we are now witnessing is a move toward the "artificialization" of humanity and, at the same time, the anthropomorphization of machines. The veil of silence surrounding such matters needs to be lifted and a debate held regarding the consequences of such an important development. In fact, the anthropological level of discourse is concerned here, with two conceptions of humanity challenging each another: the conception of the human as a person and that of the human as a machine. Great care needs to be taken to prevent the latter prevailing over the former.

The Value and Utility of the Civil Economy Paradigm

Which problematic areas of modern society can be enlightened by the salience and practical importance of the civil economy paradigm in relation to both economic reflection and praxis? Owing to the limited space available here, I shall focus on one area only, namely the worrying, systemic increase in social inequality. The question may be framed as follows: If inequality is on the increase, but this is not a result of any lack of resources or technological know-how, or of specific adversities affecting certain categories of individual or specific areas of the world, then what are its underlying causes, and, more importantly, why does it not lead to rebellion against such a state of affairs? The most plausible answer, in my view, is that this is because of the continued belief, within our societies, in the ideology of inequality. (Vilfredo Pareto indeed considered inequality to be a hard and fast law from which humankind could never escape.) There are two fundamental tenets of this ideology of inequality. The first is that society as a whole would benefit if each individual acted for their own personal gain. This is false on two accounts. First, because in order for Adam Smith's "invisible hand" argument to hold true, markets need to be close to the ideal of free competition, that is, a situation in which there are no monopolies, oligopolies, or information asymmetries. However, everyone knows that the preconditions for perfect market competition are never fulfilled in reality, and thus the invisible hand cannot operate.[3]

Moreover, people possess different abilities and skills; consequently, if the rules of the game are formulated so as to extol opportunistic, dishonest, immoral behavior, for example, then those whose moral constitution is characterized by such tendencies will invariably end up crushing all others. Likewise, avidity, that is, the passionate desire to possess things, is one of the seven deadly vices. Therefore, if systems were introduced into the workplace offering strong incentives—not rewards but incentives—then it is clear that the more avid workers would tend to dominate the less avid ones. Thus it can be said that poor people are such not by their nature but as a result of their social conditions; that is, because of the way in which economic institutions are designed. Condorcet had already made this point in his 1794 work *Esquisse d'un tableau historique des progrès de l'esprit humain*, when he stated, "It is easy to prove that fortunes naturally tend to equality, and that their extreme disproportion either could not exist, or would cease, if positive law had not introduced *fictitious means* of amassing and perpetuating them."[4] The "positive law" (civil law) of which the French Enlightenment writer spoke, refers simply to the rules of the economic game.

The second tenet of the ideology of inequality is the belief that elitism should be encouraged since it is highly effective; in other words, the wealth of the majority of the population would grow more if the abilities of the few were promoted. Therefore, greater resources, attention, incentives, and rewards should be dedicated to the most gifted members of society, since it is their endeavor and commitment that underlie society's advancement. Consequently, the exclusion of the less gifted from the economic sphere—through their employment in temporary jobs and/or their unemployment, for example—is not just normal but also necessary if the GDP growth rate is to be increased. The crisis of the ideology of equality, resulting from the fact that the application of the principle of fair distribution always requires some form of sacrifice, was clearly described by the Italian philosopher Norberto Bobbio when he wrote that the battle for equality is invariably followed by the battle for diversity:

> Human history shows that struggles for superiority alternate with struggles for equality. This alternation is natural, because the battle for superiority presupposes that certain individuals or groups have attained a certain degree of equality among themselves. The battle for equality generally precedes that for superiority . . . Before reaching the point of fighting for dominion, each social group has to achieve a certain degree of parity with its rivals.[5]

We can now understand what the adoption of a viewpoint of the common good, instead of that of the total good, would imply in terms of the solution to the problem in question. It would imply a transition from a "transcendental theory of justice," based on the anthropological premise of self-interested individualism, to a "comparative theory of justice" in the words of Amartya Sen.[6] In his radical critique of John Rawls, Sen explains (convincingly in my view) why the time has come to speak of *iustitia*, which presupposes the pursuit of the perfect, universally valid theory. In the language of the economist, this means that it is better to pursue Paretian improvements (*iustitium*) than Paretian optimality (*iustitia*). In practice, this means moving from an abstract conception of justice based on the hypothetical preferences of the agents, to the notion of benevolent justice, that is, justice aimed at the common good.[7]

A simple yet effective anecdote offers us an idea of the importance of the notion of benevolent justice. It is based on an old Arabian tale and goes like this: A father of three children, a camel driver by trade, on his deathbed decided to make out a will. After much thought, he decided to leave the oldest son one half of his wealth, the second son one quarter of his wealth, and the youngest son one sixth of his wealth. Following the death of their father, the three sons opened his will to discover that the wealth in question consisted of eleven camels, that is, all their father had managed to accumulate over the course of a lifetime. This is when the quarrelling started, since eleven is not divisible by two. The oldest son, who according to the will should have received five and a half camels, demanded that this number be rounded up to six, but the other two sons argued that since their father had favored the oldest brother by leaving him the largest share, the oldest brother should make do with five camels. A similar argument ensued with regard to the second and third sons' inheritances. The three brothers, after a heated argument, ended up fighting and would have resorted to weapons had a passing merchant atop his own camel not intervened. Curious to discover what the fighting was over, the merchant got the brothers to recount the reasons for this perilous state of affairs, and when he heard what the argument was about, he decided to donate his own camel in a gesture of pure gratuitousness. In this way, the inheritance now consisted of twelve camels: the oldest son received six, the second son three, and the third son two, making a total of eleven camels. At which point the merchant took his own camel back and continued on his way.

This moral tale contains two messages. The first is that the rules of justice—in this case represented by the testator's will—are not always sufficient to guarantee peace. History shows that numerous battles have been fought in the name of "transcendental justice." However, when the rules of justice are combined with a gratuitous gift, the desired result is guaranteed. It is important to note that what I am saying here is not that the principle of the gift can replace the rules of justice; if anything, it should supplement and complete such rules. In fact, in the case of the final division of the old man's wealth, the rules of justice were complied with, but this was achieved without any bloodshed only thanks to an act of gratuitousness. The second message that emerges from this tale is that the practice of giving never leads to impoverishment; on the contrary, it enriches the giver. In this ancient fable, not only does the merchant get his camel back but he "earns" the appreciation and gratitude of the three brothers, as well as the inner joy of having managed to prevent a serious conflict in that family. Note that the gratuitous gift—unlike the gift of a present (*munus*)—is that of a "third party." The merchant is from outside; he is not part of the three brothers' *civitas*. Thus the necessary step required to resolve the conflict is the entry of a third party. This is the meaning of the metaphor of the "perfect outsider."[8]

I believe that a successful approach to the challenges of the present age is that of returning to the principle of fraternity and placing it once again at the center of economic discourse. Antonio Genovesi's great merit was that of positing the principle of fraternity, in both institutional and economic terms, as a cornerstone of the social order. It was the Franciscan school of thought that gave the meaning to the term "fraternity" that it has preserved over the course of time. Pages of Saint Francis's "Regola" (his spiritual guidelines and practical rules) help the reader to understand clearly the true meaning of the principle of fraternity: supplementing, and at the same time going beyond, the principle of solidarity. In fact, while solidarity is the principle of social organization whereby nonequals can become equals, the principle of fraternity permits people who are equals to be diverse. Fraternity allows people who are equals from the point of view of their dignity and their fundamental human rights to express their life plan or their charisma in different ways. The 1800s and, in particular, the 1900s were centuries characterized by large-scale cultural and political battles in the name of solidarity, and this was obviously a good thing. One has only to think of the history of the trade-union movement or of the struggle for civil rights. The point is that a good society cannot be satisfied with the pursuit of solidarity alone, since a society that is only solidaristic, without being fraternal as well, is one from which all will attempt to flee sooner or later. While a fraternal society is also a solidaristic society, the opposite is not true.

The fact that a human society in which the sense of fraternity is lost, and in which everything comes down, on one hand, to improving transactions based on the trading of equivalents and, on the other hand, to increasing the transfers made by public welfare organizations, is not a sustainable society, has been forgotten. This fact explains why it is that despite the quality of the intellectual forces at play, no credible solution has yet been offered for that trade-off. A society in which the principle of fraternity fades from view is a society with no future; that is, a society is not capable of progressing if it is capable only of "giving to receive" or "giving as a duty." This is why neither the liberal-individualist vision of the world, in which everything (or nearly everything) constitutes a trade-off, nor the state-centric vision of society, in which everything (or nearly everything) is based on a sense of duty, can safely lead us out of the shallows, where the Fourth Industrial Revolution is severely testing our existing model of civilization. This is the key message contained in Antonio Genovesi's thought, which even 250 years later has maintained its originality and cogency.

The fifteenth century was the century of the first humanism, a typically European school of thought. The twenty-first century, from the very outset, powerfully demonstrates the need for a second humanism. In the fifteenth century, the shift from feudalism to modernity was the decisive factor that led in that direction. Today, it is an equally great transformation (in Polanyi's sense)—from an industrial society to a post-industrial society—that shows us the need for a new humanism. Globalization, the financialization of the economy, convergent technologies, migrations, identity conflicts, environmental challenges, and increasing inequalities are only a few of the key issues telling us of the "discontent of civilization," to cite Freud's well-known essay. In facing these new challenges, merely updating our old categories of thought or refurbishing collective decision techniques, however refined, would not be suitable for the purpose. We must have the wisdom and the courage to walk new paths.

Notes

1. For an extended discussion of this topic, see Luigino Bruni and Stefano Zamagni, *Civil Economy: Another Idea of the Market*, trans. N. Michael Brennen (Newcastle upon Tyne: Agenda, 2016).
2. Jeremy Bentham, *Introduction to the Principles of Morals and Legislation* (1823; reis., Oxford: Clarendon, 1907), chap 1, section 4.
3. Bruni and Zamagni, *Civil Economy*.
4. Marie-Jean-Antoine-Nicolas de Caritat, Marquis de Condorcet, *Outlines of an Historical View of the Progress of the Human Mind* (1796; reis., United States: Andesite, 2017). Emphasis mine.

5. Norberto Bobbio, *Destra e Sinistra* (Rome: Donizelli, 1999), 164.
6. Amartya Sen, *An Idea of Justice* (Cambridge, MA: Harvard University Press, 2009).
7. See also Michael Sandel, *Justice: What's the Right Thing to Do?* (New York: Farrar, 2009).
8. Stefano Zamagni, "Disuguaglianza e giustizia benevolente," in *L'essere che è, l'essere che accade*, ed. C. Danani and B. Giovanola (Milan: Vita e Pensiero, 2014).

PART II

RELIGIOUS TRADITIONS AND THE COMMON GOOD

CHAPTER 6

THE CONFUCIAN CONCEPTION OF THE COMMON GOOD IN CONTEMPORARY CHINA

ANNA SUN

The idea of the common good has been with us for millennia. But the calamities we have been experiencing in recent years—the refugee crisis, the COVID-19 pandemic, the unveilings of systemic racism and sexism, among others—have made the issue of the common good more urgently relevant to us than ever. It is not a surprise that Michael Sandel's reckoning of the problems of neoliberal individualism, *The Tyranny of Merit: What's Become of the Common Good*, was a best seller and received much critical praise (it was named a Book of the Year by the *Times Literary Supplement* and a Best Book About Ideas by the *Guardian*). Discussions of the common good—and its natural extension, the issue of inequality and inequity—have also been central to conversations about contemporary China.[1]

The focus of this chapter is to locate some of the main resources of the common good in contemporary Chinese life, especially in the official discourse of happiness. China has undergone a tremendous change in the past thirty to forty years in its transition into a thriving market economy, and it can indeed be argued that Confucian ethical values have grown stronger rather than weaker. One may speak of a rediscovery or renaissance of Confucian ethical life in the past twenty years.[2] However, instead of probing various Confucian ethical theories, which philosophers of ethics and moral psychology have been examining extensively with great insights, I propose that it may be important to examine the Confucian resources that are essential for ordinary Chinese people to define the meaning of the common good.

The Four Systems of Meaning and Morality in Contemporary China

In the seminal 1985 sociological study of systems of meaning in contemporary America, *Habits of the Heart: Individualism and Commitment in American Life*, the researchers—Robert Bellah, Richard Madsen, William Sullivan, Ann Swidler, and Steven Tipton—focused on the seemingly impossible questions of the social meanings of a good life. What are the major moral and ethical discourses that inform the meanings of life for ordinary Americans? After two hundred in-depth interviews with urban and suburban Americans, they did come up with an answer. It is a system that consists of four major components: the Biblical religious tradition, the civic republican legacy, the language of utilitarian individualism, and the language of expressive individualism. These are four systems of meaning that form a resilient "web of significance"; they not only inform action but also make possible the many narratives people tell about their lives: narratives about work, politics, religion, love, marriage, and other commitments and values.[3]

What are the possible "habits of the heart" in China? How do people offer an evaluation of their lives in terms of happiness? What are the moral resources of different conceptions of a good life? What major moral or ethical languages of happiness and a meaningful life exist in China today? How do people pursue the kind of happiness they deem valuable? I propose that there might be four major systems of meaning at work in contemporary urban China, together forming the "web of significance" that gives reason for action as well as narratives of action. This is the foundation of the contemporary Chinese moral imagination.

In this chapter, I focus only on the first discourse of happiness, namely the official discourse of happiness offered by the state, which in many ways defines the notion of the common good for people in everyday life in contemporary China.

The official discourse of happiness offered by the Chinese government comes from the long tradition of using mass media for propaganda purposes by the Chinese communists, which can be traced back to the years even before the founding of the People's Republic in 1949. With official national newspapers under its direct control, the Chinese government today systematically promotes a particular definition of happiness. It is a notion of happiness as something given to the people by the state, for it is the state's duty to create a "society of happiness," which in turn legitimizes its power through this achievement. This official discourse of happiness adopts a Confucian language of benevolent politics (*ren zheng*), suggesting that citizens are receivers of happiness made possible by the state, and emphasizes the importance of the common good as well as social stability. The political nature of the official discourse complicates the already multifaceted systems of meanings of happiness and a good life in China.

The second system of meaning comes from the emergence of capitalism, which could well be termed as representing "sensualists without spirit, specialists without heart," the disturbing state of being described by Max Weber in his critique of modern capitalism.[4] Scholars have been arguing whether China has become a neoliberal state, politically or economically, but what is relevant to our study is whether neoliberal—or simply raw capitalist—values have become one of the dominant ideologies affecting people's lives today in China.[5] Lisa Rofel describes it as follows: "A sea-change has swept through China in the last fifteen years: to replace socialist experimentation with the 'universal human nature' imagined as the essential ingredient of cosmopolitan worldliness. This model of human nature has the desiring subject as its core: the individual who operates through sexual, material, and affective self-interest."[6]

The third system of meaning is the ancient Confucian language of filial piety and familial devotion, which places the centrality of the family—relations between parents and children, among other familial relations, and between oneself and one's ancestors—above all other values. It is an ethical language spoken on all levels of Chinese life, followed by people from vastly different social, religious, and economic backgrounds. This Confucian ethical tradition governs the realm of family life and social relations, affecting people's conduct in everyday life.[7]

The fourth system of meaning is something we are still trying to understand in a deeper way: the role of religion in the formation of one's narrative about oneself and one's place in the world. This may be called "the religious reason to be," and it refers to the various religious values that inform people's narratives about the meaning of one's life. For instance, the Buddhist concept of *karma* may be central to a Buddhist lay believer's effort to make sense of their own life, as well as the reason behind their choices of future actions. It is important to note that having a religious system that inform one's personal values does not necessary contradict following other systems of meaning, even though contradictions or even conflicts may well be part of the daily struggle for many. For example, a Catholic whose moral imagination is formed by Christian theology may follow Confucian values such as filial love and care; and although they might not be able to stay away entirely from the neoliberal ethos, they might have a different attitude toward it compared to their non-Catholic or nonreligious friends. This complex system of religious values and multiple religious engagements, with its endless individual configurations, forms the Chinese religious moral imagination.

The Cultural Resources of the Chinese Official Discourse on Happiness

What are the historical roots of the ideas expressed in the Chinese official discourse on happiness? There are at least three sources, namely the socialist ideal of the collective class interest of the proletariats, the Confucian ideal of "benevolent politics," and the Confucian ideal of the "Great Commonwealth" (*datong*, 大同). The Confucian sources are arguably becoming increasingly significant and central in the articulation of this discourse, especially in the past two decades.[8]

There are certainly strong flavors of the Marxist ideal of the Communist Party representing the best interest of the people in the current discourse on happiness. However, the Marxian view of the future is not gradual progress but revolution, and its view of the common good is very much class bound. This can be seen clearly in Friedrich Engels's "Draft of a Communist Confession of Faith" (1847):

> Question 1: Are you a communist?
> Answer: Yes.
> Question 2: What is the aim of the communists?
> Answer: To organize society in such a way that every member of it can develop and use all his capabilities and powers in complete freedom and without thereby infringing the basic conditions of this society.
> Question 5: What are such principles?
> Answer: For example, every individual strives to be happy. The happiness of the individual is inseparable from the happiness of all, etc.[9]

Unlike the early Chinese communists, the current Chinese political regime has ceased to use the language of class conflict, such as "the proletariats versus the capitalists," which would eventually lead to a proletarian revolution. Indeed, capitalists have been allowed to join the party since 2001. As the *New York Times* puts it, the defining event of China at the beginning of the twenty-first century was not the 2008 Beijing Olympics but "the decision of the Chinese leadership to admit capitalists as members of the Communist Party. The decision raises the possibility of Communists co-opting capitalists—or of capitalists co-opting the party."[10]

Since Marxism is no longer the main foundation for the official discourse on happiness, other resources must be mobilized. Instead of using the language of "Western values" such as individual freedom and liberty, the Chinese official discourse on happiness today relies heavily on the Confucian ideal of "benevolent politics."

This is a notion that can be traced back to Confucius, as well as Mencius, who demand the state to take care of its weak and poor and to provide care to everyone in need (see, for example, sections 1A7 and 1B5 of the *Mencius*). The Confucian idea of the "mandate of heaven" implies that rulers are legitimated through their benevolent rule; the ones who are cruel or lack compassion and empathy do not carry the "mandate of heaven." Mencius states,

> The Three Dynasties won the empire through benevolence and lost it through cruelty. This is true of the rise of all, survival and collapse, of states as well. An Emperor cannot keep the Empire within the Four Seas unless he is benevolent;

> a feudal lord cannot preserve the altars to the gods of earth and grain unless he is benevolent; a minister or a counselor cannot preserve his ancestral temple unless he is benevolent.[11]

This is very much the logic followed by the constant flow of articles in *China Daily* and *People's Daily* promoting happiness and well-being, which define these terms as something shared by the nation or at least a community, rather than being achieved through individual pursuits.

The third source of the notion of happiness in today's Chinese official discourse is the ancient ideal of the "Great Commonwealth," often also translated as the "Grand Commonality" or the "Grand Unity." In "The Evolution of Rites" (*liyun*, 禮運), a chapter in the *Record of Rites* (*liji*, 禮記), an important statement regarding what an ideal society should be like is attributed to Confucius. Here is Confucius's vision of the ideal society in which the "Great Way" has been realized:

> When the Great Way was practiced, the world was shared by all alike. The worthy and the able were promoted to office and men practiced good faith and lived in affection. Therefore they did not regard as parents only their own parents, or as sons only their own sons. The aged found a fitting close to their lives, the robust their proper employment; the young were provided with an upbringing, and the widow and widower, the orphaned and the sick, with proper care. Men had their tasks and women their hearths. They hated to see goods lying about in waste, yet they did not hoard them for themselves; they disliked the thought that their energies were not fully used, yet they used them not for private ends. Therefore all evil plotting was prevented and thieves and rebels did not arise, so that people could leave their outer gates unbolted. This was the age of Grand Commonalty.[12]

But we cannot reach the age of "Grand Commonalty" before going through the age of "小康" (*xiaokang*), here translated as "Lesser Prosperity":

> Now the Great Way has become hid and the world is the possession of private families. Each regards as parents only his own parents, as sons only his own sons; goods and labor are employed for selfish ends. Hereditary offices and titles are granted by ritual law while walls and moats must provide security. Ritual and rightness are used to regulate the relationship between ruler and subject, to ensure affection between father and son, peace between brothers and harmony between husband and wife, to set up social institutions, organize the farms and villages, honor the brave and wise, and bring merit to the individual. . . . This is the period of Lesser Prosperity.[13]

As the passage makes clear, the age of "Great Commonality" and the age of "Lesser Prosperity" are two distinct stages of political and social development.

Another translation of "小康" (*xiaokang*) is "Modest Prosperity."[14] It seems "Modest Prosperity" captures better the difference between the two visions. One is an idealized stage of history, not unlike the utopian vision of many Western political philosophers, whereas the other is a realistic vision of the future, emphasizing the kind of benevolent politics necessary for the maintenance of social order.

More than two thousand years after Confucius's pronouncement, Kang Youwei (1858–1927), a key figure in the reformist movement of 1889, made use of these ideas in his notion of the "Three Ages," which is based on his interpretation of classical Confucian texts, especially the "Li Yun" chapters of the *Record of Rites*. China was at a crossroads, facing the threat of colonization by Western nations with superior military, technological, and economic power. What should China do? Kang returned to Confucius's teaching in the *Record of Rites*, looking for a solution and a map for the future:

> The course of humanity progresses according to a fixed sequence. From the clans come tribes, which in time are transformed into nations. And from nations the Grand Commonality comes about. . . . Autocracy gradually leads to constitutionalism, and constitutionalism gradually leads to republicanism. Likewise, from the individual the relationship between husband and wife gradually comes into being, and from this the relationship between parents and child is defined. This relationship of parent and child leads to the loving care of the entire race, which in turn leads gradually to the Grand Commonality, in which there is a reversion to individuality. . . . Confucius was born in the Age of Disorder. Now that communications extend through the great earth and changes have taken place in Europe and America, the world is evolving towards the Age of Order.[15]

Kang's global consciousness allowed him to see Europe and America as parts of the world "evolving toward the Age of Order" and the Confucian vision becoming a universal vision for the future:

> There will be a day when everything throughout the earth, large or small, far or near, will be like one. There will no longer be any nations, no more racial distinctions, and customs will be everywhere the same. With this uniformity will come the Age of Great Peace. Confucius knew all this in advance.[16]

It is clear that, in Kang's mind, the Confucian ideal of the "Great Commonalty" is the final goal of world history, not merely an ideal for China. The "Age of Great Peace" is for everyone on Earth.

As Barbara Hendrischke remarks, Confucius's vision of a universal "Age of Great Peace" coming after an age of disorder and "a period of approaching peace" is not dissimilar to the early Daoist vision of "great peace."[17] This is a reading that echoes Fung Yu-lan's understanding of Confucius's idea of the Great Commonwealth:

> This [Confucius's idea of the Great Commonality or Great Unity] says that the government and society so striven for by some of the Confucians is, in the final analysis, only that of the Small Tranquility [or Modest Prosperity], above which there is the government of Great Unity. This idea is one plainly borrowed from the social and political philosophy of the Taoists [Daoists]. In recent times the philosophy of the Confucian school exemplified here has been much exalted by certain Chinese political leaders, such as the reformer, K'ang Yu-wei [Kang Youwei], and Sun Yat-sen.[18]

It seems inevitable that the ancient ideal of the "Great Commonwealth" will return to the foreground of contemporary Chinese political discussions. In the Western philosophical and political tradition, the notion of the common good can be traced back to its historical roots in the works of Aristotle and Aquinas.[19] There are also religion-minded thinkers today who wish to bring back the notion of the common good as central to human flourishing, as well as an essential source of social and political action. José Casanova is one of those thinkers:

> Some religious traditions, most prominently for instance, Catholic Social Teaching[,] insist that it is both empirically and morally wrong to view the common good as merely the sum of individual goods as if the common good could result simply from individuals pursuing on their own self-interests without regard for the interests of others or for the general interest. The Catholic tradition insists that all individuals and all social institutions have a moral obligation to pursue the common good, and not just their own self-interest, their well-being or their own happiness. Within the Catholic tradition three principles play a crucial role in regulating the pursuit of the common good: catholicity, solidarity, and subsidiarity.[20]

In the Catholic tradition, as Casanova describes it, a map of moral principles is given for the pursuit of the common good, with moral obligations articulated for both institutions and individuals. In China today, when the spirit of neoliberalism is eroding many other values, are Confucian ethical teachings strong enough to balance its power? It can be argued that the continuing promotion of the common good in public discourse, incorporating the ancient ideal of the "Great Commonwealth," is exactly what China needs to solidify its current growth and to assure its future prosperity. And this future prosperity should be not only for China but also for all on our shared Earth.

Notes

1. Richard Madsen, "Inequality, Culture War, and Imperiled Common Good: America and China" (paper presented at the Asia/Pacific Studies Institute, 2020); Perry Link, Richard P. Madsen, and Paul G. Pickowicz, eds., *Restless China* (New York: Rowman & Littlefield, 2013); Martin King Whyte, *Myth of the Social Volcano: Perceptions of Inequality and Social Injustice in Contemporary China* (Stanford, CA: Stanford University Press, 2010).
2. Sébastien Billioud and Joël Thoraval, *Le sage et le peuple. Enquête sur le renouveau confucéen en Chine* [*The Sage and the People: Exploring the Confucian Revival in China*] (Paris: CNRS, 2013); Anna Sun, *Confucianism as a World Religion: Contested Histories and Contemporary Realities* (Princeton, NJ: Princeton University Press, 2013); Daniel Bell, *China's New Confucianism: Politics and Everyday Life in a Changing Society* (Princeton, NJ: Princeton University Press, 2008).
3. Robert N. Bellah, Richard Madsen, William M. Sullivan, Ann Swidler, and Steven M. Tipton, *Habits of the Heart: Individualism and Commitment in American Life* (Berkeley: University of California Press, 1985).
4. Max Weber, *Protestant Ethics and the Spirit of Capitalism* (New York: Routledge, 1992), 124.
5. Hai Ren, *The Middle Class in Neoliberal China: Governing Risk, Life-Building, and Themed Spaces* (London: Routledge, 2012); Donald Nonini, "Is China Becoming Neoliberal?" *Critique of Anthropology* 28, no. 2 (2008): 145–76.
6. Lisa Rofel, *Desiring China: Experiments in Neoliberalism, Sexuality, and Public Culture* (Durham, NC: Duke University Press, 2007), 3.
7. Chi-Ming Lam, ed., *Philosophy for Children in Confucian Societies: In Theory and Practice* (Oxford and New York: Routledge, 2020); Sun, *Confucianism as a World Religion*; Bell, *China's New Confucianism*; Wei-ming Tu, ed., *Confucian Traditions in East Asian Modernity: Moral Education and Economic Culture in Japan and the Four Mini-Dragons* (Cambridge, MA: Harvard University Press, 1996); Philip J. Ivanhoe, *Confucian Moral Self-Cultivation* (New York: Peter Lang, 1993).
8. See, for instance, David Solomon and P. C. Lo, eds., *The Common Good: Chinese and American Perspectives* (Dordrecht and New York: Springer, 2014).
9. Friedrich Engels, "Draft of a Communist Confession of Faith," in *Gründungsdokumente des Bundes der Kommunisten (1847)*.
10. Charles Wolf, "China's Capitalists Join the Party," *New York Times*, August 13, 2001.
11. Mencius, 4A3, in *Mencius: Revised Edition*, trans. D. C. Lau (Hong Kong: Chinese University Press, 1979), 153.
12. Cited in *Liji zhengyi* ("Explications of the *Record of Rites*"), 21:1a–3a, one of the "Thirteen Classics" of the Confucian canon. In W. M. Theodore de Bary and Irene Bloom, eds., *Sources of Chinese Tradition*, vol. 1, *From Earliest Times to 1600* (New York: Columbia University Press, 1999), 343.
13. *Liji zhengyi* 21:1a–3a.
14. Michael David Kaulana Ing, *The Dysfunction of Ritual in Early Confucianism* (Oxford: Oxford University Press, 2012), 105.
15. From Kang Youwei's *Lunyu zhu* ("Commentaries on the *Analects*"), 2.11a–12b, first published in 1917. In W. M. Theodore de Bary and Irene Bloom, eds., *Sources of Chinese Tradition*, vol. 2, *From 1600 Through the Twentieth Century* (New York: Columbia University Press, 2000), 268.

16. Kang Youwei, *Lunyu zhu*, 2.11a–12b.
17. Barbara Hendrischke, *The Scripture on Great Peace: The Taiping Jing and the Beginning of Daoism* (Berkeley: University of California Press, 2006), 11.
18. Fung Yu-lan, *A History of Chinese Philosophy*, vol. 1 (Princeton, NJ: Princeton University Press, 1952), 378.
19. Dennis P. McCann and Patrick D. Miller, eds., *In Search of the Common Good* (New York and London: T&T Clark, 2005); M. S. Kempshall, *The Common Good in Late Medieval Political Thought* (Oxford: Clarendon, 1999).
20. José Casanova, "The Pursuit of Happiness, Religion, and Globalization" (lecture given as part of "The Bridges to the Future" lecture series, University of Denver, Denver, CO, 2008), 7.

CHAPTER 7

HINDUISM

"Consider the common good in all actions"

ANANTANAND RAMBACHAN

Hinduism is an astonishingly diverse tradition, suggested by the word "Hindu" itself. This term has been used at different times, and even at the same time, to signify geographical, religious, cultural, and, in more recent times, national realities. "Hindu" is the Iranian variation for the name of a river that the Indo-Europeans referred to as the "Sindhu," the Greeks as the "Indos," and the British as the "Indus." Those who inhabited the regions drained by the Indus river system were derivatively called Hindus. They did not share a homogeneous religious culture, and today the Hindu tradition reflects the astonishing variation in geography, language, and culture across the Indian subcontinent. It helps to think of Hinduism as a large, ancient, and extended family, recognizable through common features but also reflecting the uniqueness of its individual members.

Though generalizations are hazardous, scholars have identified some common features and themes.[1] One common feature is a recognition of the significance of the four Vedas (*Rg*, *Sama*, *Yajur*, and *Atharva*) as sources of authoritative teaching. The four Vedas are widely acknowledged by Hindus to contain revealed teachings, and acceptance of the authority of the Vedas is commonly regarded as necessary for Hindu orthodoxy, even though such acceptance may be merely formal and nominal. In reality, Hindus look to a wide variety of sacred sources, including the *Bhagavadgita*, the *Ramayana*, and the *Mahabharata*. A second common feature is a belief in a moral order and rebirth (*karma* and *samsara*), emphasizing human responsibility and the short- and long-term consequences of actions. The idea here is that every volitional action produces a result that is determined by the nature of the action and the underlying motive. The consequences of these actions may stretch into lives beyond the present one. A third common feature is the affirmation of an ultimate reality, *brahman*, that is regarded as both transcendent and immanent; it is called by many names and imagined in diverse forms. A fourth common feature is the belief that ignorance (*avidya*) of the nature of this reality is the fundamental human problem and the primary cause of suffering. "Ignorance" is defined as the assertion of a separate selfhood, our failure to understand ourselves as identical with, or as part of an unlimited, interrelated reality. Ignorance expresses itself as self-insecurity, a persistent sense of want and lack, which in turn leads to greed, an insatiable grasping after wealth and power with no thought of consequences for the common good. A fifth common feature is an optimistic outlook flowing from a belief in the possibility of liberation (*moksha*). Liberation awakens us to the unity of life, frees us from self-centeredness, and inspires us to live lives committed to the common good. These common orientations provide the fabric, as it were, out of which a rich theological diversity is woven.

The Common Good (*Lokasangraha*)

The *Bhagavadgita*, regarded as one of the pillars of the Hindu tradition, commends, on two occasions, a commitment to the common good.[2] In the first reference (3:20), the teacher Krishna commends ancient political leaders, like King Janaka, who were active in the world and who attained the highest goal of the religious life. Krishna encourages his student Arjuna to work as these leaders did and to consider the common good in all undertakings. In the second reference (3:25), commitment to the common good is commended and praised as superior to actions that are selfish and self-centered.

The Sanskrit equivalent in the *Bhagavadgita* for "the common good" is *lokasangraha*. *Lokasangraha* is inclusive. It includes all human beings and the world of nature. It does not allow us to privilege unjustly the interests of a particular institution, nation, religion, race, or gender. It excludes the pursuit of personal and institutional interests that violently impede the flourishing of all. It excludes trying to lift oneself or one's nation by crushing others. *Lokasangraha* ensures that virtues are not privatized but applied publicly for peace, justice, and flourishing of all. The common good is not served by an economic system that depletes our natural resources, eradicates our biodiversity, and adversely affects our climate. The common good becomes the measure of the meaning of all that we do. In the Hindu tradition, this concern for the common good is not limited to the actions of individual human beings. It must also become normative for the policies and practices of corporations, institutions, and states.

The Hindu tradition that commends this commitment to the common good also provides its theological justification. At the heart of this theology is an understanding of the unity and interdependent character of existence. One of Hinduism's important creation narratives occurs in the *Taittiriya Upanishad* (2.2.1), which describes a sequence of the emergence of the great elements and life from an infinite source (*brahman*). The first to emerge is space; from space comes air; from air comes heat or fire; from fire comes water; from water comes the earth; from the earth comes vegetation; from vegetation comes food; and from food come living beings.[3] The text describes human beings as formed from food. The point is that we are formed of the universe, represent it in our bodies, and are always organically connected with it. The earth is in everything solid in our bodies, water is in all that is liquid, fire is in the body's warmth and energy, and air is in the breath. Dualistic language that speaks of human beings and nature as separate realities is both false and dangerous. It is false because we are we an organic and integral part of the natural world, made from and embodying space, air, fire, water, and earth. We do not exist independently from the natural world. It is dangerous because it objectifies the natural world and contributes to the belief that we can destroy the natural world without destroying ourselves.

The interdependent character of existence that is the foundation of the Hindu commitment to the common good includes our relationships with the human community. The Hindu tradition describes every human being as born into debt (*rnam*). To be human is to be indebted; one does not develop one's human potential outside of a human community. The tradition emphasizes our indebtedness to, among others, the divine, parents and ancestors, teachers, other human caregivers, and nature. We exist and have the potential to flourish because we are a part of a generous and supportive interdependent whole. The proper response to this understanding of our unity, indebtedness, and interdependence is a commitment to the common good. If we selfishly receive and exploit without generous self-giving, we deplete the resources of our world and become agents of suffering. *Yajña* is the term used in the *Bhagavadgita* to describe this generosity. The *Bhagavadgita* (3:12) describes those who benefit from others' giving without responding with generosity as thieves. As noted earlier, we ought to apply this description to individuals, corporations, and states. The text (3:13) likens those who are not committed to the common good to people who selfishly cook only for themselves; virtuous people committed to the common good cook both for themselves and for others. The food of the selfish is impure; the food of the virtuous is pure. Mutual commitment to the common good, teaches the *Bhagavadgita* (3:12), is the way ordained by the creator for human prosperity. It is the *kamadhuk*, that is the wish-fulfilling cow that provides for all human needs. We attain the highest good by mutually caring for each other (*parasparam bhavayantah sreyah param avapsyatha* 3:11).

In the discussion that follows, I selectively identify and discuss significant dimensions of a commitment to the common good that the Hindu tradition regards as essential for the flourishing of all. I take as my inspiration the description of a flourishing community offered by the sixteenth-century poet-saint Tulasidasa in his retelling of the life story of Rama in the *Ramacaritamanas*.[4] Tulasidasa speaks of this utopian community as *Ramarajya* (the kingdom of Rama). In this community, "there is no premature death or suffering of any kind; everyone enjoys beauty and health. No one is poor, sorrowful or in want; no one is ignorant." The community is free from hate and violence, and nature flourishes. "The trees in the forests," wrote Tulasidasa, "bloom and bear fruit throughout the year; the elephant and lion live together as friends; birds and beasts of every kind are no longer hostile and live in harmony with one another."[5]

Human Dignity and the Common Good

Commitment to the common good, as discussed earlier, is rooted theologically in the unity of life and in the nature of the world as an interdependent reality. In this reality, we thrive because we are receivers of the generous gifts of others. It is our obligation, in Hindu teaching, to act reciprocally by being willing givers and to consider the common good in all decisions. A community prospers when human beings are both receivers and givers. To be a receiver is not a choice, but to be a giver is a moral obligation.

At its heart, a commitment to the common good is the expression of care and concern for the flourishing of all living beings and the planet that is our common home. It is an expression of our value and reverence for life and for the happiness of all. "May all be happy (*sarve bhavantu sukhinah*)" is the opening line of a well-known Hindu prayer. The prayer concludes with the hope that all may be free from suffering (*ma kascit duhkha bhagbhavet*). Care for the well-being of others and delight in their happiness originate in a theological understanding that enriches the argument for interdependence.

The *Upanishads* define the divine as "that from which all beings originate, by which they are sustained and to which they return" (*Taittiriya Upanishad* 3.1.1). The *Bhagavadgita* (9:17–18) speaks of the divine as father and mother of the universe and as its nourisher, lord, goal, and friend. God is not the national or tribal deity of a particular religious or ethnic community but the source of all life and existence. For Mahatma Gandhi, the oneness of the divine implied the unity of humanity. "I believe," wrote Gandhi, "in the absolute oneness of God and, therefore, of humanity. What though we have many bodies? We have but one soul. The rays of the sun are many through refraction. But they have the same source. I cannot, therefore, detach myself from the wickedest soul nor may I be denied identity with the most virtuous."[6]

In addition, Hindu traditions affirm, without qualification, the equal existence of God in all. The *Bhagavadgita* (18:61) teaches that "God dwells in the heart of all beings." There is no life outside of God, and there is nothing that exists that is not sustained by God. God is the unifying reality present in the entire creation, the singular truth permeating, uniting, and sustaining all existence. The equal presence of God in all beings is the source of the inherent dignity and equal worth of every human being. It invites us to respond with reverence toward all beings. The value of human beings is not derived from the state and is not reducible to economic or political considerations. We cannot value the divine and devalue human beings or be indifferent to the conditions of their lives. We cannot give our assent or support to any social or cultural system that is founded on human inequality and indignity. To see women as inferior to men, to prefer the male child, to mistreat the elderly, to ascribe unequal worth to and demean people on the basis of birth, and to discriminate and practice violence against any group are all in fundamental contradiction to the deepest Hindu teachings.

The ethical value that most eloquently expresses this reverence for life is *ahimsa* (nonviolence), regarded in the Hindu tradition as the foremost of virtues. In his understanding of the meaning of *ahimsa*, Gandhi explained that in its negative form it means abstention from injury to living beings. In its positive form, *ahimsa* is the practice of love and compassion (*daya*) for all. For Gandhi, *ahimsa* also means justice for everyone and abstention from all forms of exploitation. In other words, *ahimsa* is consistent with a commitment to the common good. Tulasidasa, cited earlier, emphasized caring for others as the highest expression of the moral life and described it as empathy or identifying with the other in suffering and in happiness (*para duḥkha duḥkha sukha sukha*). He equated the moral life with working for the well-being of others and immorality with being an agent of oppression and suffering.

The intrinsic dignity of living beings, the unity of life, noninjury, compassion, and empathy, when translated into active virtues, require that we properly inquire into the causes of suffering with the aim of overcoming them. We cannot ignore the suffering of human beings when they lack opportunities to attain the necessities for dignified and decent living or when suffering is inflicted through oppression and injustice based on gender, birth, or race. Working to overcome suffering requires identifying those political, social, and economic structures that cause and perpetuate suffering. The unmistakable call to be one with the suffering other requires nothing less. "No man," wrote Gandhi, "could be actively nonviolent and not rise against social injustice no matter where it occurred."[7] The practice of nonviolence, in other words, requires a commitment to the common good.

Prosperity and the Common Good

The Hindu tradition identifies four goals for a good human life: virtue (*dharma*), wealth (*artha*), pleasure (*kama*), and liberation (*moksha*). Contrary to popular impressions, the Hindu tradition is neither life denying nor narrowly other-worldly. Hinduism has never given its blessings to involuntary poverty. It recognizes poverty to be a great cause of suffering. By including wealth, *artha*, as one of life's four goals—along with pleasure, virtue, and liberation—Hinduism recognizes the need of every human being for access to those material necessities, such as food, health care, shelter, and clothing, that make life possible and enable human beings to live with dignity.

The goal of virtue (*dharma*) emphasizes the social and relational context in which we live our lives and the need to regulate our pursuit or wealth and pleasure in the interest of the well-being of others. The attainment of wealth and pleasure by inflicting pain and suffering on others or by denying them the freedom to pursue these ends themselves would be opposed to *dharma*. The pursuit of wealth and pleasure must be consistent with the requirements of the common good. There is no good reason why the requirement to adhere to the demands of *dharma* should be limited to individuals. Today, we must extend the ethics of *dharma* to corporations and nations because of their power to affect the common good.

The growth of corporate power and wealth has been phenomenal. The domination of the world economy by corporations is unprecedented. The revenues of Walmart, for example, surpass those of 157 small countries. Sixty-nine of the richest 100 entities in the world are global corporations.[8] Although enjoying many of the rights and privileges of individual people, as well as wealth and power that surpass those of nations, corporations have not consistently concerned themselves with the common good. Because of the global nature of corporations, the rights and powers they enjoy are often exercised by shareholders who have no concern for the good of the local communities where these businesses establish operations. The sole measure of success of many of these corporations is the maximization of profit and market share. Toward this end, consumerism is encouraged, and relentless advertising generates artificial desires for more and more products to the long-term detriment of the common good.

Since corporations, at least as legal entities, did not exist during the early history of most religious traditions, teachings and teachers do not speak directly about their responsibilities or offer specific perspectives. Religious traditions, however, are eloquent about the fundamental causes of human suffering and offer visions of the good life for all. In the case of the Hindu traditions, the root causes of suffering are ignorance, greed, and greedful actions (*avidya-kma-karma*), spoken of as the three knots of the heart.

When individual choices and actions spring from ignorance and greed, the consequence is inevitable suffering for oneself and harm to the common good. If ignorance and greed are conditions that afflict us as individuals, it is true also that these afflictions extend to the group or collective identities that we profess. The corporation is another more recent expression of our collective identity. Here the knots of the heart, as we see in so many cases, become institutionalized in corporate personhood and identity. The maximization of profits through growth becomes the sole concern, with little or no concern for the common good. Where there is greed, individual or institutionalized, it is believed that there is never enough, and there is little concern about outcomes beyond one's own gains.

Religious analysis and critique of the causes of suffering in ignorance and greed must not be limited to the individual. It is urgent that such analysis be extended to corporations. Because of the global reach of corporations and their growing power and wealth, they cause suffering that outweighs anything that individuals can do. Religious traditions concerned with the common good and overcoming suffering cannot ignore the role and impact of corporations. In a world where the prevailing voices speak for a limited good, the nation, a specific community, or the corporation, our traditions have the vision and potential to speak powerfully for the universal common good.

Gandhi emphasized the need for economics to be infused with an ethical concern and focus. "True economics," said Gandhi, "never militates against the highest ethical standard, just as all true ethics to be worth its name at the same time must also be good economics."[9] The *Mahabharata* emphasizes that one should seek *artha* in ways that avoid harm, deception, and cheating. One's gain must not be unlawful or result in another's loss; such gains, ignoring *dharma* and driven by greed, are destructive of self and others.[10] The text is clearly referring to corruption in its various forms.

Corruption can become and is often systemic and structural, leading to a concentration of power and privilege in the hands of a few and disempowering and limiting opportunities for the many. Systemic corruption benefits a minority who are deeply invested in the perpetuation of unjust structures. Because corruption is structural, structural change is necessary to overcome it. We need to give attention not only to the corruption of the human heart but also of social, political, and economic structures that harm the common good.

Our Planet and the Common Good

Lokasangraha, the Hindu equivalent of "the common good," is not anthropocentric. It is concerned with the flourishing of the whole, which includes all living beings and the earth that is our common home. The earth community includes every creature depending on her for sustenance—everyone. The earth is not a passive and inert field, dualistically separate from us, that we may thoughtlessly and inconsequentially exploit for our purposes. Our lives are inextricably bound together and our well-being inseparably linked. We do not hurt the planet without hurting ourselves, as the devastating effects of climate change show.

Human self-interest is a valid cause for prudent care for the world, and it is a prominent argument. From a place of religious concern, however, it is not the highest or only argument. A philosophy of environmental justice that is justified only by an appeal to human self-interest still instrumentalizes the world and makes human needs the norm for measuring the value of everything else. From a Hindu theological perspective, the world has integrity and intrinsic value, and our care for it ought to express this value. Its value cannot be measured only by what it contributes to our existence. We are not at liberty to destroy and recklessly exploit everything that we perceive as unnecessary for our own prosperity.

To emphasize this point, allow me to share the famous opening verse of the *Isa Upanishad*:

> This entire universe, moving and unmoving, is enfolded in God. Renounce and enjoy. Do not covet the wealth of others (*Isavasyam idam sarvam yat kim ca jagatyam jagat, tena tyaktena bhunjitha, ma gridhah kasyasvid dhanam*).

The first line affirms the truth of God, named here as *isa*, meaning "ruler" or "lord," and describes the universe as enfolded or wrapped in the divine (*ishavasyam*). No thing and no one is outside of God; nothing exists separately from God. The text emphasizes this divine inclusivity.

Enfolding is a form of embracing and suggests care and love. It points also to the precious worth of the universe, which is embraced by God as a child in the hands of a mother or father. The text invites us to see the universe as a sacred reality that exists within God, is pervaded by God, and is God's gift to us. Such "seeing" in the Hindu tradition is spoken of as sacred seeing (*darshan*). *Darshan* is seeing the universe but also seeing more, seeing the invisible divine that penetrates and is the ontological foundation and ground of the universe. To see the world existing independently of or separate from God, is to see in error; it is false seeing.

The *Bhagavadgita* (7:7) offers a similar vision of the universe:

> Nothing higher than God exists; everything exists in God like jewels on a string (*mattahparataram nanyat kimcid asti dhananjaya/mayi sarvam idam protam sutre manigana iva*).

When we look at a beautiful necklace, we may notice the jewels, but we do not often see or think about the string that is concealed within each jewel giving it order and sustenance. The invisible string links and unites the jewels with each other, however different each one may be. In an analogous way, the divine is the common and unifying reality in all creation. The divine sustains the planet not from a remote distance but by being intimately present in everyone and in everything. The *Bhagavadgita* (7:8–9) invites us to experience the divine in the taste of water, the brilliance of the moon and sun, the sound in air, the pure fragrance of the earth, the radiance of fire, and the life in all beings.

In the second line of the opening verse of the *Isa Upanishad*, seeing the planet in its connectedness to God and as God's gift is presented as a mode of renunciation. The renunciation commended here is not a negativization of the world, a withdrawal or escape from it. Here, in a profound insight, renunciation is paired with enjoyment (*tyaktena bhunjitha*; "renounce and enjoy"). Seeing the world wrapped in and inseparable from the divine transforms our relationship with it. This seeing liberates us from the urge to possess, to claim, to dominate, and to see the universe as existing solely to serve our human needs.

As theologians of various traditions have noted, the desacralization of the planet helps to create the condition for its reckless exploitation. Its sacralization, on the other hand, moves us from greed to gratitude. "Renounce and enjoy," says the *Isa Upanishad*. We are invited to enjoy the universe, to celebrate its existence, and such joy is, in fact, enhanced, not diminished, by renunciation. We can delight in and enjoy the world as divine generosity, spoken of in the Hindu tradition as *prasadam*. This includes appreciating the air we breathe, the water we drink, the food we eat, and the clothing we wear. The renunciation commended in this verse is not a rejection of or turning away from the world but our own loving embrace, and care for and enjoyment, of all that exists.

The final line of the *Isa Upanishad* text, "Do not covet the wealth of others," speaks to us in a special way about our human relationships and the need to cultivate and promote communities and institutions, at all levels, that are motivated not by greed but that reflect the joy of justice, generosity, and commitment to the common good. It is the call to value other human beings intrinsically, not on the basis of whether they are useful in achieving our own desires. A greedful way of being in relation with other human beings or the planet does not express the vision of the world enfolded in God. A greedful way of being is the antithesis of the beauty and the joy in creation that the *Isa Upanishad* offers us. It is the antithesis of valuing the common good. What we are called upon to renounce is greed; what we rejoice in is the common good, the well-being of all. The *Bhagavadgita* condemns and describes as thieves those who selfishly exploit the planet's resources without regard for its sustainability. It commends a life of moderation in consumption and mutuality in receiving and giving.

At the heart of the Hindu understanding of the common good is a vision of life's unity that flows from the existence of all in the one enfolding divine reality. This vision is the source of a reverential attitude to all life and to a way of being that is infused with joy, compassion, generosity, and delight in the flourishing of all. Living out of this truth requires work to overcome structures of inequality, indignity, and injustice.

Notes

1. See Ramakrishna Puligandla, *Fundamentals of Indian Philosophy* (Nashville, TN: Abingdon, 1975), 25–26; Troy Wilson Organ, *Hinduism: Its Historical Development* (Woodbury, NY: Barron's Educational Series, 1974), 28–34.
2. Winthrop Sargeant, trans., *Shri Bhagavad Gita* (Albany: State University of New York Press, 1993). The text dates from about 150 BCE to 250 CE. Translations modified.
3. Patrick Olivelle, trans., *Upanisads* (New York: Oxford University Press, 1996).
4. Rama is regarded by most Hindus as an *avatara* (a descent of the divine into our world). See R. C. Prasad, trans., *Tulasidasa's Sri Ramacaritamanasa: The Holy Lake of the Acts of Rama* (Delhi: Motilal Banarsidass, 1991).
5. See Prasad, *Tulasidasa's Sri Ramacaritamanasa*, 704–7 (*Uttarakanda*), for a full description of *Ramarajya*.
6. Mahatma Gandhi, *All Men Are Brothers: Life and Thoughts of Mahatma Gandhi, as Told in His Own Words* (New York: Continuum, 1980), 75.
7. Gandhi, *All Men Are Brothers*, 85.
8. See Global Justice Now, "69 of the richest 100 entities on the planet are corporations, not government, figures show," October 17, 2018, https://www.globaljustice.org.uk/news/69-richest-100-entities-planet-are-corporations-not-governments-figures-show/. Accessed August 7, 2017.
9. M. K. Gandhi, *The Voice of Truth*, trans. Valji Govindji Desai (Ahmedabad: Navajivan, 1969), 321–22.
10. See Y. Krishan, "The Meaning of the Purusarthas in the *Mahabharata*," in *Moral Dilemmas in the Mahabharata*, ed. Bimal Krishna Matilal (Shimla: Indian Institute of Advanced Studies, 1989), 57–58. The *Mahabharata* (400 BCE–100 CE) is traditionally attributed the author, Vyasa, but the name means simply "compiler." It is considered the to be the longest work in Indian literary history and consists of one hundred thousand verses. The content is diverse, but at its core is the story of the rivalry between two sets of cousins culminating in a climactic battle that sets the scene for the *Bhagavadgita* dialogue.

CHAPTER 8

JUDAISM AND THE COMMON GOOD

DAVID ROSEN

The idea that the human person is created in the Divine Image, as described in the Creation story at the beginning of the book of Genesis, serves as the foundation for Judaism's teaching regarding the sanctity of human life and the importance of freedom and dignity for all people. Moreover, the Creator and Guide of the Universe is presented in the Hebrew Bible as a moral deity of justice, righteousness, love, and compassion; accordingly, the idea of being created in the Divine Image is called upon as the basis for the expectation of human moral behavior.

Indeed, the idea of the Divine Image in all people is described by the second-century sage Ben Azzai as the most important of all Biblical ethical principles.[1] Rabbi Tanhuma adds that this idea serves as a constant reminder not to treat any person with disrespect. A century and a half earlier, the great sage Hillel affirmed that the essence of Judaism is "do not do to others what is hateful to you."[2]

Hillel also taught the concept that taking or saving one human life is comparable to destroying or saving the whole world.[3] The Mishnah indicates that this is precisely the message behind the Genesis narrative describing the first human being as a single creature (as opposed to all other beings, which are created male and female): "In order to teach you that he who destroys one person's life, it is considered as if he destroyed a whole world; and he who preserves one person's life, it is as if he has preserved a whole world."[4] Furthermore, while self-defense and the defense of others whose lives are threatened are duties,[5] murder is seen as the most heinous of sins; it diminishes the Divine in the world precisely because each person is created in the Divine Image.[6]

Thus, rather than be party to killing an innocent person, one must be willing to suffer martyrdom, and one may not claim that "one's blood is redder than that of another person."[7]

The Mishnah indicates another moral message behind the narrative regarding the Divine creation of a single first human, which emphasizes that we are all descended from a common ancestor: "Therefore none should say, [my] Father is greater than yours."

The vast majority of the Biblical commandments concern human conduct and responsibility toward one another, summed up in the words in Leviticus: "And you shall love your neighbor as yourself, I am the Lord,"[8] and in the words of the prophet Micah: "It has been told to you O human being, what is good and what the Lord requires of you; but to deal righteously, love acts of kindness, and walk humbly with your God."[9]

As indicated, the Hebrew Bible views the supreme ethical goal as being the emulation of the Divine moral qualities. In addition to the injunction to "be holy as . . . the Lord your God is Holy,"[10] rabbis understand the idea of *imitatio dei*, to be the intent of the commandment to cleave to God.[11]

Accordingly, the sages of the Talmud declare, "Just as He is gracious and compassionate, so you be gracious and compassionate."[12]

And because, in the words of the Psalmist, God's "mercies extend to all His creatures,"[13] the Bible makes it clear that this merciful nature means a special care for the weakest members of society,[14] whom we are accordingly obliged to reflect in our own actions.

Thus the Hebrew Bible is replete with instructions to pay special attention and concern to those whose dignity is vulnerable and who are marginalized—the poor, the stranger, the widow, and the orphan. For example: "And when you reap the harvest of your land, you shall not reap the corners of your field, nor shall you gather the gleaning of your harvest. And you shall not glean your vineyard, neither shall you gather the fallen fruit of your vineyard; you shall leave them for the poor and for the stranger."[15]

The Talmud notes that the obligation to care for the stranger or sojourner is repeated thirty-six times throughout the Pentateuch, more than any other injunction. Notable in this regard are the following instructions: "There shall be one law [judgment] for you; it shall be for the sojourner as well as the native, I am the Lord your God,"[16] and "If your brother becomes poor and cannot maintain himself with you, you shall support him, similarly the stranger and sojourner, he shall live with you."[17] Jewish tradition understands these verses as requiring the Jewish community to guarantee to the non-Jewish resident who dwell among them the same civil rights and benefits that they enjoy.

While the Hebrew Bible envisions the people of Israel as a particularly covenanted people, requiring of them a rigorous ethical and ritual life as "a kingdom of priests and a holy nation,"[18] rabbinic Judaism understands the covenant made with the children of Noah after the flood[19] to have been an original universal covenant reflecting Divine care for all human beings and requiring human observance of seven universal commandments summarizing the essence of universal morality.[20] Yet the Jewish community is called upon to care also for those who do not observe even these basic universal precepts.

On the basis of the guiding statement in the Talmud "that the whole of Judaism is for the sake of Peace," the sages demand charity for the poor, care for the sick, and respect for the dead, even for idolaters.[21]

Maimonides codifies these responsibilities and places them in their scriptural moral context:

> We are obliged to maintain even the poor of idolaters, attend to their sick and bury their dead, as we do with those of our own community, for the sake of the Ways of Peace. Behold it is said: "Her ways are pleasant ways and all her paths are Peace" (Proverbs 3:17); and it is written, "God is good to all and His mercy extends to all His creatures" (Psalms 145:9).[22]

While this demand goes beyond the minimal letter of the law, it reflects the view of the Torah as aspiring for an ethic—the ways of Peace—to advance the well-being of all people.[23]

While the political system envisaged in the Hebrew Bible is a monarchic one, there is debate among medieval rabbinic authorities as to whether this was an ideal or a necessity. Regardless, the monarch is explicitly subject to the law, which not only requires him to treat all subjects equally with justice and dignity but also subjects the king himself to the same moral standards.[24] Additionally, although Judaism has nothing analogous to the Greek and Roman concept of "state," it has a profound sense of community and its legal obligations. Thus it was the duty of the Sanhedrin in ancient Israel to appoint qualified courts[25]; further, the ancient rabbis instituted the establishment of schools throughout the land to guarantee widespread educational opportunities.[26] It appears that municipal governance had also been instituted on a national scale by Talmudic times,[27] and the Talmud clarifies the rights of the residents of a city to establish by mutual consent standards of measurement, market prices, wages, and so on, as well as the right to apply sanctions against those who violate these.[28]

Communal interests, however, must still respect the inalienable rights and freedoms of the individual. In the words of Samuel Belkin,

> The belief in the sacredness of the human personality not only governs the relations of one individual to another: it defines [the individual's] relation to society as a whole. . . . (While) each individual shares in the responsibilities of the social order, [nevertheless it] is guaranteed that just as no individual can acquire ownership [of another], so the group will never be given unlimited authority over his person.[29]

The Hebrew Bible takes the concept of collective identities seriously, notably the identities of nations.[30] Indeed, the ideal human society, the Messianic vision, is not one of a denationalized society, but a truly international one:

> And many peoples shall go and say let us go up unto the mountain of the Lord to the house of the God of Jacob that He will teach us His Ways and we will walk in His Paths, for from Zion shall go forth Instruction and the Word of the Lord from Jerusalem. And He will judge among nations and decide for many peoples; and they shall beat their swords into ploughshares and their spears into pruning hooks; nation shall not lift up sword against nation, nor shall they learn war anymore.[31]

This Biblical aspiration for global peace, however, is not something that may be left to pious expectation. Commenting on the words of the Psalmist, "search for peace and pursue it,"[32] the sages point out that the obligation to seek peace is of a much higher order than other Biblical injunctions. For whereas many of the Torah's commandments are phrased in conditional terms such as "if you see," "if you meet," and "if you come across," which indicate that they are operative only in specific situations, the imperative of peace requires us to go out of our way, to seek it out and pursue it everywhere to the best of our ability.[33]

Furthermore, we are told that the very goal of Divine creation was in order "that there be peace between all human beings"[34] Similarly, the midrashic work Seder Eliahu Rabbah teaches that Divine peace is generated "in seventy languages for all persons whom He created and brought into the world"[35]; that is, peace is the goal of all people and peoples, requiring behavior toward all accordingly.

As mentioned, the promotion of the common good concerns not only human social behavior but also ensuring that basic necessary resources are available to all (as reflected in the previously mentioned legislation requiring that corners of fields and gleanings be left for the poor and in the general tithes for the poor and provisions for the orphan, widow, and stranger). Accordingly, integral human development necessitates a sustainable ecology to provide for such. Moreover, the very Biblical foundational view of our world as a Divine Creation demands special care and protection of the environment as indicated in Genesis: "And He placed him in the Garden of Eden to cultivate it and to preserve it"[36]—what we would describe today as sustainable development.

The perception of our ecosystem as a Divine Creation leads to a particular perspective on the human relationship with natural resources. First and foremost in this regard is the recognition that "the earth is the Lord's and all that is in it."[37] Leviticus presents God as saying, "For the earth is mine and you are sojourners and temporary residents in it with me."[38] Fundamental to the Biblical description of the world in which we live and the way in which we should conduct ourselves in relation to it, is the understanding that we are but tenants in a world that belongs to its Creator.

Above all, Judaism derives the imperative of environmental responsibility from the prohibition in Deuteronomy against cutting down fruit trees when laying siege to a city in a context of war.[39] The sages of the Talmud draw an *a fortiori* conclusion that if in a situation of war in which human life is in danger it is prohibited to cut down a fruit tree, under normal conditions the prohibition against destroying anything that provides sustenance is even greater. Indeed this prohibition is extended to anything that can be of use and of value; it extends to waste[40] and even to ostentation and overindulgence.[41]

The Talmud also requires certain businesses—notably threshing floors and tanneries—to be kept at a distance from human domiciles so that they do not cause harm through physical pollution or even pollution of the senses through, for example, unpleasant odors.[42] Moreover, the prohibition against pollution is contingent not only on private ownership but extends also to public spaces and thus the environment as a whole.[43] Rabbi Samson Raphael Hirsch, one of the great Orthodox rabbis of the nineteenth century, considered the concept of *bal tashchit*—the prohibition against waste and wanton destruction—to be the most basic of Jewish ethical precepts, acknowledging the sovereignty of God as the Source and Owner of all, and requiring moral discipline of the human ego and unbridled impulses. When we preserve the world around us, he explains, we act with an understanding that God owns everything. When we destroy it, however, we are worshipping the idols of our own desires, indulging only in self-gratification and forgetting, if not denying, the One Source of all.

By observing the discipline of this prohibition, we restore harmony between ourselves and the world around us and above all consciously respect the transcendent Divine Will, which we place above our own selfish interests. Failure to do so, declares Hirsch, is in effect idolatry:

> If you destroy, if you ruin, at that moment you are no longer truly human and you have no right to the things around you. I, God, lent them to you for wise use only, never forget that I entrusted them to you.
>
> As soon as you use them unwisely, be it the greatest or the smallest, you commit treachery against My world, you commit murder and robbery against My property, you sin against Me.
>
> This is what God calls unto you and with His call God represents the greatest and the smallest against you and grants the smallest as well as the greatest, a right against your presumptuousness. In truth there is no one nearer to idolatry than one who can disregard the fact that all things are the property of God; and who then presumes also to have the right because he has the might to destroy them according to a presumptuous act of will.
>
> Indeed such a person is already serving the most powerful of idols, anger, pride and above all ego, which in its passion regards itself as the master of all things.[44]

The Sabbath is arguably the most central precept of ritual observance in Judaism and serves as an ecological paradigm, providing for a day on which the natural ecosystem, as well as human society, are able to rest, regardless of position or authority. In his commentary on the Pentateuch, Rabbi Hirsch explains:

> The Sabbath was given so that we should not grow arrogant in our dominion in God's Creation . . . (to) refrain on this day from exercising our human sway over the things of the earth, and not lend our hands to any object for the purpose of human dominion . . . the borrowed world is, as it were, returned to its Divine owner in order to realize that it is but lent (to us). On the Sabbath you divest yourself of your glorious mastery over the matter of the world and lay yourself and your world in acknowledgment at the feet of the Eternal, your God.[45]

The Sabbath is accordingly seen as a weekly restoration of the natural relationship both in relation to the Divine and in relation to our environment: social and ecological. It is not that the work and material development of our weekday activity are unnatural—on the contrary. In the words of Hirsh, this is our "glorious mastery," and we are commanded, "Six days shall you work."[46] But there is a real danger that our creative labor can take us over, subjugating and even stifling our social and spiritual potential. Indeed, there is a danger that our technological capacities can become the be-all and end-all in what has often become a kind of modern idolatry. To quote Hirsch again,

> To cease for a whole day from all business, from all work, in the frenzied hustle and bustle of our time? To close the exchanges, the workplaces and factories; to stop all railway services? [We might add, "to switch off our computers, to do without our smartphones."] Great Heavens, how would it be possible—the pulse of life would stop beating and the world would perish! The world would perish? On the contrary, it would be saved.[47]

The sabbatical year in many respects echoes the Sabbath day and seeks the social balance and restoration of society at large. Three essential components are involved. First and foremost, as stated in Exodus, the land is to lie fallow, untilled, and unpossessed, serving as the most eloquent testimony that "the earth is the Lord's."[48] In an agricultural society, land is also the source of status. Thus, in requiring that every seven years the land is, as it were, returned back to its original owner, to God, an important social ethical statement is being made with regard to the equality of all before God. This restoration of social equilibrium is further reinforced by the other two precepts. The first is the cancellation and annulment of debts. The moral significance of this concept can be understood only within the agrarian context in which it functioned.

Loans (and consequent debts), which for us are a normal part of commercial life, were not part of a normal, healthy agricultural society in the ancient past. A loan was taken only when a farmer fell upon unusual hardship through diseased crops, drought, or suchlike and did not have the resources to enable them to harvest once again. Thus a loan was an exceptional but essential means for the restoration of a healthy, productive agricultural cycle.

However, taking a loan posed the threat of a poverty trap. If the following year's harvest is not successful enough, one may not be able to repay one's debt. This may continue year after year, the debt compounding and the farmer becoming economically ensnared. Accordingly, the cancellation of debts in the sabbatical year ensured that nobody would be caught in a poverty trap for very long. Similarly, the third component of the sabbatical year is the manumission of servants who either sold themselves into the employment of others in order to escape poverty or to pay off debts (or punishments administered by the courts).

The most extensive passage in the Bible dealing with the sabbatical year, Leviticus 25, is followed in the next chapter by the promise of good rains and harvests, prolonging our days on the earth and guaranteeing peace. This, the Bible explains, is the consequence of observing the Divine commandments, but if we disregard these, we face ecological disaster, failed harvests, war, and devastation.

Many medieval Jewish commentators on the Bible used the Talmudic phrase "the Torah speaks in the language of humanity"[49] to explain Biblical imagery in metaphoric terms, in this case seeking to convey the higher idea of the spiritual consequences of our actions in a manner that even the most simple might be able to grasp.[50] However, today we can interpret these texts more literally than ever before, because the consequences of human conduct toward our environment are so strikingly evident. Human avarice, unbridled hubris, insensitivity, and lack of responsibility toward others and our environment have led to the pollution and destruction of much of our natural resources and interfered with the climate as a whole, jeopardizing our rains and harvests and threatening the very future of sentient life on the planet.[51]

Moreover, unrestrained, irresponsible indulgence in modern society has led not only to far greater cruelty toward animal life, abused on a massive scale for the sake of human indulgence, but also to the exploitation of large sections of humanity to serve a much smaller segment. Indeed, shocking numbers of human beings languish in hunger while others overindulge. The need to restore our society to one that is in tune with the values reflected in these Jewish teachings that seek the common good is more critical than ever. Our world today demonstrates not only how much moral conduct or lack thereof impacts upon human society as a whole but also how it impacts upon our very capacity to live on the earth that has been entrusted to us.

Notes

1. Jerusalem Talmud, Nedarim 9:4; Bereshit Rabbah 24:7; Sifre, Kedoshim 4:12.
2. Babylonian Talmud, Shabbat 31a.
3. Jerusalem Talmud, Sanhedrin 4:1.
4. Mishnah, Sanhedrin 4:5.
5. Leviticus 19:16.
6. Tosefta, Yevamot 8:4.
7. Babylonian Talmud, Sanhedrin 74a.
8. Leviticus 19:18.
9. Micah 6:8.
10. Leviticus 19:1.
11. Deuteronomy 10:20, 11:22.
12. Babylonian Talmud, Shabbat 133b. See also Babylonian Talmud, Sotah 14a.
13. Psalm 145:9.
14. See Exodus 22:22,26.
15. Leviticus 19:9–10.
16. Leviticus 24:22; cf. Exodus 12:49, Numbers 5:15,16.
17. Leviticus 25:35.
18. Exodus 19:6.
19. Genesis 9:8–17.
20. Babylonian Talmud, Sanhedrin 56a; Maimonides, *Mishneh Torah*, Melakhim 9:1.
21. Babylonian Talmud, Gittin 59b.
22. Mishneh Torah, Melakhim 10:12.
23. Babylonian Talmud, Shabbat 133b.
24. Deuteronomy 17:16–20.
25. See Sifrei, Deuteronomy 144.
26. Babylonian Talmud, Bava Bathra 21a.
27. Babylonian Talmud, Megillah, 26a.
28. Babylonian Talmud, Bava Bathra, 8b.
29. Samuel Belkin, *In His Image: The Jewish Philosophy of Man as Expressed in Rabbinic Tradition* (London and New York: Abelard-Shuman, 1960), 117.
30. See Genesis 10.
31. Isaiah 2:3–4.
32. Psalm 34:15.
33. Jerusalem Talmud, Peah 1a; Yalkut Shimoni 711.
34. Bamidbar Rabbah 13:1.
35. Seder Eliyahu Rabbah.
36. Genesis 2:15.
37. Psalm 24:1.
38. Leviticus 25:23.
39. Leviticus 20:19
40. Babylonian Talmud, Berachot 52b.
41. Babylonian Talmud, Hullin 7b; Shabbat 140b. See also Maimonides, *Mishneh Torah*, Melakhim 6:10.
42. Mishnah, Bava Bathra 2:8,9.
43. Shulhan Arukh Harav, Shemirat Haguf v'hanefesh, chapter 14.
44. Samson Raphael Hirsch, *Horev* (London: Soncino, 1962), 397–98.

45. Samson Raphael Hirsch, *The Pentateuch*, trans. Isaac Levy (Gateshead: Judaica, 1982).
46. Exodus 20:9.
47. Samson Raphael Hirsch, *The Jewish Sabbath*, trans. Joseph Ussoro (independently published, CreateSpace, 2013).
48. Exodus 23:10.
49. See Babylonian Talmud, Nedarim 3a.
50. For example, see Maimonides, *Guide for the Perplexed*, 1:26.
51. See the reports of the Intergovernmental Panel on Climate Change, available at http://www.ipcc.ch/.

CHAPTER 9

BUDDHISM AND THE COMMON GOOD

REVEREND KYOICHI SUGINO

> *[We] are all* relational beings. . . . *A lively awareness of our relatedness helps us to look upon and treat each person as a true sister or brother. . . . Inner peace is closely related to care for ecology and for the common good. Because, lived out authentically, it is reflected in a* balanced lifestyle *together with a capacity for wonder which takes us to a deeper understanding of life. . . . It is a return to that* simplicity *which allows us to stop and appreciate the small things, to be grateful for the opportunities which life affords us, to be* spiritually detached from what we possess, *and not to succumb to sadness for what we lack.*[1]

Some might consider the above quote Buddhist teachings of interrelatedness, inner peace, the middle path, simplicity, contentment, and detachment. In fact, these are the messages of Pope Francis in his encyclical *Laudato si'*. Buddhist leaders from across the globe expressed their strong solidarity and support for *Laudato si'* as the world's religious communities advocated individually and collectively for the Paris Agreement on climate change.[2] Such considerable overlap between Christian and Buddhist expressions of shared values and commitments demonstrates an emerging global interreligious consensus on fundamental values and virtues for the advancement of integral human development.

Paṭiccasamuppāda and Nonself

The Buddhist vision of integral human development derives from the ontological awareness of *Paṭiccasamuppāda* (dependent origination). Everything is interrelated. Everything affects everything else. Everything that is is because other things are. This notion makes clear that what we experience as "self" arises out of interaction with others. Self and other are mutually implicated. Thus, the notion that one could somehow attain enlightenment outside of relationship with others is an illusion. So, too, is it clear that only through helping all others attain their true Buddha-nature can one realize one's own truth.[3]

The Buddhist awareness of dependent origination and the interrelated reality of human beings and their relationship with nature offers spiritual and moral imperatives of human dignity for all, shared well-being, inclusivity and equality, and compassion for others. These are also foundational values for the realization of the Sustainable Development Goals. Such an interrelated perspective also invites Buddhists to nurture a holistic and integral understanding of peace and development that relates the SDGs to one another as interrelated goals to achieve holistic human flourishing and shared well-being.

Nonself or self-emptying in Buddhism and kenosis[4] in Christianity are key dynamics in our interactions with the "other." Nonself in Buddhism is a mystic notion. Human subjects exist, but they have no fixed essence that we would ordinarily call a substantial self. The doctrine of nonself is paradoxically designed on the one hand to protect us from a dangerous illusion of a false self and, on the other, invite us into the mysterious and always essenceless experience of being a subject intimately connected in the endless web of dependent origination. Self-emptying creates the space for the other to be the other, offers a form of self-donation, and expresses the true meaning of compassion.

Clinging to an imaginary self can produce harmful selfish desires, cravings, attachments, hatreds, ill will, conceit, pride, egoism, and other defilements, impurities, and problems. Liberating ourselves from *dukkha* (suffering, anguish, pain, or unsatisfactoriness) by seeing things as they are (*sammā-diṭṭhi*—right view) and putting wisdom and compassion in action are considered the essence of Buddhism. As an embodiment of wisdom and compassion in action and as an ideal of the fulfillment of "true self," the notion of the *bodhisattva* is introduced in Mahayana Buddhist scriptures. At its core, the notion links benefitting oneself to benefitting others. While the notion of the *bodhisattva* is intrinsically altruistic, it radically expresses the Buddhist notion of dependent origination.

The Buddhist Notion of Human Flourishing and Shared Well-Being

The Mahayana Buddhist experience of human flourishing and shared well-being originates from a holistic and harmonious awareness of the following three fundamental visions and values found in its major scripture, *Saddharma Puṇḍarīka Sūtra* (the *Lotus Sutra*): Buddha-nature, the Eternal Buddha (universal saving compassion), and the *bodhisattva*.

Buddha-Nature

All living beings can attain enlightenment, that is, they all have (essence-less) Buddha-nature. The aspirations expressed in "One Great Vehicle"[5] in the *Lotus Sutra* are to help all living beings become aware of their own Buddha-nature. This passage expresses the ultimate compassion of the Buddha in helping all beings awaken to their Buddha-nature.

The Mahayana Buddhist understanding of the essential components of Buddha-nature, or of a good person, is found in the teaching of the four "immeasurables" of Buddha-nature: an immeasurable mind of lovingkindness (*maitrī*, 慈無量心) and bestowing of joy or happiness; an immeasurable mind of compassion (*karuṇā*, 悲無量心), empathy, and the desire to eradicate the suffering of others; an immeasurable mind of sympathetic joy (*muditā*, 喜無量心), rejoicing in others' happiness; and an immeasurable mind of equanimity (*upekṣa*, 捨無量心), in which all distinctions between friends and foes disappear.

The Eternal Buddha

The Mahayana Buddhist experience of the Buddha's universal compassion culminates in the notion of the Eternal Buddha (described in chapter 16 of the *Lotus Sutra*). The everlasting Eternal Buddha[6] is omnipresent in the universe and is the life force that sustains and guides all sentient beings. The ultimate substance of the Buddha is the eternal and imperishable life force, which abounds within and about us all.

The everlasting universal compassion embodied in the Eternal Buddha is also believed to employ *upāya* (skillful means),[7] taking the form most appropriate to a person's capacity, as well as the time and place, and guiding them to enlightenment as swiftly as possible. The Eternal Buddha saves people in such a way that all can fully be enlightened to their Buddha-nature and flourish, achieving all their potential as human beings.

The *Bodhisattva*

A *bodhisattva* has been described as one who seeks upward for *bodhi* (wisdom) and teaches downward to all beings, that is, one who perfects oneself by aiming at the attainment of enlightenment while also descending to the level of the unenlightened in order to save them. In the simplest Mahayana Buddhist terms, a *bodhisattva* is one who devotes oneself to attaining enlightenment not only for oneself but for all sentient beings. The *bodhisattva* derives from the Buddhist notion of *Paṭiccasamuppāda* (dependent origination) and represents the ideal of the fulfillment of one's true self.

Causes of Suffering and Paths to Shared Well-Being

A Buddhist approach to problem-solving is elucidated in the teaching of *Catvāri āryasatyāni* (the Four Noble Truths):

(1) Truth of Suffering: All existence is *dukkha*. The word *dukkha* has been variously translated as "suffering," "anguish," "pain," or "unsatisfactoriness."
(2) Truth of Cause: The cause of *dukkha* is craving. The natural human tendency is to blame our difficulties on things outside ourselves. But the Buddha says that their actual root is to be found in the mind.
(3) Truth of Extinction: The cessation of *dukkha* comes with the cessation of craving. As we are the ultimate cause of our difficulties, we are also the solution. We cannot change the things that happen to us, but we can change our responses.
(4) Truth of the Way: There is a path that liberates us from *dukkha*, which is the Eightfold Path.

The *ariyo aṭṭhaṅgiko maggo* or ārya̅ṣṭāṅgamārga (the Eightfold Path) outlines the path from *dukkha* to liberation and wisdom. It consists of the following components:

(1) Right view (*sammā-diṭṭhi*)
(2) Right thought (*sammā-saṃkappa*)
(3) Right speech (*sammā-vācā*)
(4) Right conduct (*sammā-kammanta*)
(5) Right livelihood (*sammā-ājīva*)
(6) Right effort (*sammā-vāyāma*)
(7) Right mindfulness (*sammā-sati*)
(8) Right concentration (*sammā-samādhi*)

These eight factors promote and perfect the three essential elements of Buddhist training and discipline: *sila* (ethical conduct), *samadhi* (mental discipline), and *panna* (wisdom). Mental and ethical training and the cultivation of virtues at both the personal and social levels are aimed at overcoming thoughts of selfish desire, ill will, hatred, and violence, leading us to true wisdom.[8]

Personal and Societal Virtue Cultivation in Buddhism

A comparative survey of Buddhist and Christian entrepreneurs' value orientations suggests both similarities and differences in nuances and approaches in personal and social values and in the cultivation of related virtues (table 9.1).[9]

Buddhist approaches to problem-solving (the Four Noble Truths and the Eightfold Path) and notions of human dignity and shared well-being (Buddha-nature and the universal compassion of the Eternal Buddha) have both personal and structural or institutional dimensions, and these are interrelated.

Chapter 28 of the *Lotus Sutra*, "Encouragement of the Bodhisattva Universal Virtue," specifies four conditions related to the cultivation of virtues as follows: "The first is to be protected and kept in mind by the *buddhas*; the second is to plant roots of virtue; the third is to join those who are headed for awakening; and the fourth is to be determined to save all the living. After the extinction of the *Tathagata*, any good son or good daughter who meet these four conditions will be certain to acquire this sutra."[10]

The Four Conditions of Buddhist Virtue Practice demonstrate interrelated phases of virtue cultivation, from faith in the Eternal Buddha's unlimited universal compassion, to the cultivation of personal virtues, to the aspiration to become enlightened through virtue practice, together with other members in the *sangha* (community),[11] and finally to embodiment of the Buddha's universal compassion to save all the living. Personal, societal, structural, and institutional virtue cultivation are integrated to express the Buddhist notion of human dignity and the common good. The notion of the "One Great Vehicle" in the *Lotus Sutra* is an

TABLE 9.1 A comparison of core Christian and Buddhist values

Value category	Christian value	Buddhist value(s)
Ontological conceptions	Human dignity	Interconnectedness
Procedural values	Justice	Moderation, mindfulness, the middle path
Other-directedness	Solidarity	Compassion

expression of holistic development and the flourishing of humans as relational beings, contributing to the wholeness of the human family and the universe.

Greed is manifested at the individual and social levels, as well as structurally in political, economic, social, cultural, and other domains. Personal and structural greed have resulted in unparalleled disparities between the super rich and those who go hungry every day and in the accelerated degradation of the environment. More people have become comfortable with greed and have begun to believe that unregulated greed is acceptable and that unbridled competition and the accumulation of wealth are necessary for human progress.[12]

Addressing greed in each realm requires transformation and a variety of strategies and skillful means to achieve positive transformation. Strategies for addressing greed at the personal and social levels need to begin with cultivating compassion for others. Counteracting the structural greed embodied in political and economic power structures requires additional anti-greed measures, such as developing and enforcing adequate regulation of financial transactions and policies that promote the equitable distribution of wealth.[13]

Buddhist Economics: Practical Application of the Buddhist Notion of Integral Development and Shared Well-Being

It is in the light of both immediate experience and long-term prospects that the study of Buddhist economics could be recommended even to those who believe that economic growth is more important than any spiritual or religious values. For it is not a question of choosing between modern growth and traditional stagnation. Rather, it is a question of finding the right path of development, the Middle Way between materialistic heedlessness and traditionalist immobility, of finding "Right Livelihood."

The German-born British economist Ernst Friedrich Schumacher applied the Buddhist Eightfold Path to develop Buddhist economics. In contrast to modern economics, which is centered on profit, wealth, desire, and maximizing consumption, Schumacher proposed an alternative economic policy framework by reintroducing Buddhist notions of interrelatedness, nonself, minimizing suffering, and maximizing the well-being of all.

In Buddhist economics, consumption is considered merely a means of achieving human well-being, the aim being to "obtain the maximum of well-being with the minimum of consumption." Modern economics, on the other hand, considers consumption to be the sole end and purpose of all economic activity, taking the factors of production—and labor and capital—as the means. The former tries to maximize human satisfaction through an optimal pattern of consumption, whereas the latter tries to maximize consumption through an optimal pattern of productive effort (table 9.2).

TABLE 9.2 A comparison of the features of modern and Buddhist economics

Modern economics	Buddhist economics
Self-interest	*Anatta* (nonself)
Maximize profits and individual gains	Minimize suffering (losses) for all living and nonliving things
Encourage material wealth and desire	Simplify one's desires
Maximize markets to the point of saturation	Minimize violence
Bigger is better, and more is more	Small is beautiful, and less is more

In Buddhist economics, simplicity and nonviolence are closely related. Producing a high degree of human satisfaction by means of a relatively low rate of consumption allows people to live without great pressure and strain and to fulfill the primary injunction of Buddhist teaching: "Cease to do evil; try to do good." As physical resources are everywhere limited, people satisfying their needs by means of a modest use of resources and living in self-sufficient local communities are less likely to get involved in large-scale violence than people whose existence depends on worldwide systems of trade. From the point of view of Buddhist economics, therefore, production from local resources for local needs is the most rational way of economic life. As market-driven global economies have become harmful to small businesses and devastating to local communities, efforts to create alternative economies at the local level must be encouraged. Local exchange and trading systems, in which trading is done in local and regional currencies; cooperative banking; decentralized energy; and localizing the production and exchange of basic commodities such as water and food, should be further encouraged and promoted.

Schumacher also referred to the Buddhist tradition of tree planting for the preservation of forests, noting the following:

> The teaching of the Buddha . . . enjoins a reverent and non-violent attitude not only to all sentient beings but also, with great emphasis, to trees. Every follower of the Buddha ought to plant a tree every few years and look after it until it is safely established, and the Buddhist economist can demonstrate without difficulty that the universal observation of this rule would result in a high rate of genuine economic development.[14]

Buddhist economics is an attempt to harmonize personal and structural approaches and to transform self-interest and profit maximization into the alleviation of suffering and enlightened simplicity. Further refining Buddhist economic theory would have a substantive and concrete impact on the likelihood of achieving the SDGs. Various efforts have been made to put into practice key elements of Buddhist economics, including the Eco-Temple Community Development Project,[15] which is a prototype of Buddhist integral human development. This experiment of Buddhist holistic and integral human and ecological development advances the notion of following interrelated components simultaneously: an ecological temple structure and energy system, economic sustainability, integration with the surrounding environment, engagement with the community, and the development of spiritual values and teachings on the environment ("eco-dharma"), which fosters the cultivation of good habits. The project has been advanced by socially engaged Buddhist communities and leaders in various Asian countries.

Conclusion

The Buddhist understanding of human dignity and shared well-being rooted in its fundamental teachings and practice supports the notion of the common good. The teaching of the "One Great Vehicle" in the *Lotus Sutra* is an expression of holistic development and the flourishing of humans as relational beings, contributing to the wholeness of the human family and the universe.

Today's developmental and environmental challenges further nuance Buddhist notion of the common good. In an evolving global Buddhism that cuts across major Buddhist streams such as Theravada and Mahayana, reorienting Buddhist practice from the cultivation of personal virtues to socially engaged Buddhism has been a critical development as Buddhists respond to structural problems. A number of socially engaged Buddhist leaders and practitioners are revisiting and refining Buddhist economics, developed in the 1970s, to challenge modern economics and discern alternative sustainable and integral development strategies and approaches.

Religious communities normally work at two levels: normative and pastoral. It is important for religious leaders to reaffirm the authentic teachings of their religions and to establish a primary narrative of peace and a vision of integral human development. Such normative-level interventions include Buddhist leaders from across traditions gathering to proclaim and reconfirm basic Buddhist teachings that support the spiritual and moral imperatives of the SDGs, such as peaceful and inclusive societies, equality, respect for the dignity for all, mercy and compassion for others, and the preservation of forests and the environment. Religious and moral norm-setting should be part of religious communities' ongoing efforts.

At the pastoral level, religious leaders work with individual followers and believers who are facing challenges, suffering, and ambiguity in their daily lives.

Religious leaders offer pastoral support, accompanying and guiding their followers to salvation and liberation. In places such as Myanmar, where ordinary people are experiencing political manipulation of religions and prejudice against minority groups, people need to be reminded again and again of the shared values and authentic teachings of their own faith and practice. This "again and again" need is indeed the essence of virtue ethics. Virtue is a habitual orientation to values.[16] Virtue ethics based on the values of religious traditions has to be cultivated every day with religious believers and practitioners.

For Buddhists, wisdom and compassion are the two fundamental virtues to be cultivated. Wisdom arises from our own conscious effort to see things as they are, to discern challenges, problems, and suffering as they are, without self-centered or distorted perspectives. This wisdom is cultivated through the practice of right view in the Eightfold Path and of the truths of suffering and of cause in the Four Noble Truths. With the wisdom obtained through these practices, Buddhists are called to walk "the *Bodhisattva* way," taking concrete action to serve others, not for one's own desire or interest but to serve the common good. Wisdom and compassion in action are the essence of Buddhism.

Notes

1. Francis, *Laudato si'* (Vatican City: Libreria Editrice Vaticana, 2015), sec. 222–225. Emphasis mine. Interrelatedness is also mentioned in sec. 120, 137, 138, 139, 142, and 164.
2. "Buddhist Climate Change Statement to World Leaders 2015," Plum Village, October 31, 2015, https://plumvillage.org/articles/buddhist-climate-change-statement-to-world-leaders-2015/.
3. The interpretations of Buddhist teachings and comparative studies between Christianity and Buddhism in this chapter largely derive from the author's presentation at the Christian Buddhist Dialogue conference in June 2015 hosted by His Holiness Pope Francis. Kyoichi Sugino, "Fraternity as the Way Forward," *Claritas: Journal of Dialogue and Culture* 4, no. 2 (2015), https://docs.lib.purdue.edu/claritas/vol4/iss2/24.
4. Steve Odin, "Kenōsis" as a Foundation for Buddhist-Christian Dialogue: The Kenotic Buddhology of Nishida and Nishitani of the Kyoto School in Relation to the Kenotic Christology of Thomas J. J. Altizer," *Eastern Buddhist New Series* 20, no. 1 (1987): 34–61, www.jstor.org/stable/44361804.
5. The word *maha* means "great," and *yana* means "vehicle" or "raft," which evokes the image of Buddhist teachings as a raft or vehicle that can help one cross the river of suffering to reach the other shore.
6. The Eternal Buddha is an evolution of the notion of the Buddha in Mahayana Buddhism. It is distinct from the historical Buddha Shakyamuni.
7. Chapter 2 of the *Lotus Sutra* is titled "Skillful Means," which, together with chapter 16, "Revelation of the [Eternal] Life of the Tathagata," has long been regarded as the heart of the *Lotus Sutra*. Nikkyo Niwano, *A Guide to the Threefold Lotus Sutra* (Tokyo: Kosei, 1989).
8. Walpola Sri Rahula, "The Noble Eightfold Path: The Buddha's Practical Instructions to Reach the End of Suffering," *Tricycle Journal of Buddhist Review*, accessed March 31, 2021, https://tricycle.org/magazine/noble-eightfold-path.

9. Gábor Kovács, *The Value Orientations of Buddhist and Christian Entrepreneurs: A Comparative Perspective on Spirituality and Business Ethics* (Cham: Palgrave Macmillan, 2020).
10. For the English translation of the *Lotus Sutra*, the following books were referred to: Nikkyo Niwano, *Buddhism for Today: A Modern Interpretation of the Threefold Lotus Sutra* (New York and Tokyo: Weatherhill/Kosei, 1976); Gene Reeves, *The Lotus Sutra: A Contemporary Translation of a Buddhist Classic* (Boston: Wisdom, 2008); and Burton Watson, *The Lotus Sutra* (New York: Columbia University Press, 1993).
11. In the Theravada tradition, *sangha* (community) refers to a monastic community, whereas in the Mahayana tradition, it refers to a broader community of lay followers and practitioners.
12. "A Buddhist-Christian Common Word on Structural Greed," International Network of Engaged Buddhists, December 21, 2010, https://www.inebnetwork.org/a-buddhist-christian-common-word-on-structural-greed/.
13. Martin L. Sinaga, "Transforming Greed: An Interfaith Common Word," *Dharma World* 40 (2013).
14. E. F. Schumacher, "Buddhist Economics," Schumacher Center for a New Economics, accessed March 31, 2021, https://centerforneweconomics.org/publications/buddhist-economics/. This essay is also included in E. F. Schumacher, *Small Is Beautiful: Economics as if People Mattered—25 Years Later . . . with Commentaries* (Vancouver: Hartley and Marks, 1999).
15. "Eco-Temple Community Development Project," Japan Network of Engaged Buddhists, last updated May 2020, https://jneb.net/activities/buddhistenergy/eco-templeproject/.
16. "Declaration of the 10th World Assembly of Religions for Peace," Religions for Peace, August 23, 2019, https://rfp.org/declaration-of-the-10th-world-assembly-of-religions-for-peace.

CHAPTER 10

GREEK ORTHODOXY AND THE COMMON GOOD

JOHN D. ZIZIOULAS

ADAPTED BY AND WITH CONTRIBUTIONS FROM JESSE THORSON

> *The life of the Orthodox Church is applied ecology, a tangible and inviolable respect for the natural environment. The Church is an event of communion, a victory over sin and death, as well as over self-righteousness and self-centeredness—all of which constitute the very cause of ecological devastation. The Orthodox believer cannot remain indifferent to the ecological crisis. Creation care and environmental protection are the ramification and articulation of our Orthodox faith and Eucharistic ethos.*[1]

In 1989, Ecumenical Patriarch Dimitrios decided to devote the first day of September each year to praying for the environment. According to the Orthodox liturgical calendar going back to Byzantine times, this date is the first date of the ecclesiastical year.

Why would the Orthodox devote such an important day of the church calendar to praying for the environment? In his "Message Upon the Day of Prayer for the Protection of Creation," His All Holiness Ecumenical Patriarch Bartholomew explains:

> The Ecumenical Patriarchate's ecological initiatives provided a stimulus for theology to showcase the environmentally friendly principles of Christian anthropology and cosmology as well as to promote the truth that no vision for humanity's journey through history has any value if it does not also

> include the expectation of a world that functions as a real "home" (*oikos*) for humanity, particularly at a time when the ongoing and increasing threat against the natural environment is fraught with the possibility of worldwide ecological destruction.[2]

Some theologies describe human beings as temporary residents of a foreign land, able to do whatever they please with the vast resources of that land before its eventual destruction. Orthodox Christianity tells a different story, one that is relational, eucharistic, and sacramental, and this generates important implications regarding the relationship between human beings and the natural world. This chapter will briefly contextualize the current ecological crisis in a relational, theological framework and highlight aspects of Orthodox Christianity that can serve as motivation for proper environmental care, stewardship, and integral human development.

Forgetting "Being as Communion"

First, we must recognize—even celebrate—that nothing in existence is conceivable in itself. As the Holy Trinity, God is relational. God is a communion of people, Father, Son, and Holy Spirit. Like God, our human identity is relational. As relational creatures, our personhood is bound up in our relationships with God, with other people, and with the created world.[3]

The ecological crisis—although certainly multidimensional—is fundamentally spiritual: it arises due to the imbalance between humanity's relationship with God, other people, and the natural world. The proper relationship between humanity and the earth was broken with the fall. Indeed, we forget that we depend on the earth for our very being. We cannot be human beings outside of our dependence on water cycles, geological events, ecosystem services, and predator-prey relationships, to name only several of a perhaps infinite number of natural phenomena that sustain our being, none of which can be named in isolation. And it is precisely these life-giving, human-constituting phenomena that are impacted by the distortion of our relationship with the environment. Indeed, our forgetting of the relational character of being leads to all ecological, environmental, and even sociopolitical challenges.

In our time, this rupture of the relationship between humans and the environment—what we could call environmental sin, as seen in such environmental crises as climate change, ocean acidification, and loss of biodiversity—can be attributed to the rise of individualism in our culture. The pursuit of individual happiness has become an ideal in today's culture. Human greed blinds people to the point of ignoring and disregarding the basic truth that the flourishing of the individual depends on its relationship with the rest of humanity and the created world, whose sacredness has been affirmed by Christ's assumption of materiality in the Incarnation.

Importantly, however, ecological sin is a sin not only against God but also against our neighbor. Our common flourishing depends greatly on our proper stewardship of our environment and natural resources. The poor, the Indigenous, and other socioeconomically marginalized peoples are especially liable to suffer as a result of ecological crises, including anthropogenic climate change. Environmental sin is not only a sin against the *other* of our own time but also against the *other* of future generations. By destroying our planet in order to satisfy our greed for happiness, we bequeath to future generations a world damaged beyond repair with all the negative consequences this will have for their lives.

The Gift and Sacrament of Nature

Creation is a gift. The characteristic of biblical religion as distinct from ancient Greek cosmology is that for the former, creation has come out of nothing. It is—in Scholastic terminology—contingent. Creation's sacredness is not intrinsic but dependent on its relation with its creator. Importantly, nature cannot be possessed by human beings, for it is a gift. Human beings can only receive nature, enjoy it with thanksgiving, and pass it along to others in love and gratitude. This conception of creation as a gift is present particularly in the Eucharistic liturgy, in which nature is given in order to be given, not to be possessed.

The sacramentality of nature reminds us of the ontological link between humanity and the rest of creation. Humans can have communion with God and with other human beings only via nature. In the Christian tradition, the sacraments remind us that no communication or relationship is possible outside of nature or natural material. It is not without significance that we are baptized with water, offer the Eucharist in the form of bread and wine, anoint with oil, and exchange the kiss of peace by embracing one another with our bodies. In this way, the world becomes a sacrament that is a material or natural reality that carries the capacity of communion. And, importantly, our disregard for and abuse of nature thwarts the possibility of communion with God and each other, thus obstructing the common good.

The Gospel and the Common Good

For a long time, Christian theology in both the West and the East has tended to exhaust the meaning of salvation with the redemption of the human being. A simple glance at the manuals of Christian dogmatics (whether Catholic, Orthodox, or Protestant) would suffice to notice the almost total absence of any reference to nature in Christ's salvific work, unfortunately in contrast to both the biblical evidence[4] and patristic teaching.[5] In a now famous *Science* article, the American historian Lynn White accused Christian theology of being responsible for the modern ecological crisis.[6] In much of Christian theology, the human being has been so exalted above the material creation as to allow humans to treat it as material for the satisfaction of their needs and desires. The human being has been denaturalized, and in its abuse and misuse of the biblical command to the first human couple, "Increase and multiply and subdue the earth,"[7] humanity has been encouraged to exploit the material creation unrestrictedly, with no respect for its integrity or sacredness. So too have many other religions contributed to a disjunction between anthropology and cosmology. Nature has become something quite different from the human being, as if the human being were not part of nature. The human being, in this telling, becomes the master of creation through the advancement of science and technology.

This exploitative story has gripped the imagination of many *not because of, but in spite of, the fundamental principles of Christian faith.* Rightly construed, the Gospel reminds us of our common good. In his Epistle to the Romans, St. Paul speaks clearly of the inseparable link between humanity and nature in the good news of the Gospel: "The material creation awaits with eagerness the salvation of the human beings, so that it may share itself in the glorious future offered by God in Christ."[8]

Indeed, the destiny of humanity is tied up with that of nature, but what is this future that humanity and nature are destined to share? In the patristic tradition, the link between nature and human is often presented in the image of microcosm and mediator. The human being is the priest of creation who takes nature in their hands in order to refer it with gratitude to God (the giver) and then offer it to others as a gift. Nature is in constant motion and circulation. It is given and taken in order to be given and taken and given again. It is never contained anywhere or by anyone. It moves endlessly from one person to another until the coming of the kingdom.

In the celebration of the Eucharist, the Church offers to God the material world in the form of bread and the wine. In this sacrament, space, time, and matter are sanctified. They are lifted up to the Creator with thankfulness as his gifts to us. Creation is solemnly declared as God's gift, and human beings, instead of as proprietors of creation, act in the Eucharist as its priests, who lift it up to the holiness of the divine life. As the psalmists say, "Everything declares God's glory," and the human being leads this cosmic chorus of glorification to the creator as the priest of creation.

Nature declares the good news that life is communion. Nature declares the good news that the human being fulfills itself when one takes in one's hands the natural world and transforms it into a vehicle of communion. The Gospel is good news not only for humanity but also for nature, and this way of understanding the place and mission of humanity in creation is common to both Eastern and Western traditions and is of particular importance for the cultivation of an ecological ethos.

On Personhood and Ethics

Although a proper ecological ethics must emphasize relationality, this must be complemented by the idea of "otherness" or "uniqueness." Relationality can ignore the specificity of beings and specificity in nature or elsewhere in human society. Relationality and uniqueness, properly combined, constitute the idea of "person," which must become and remain a key term in an ecological ethics.

"Person" is not understood in the Western tradition as it is in the Eastern tradition. Of course, according to Saint Augustine "person" is a relational notion. But according to the Greek Fathers, the person—in the Greek, *hypostasis*—is a unique particularity. A focus on relationship should not lead us to conceive of a universal communion in which the particular consistence of this communion is sacrificed for the sake of the general.

There is a danger in our culture today because we live in a global situation in which we tend to forget that each particular entity in the world, in nature, as well as in human society, is unique. We have to respect the *uniqueness of beings*, and ecological behavior must take that into account. It is the uniqueness of beings that grounds our concern for particular animal or plant species in creation, yet we continue to destroy many. It is the uniqueness of our earth that should motivate our stewardship of the planet, not some vague relational concept. And, most importantly, it is the particularity of people—this person, that person—that should prompt us to act on behalf of the poor and the marginalized. An ecological ethic must be a personal ethic, though not an individualistic one.

Ecological Eschatology, Asceticism, and Prayer

The Christian religion provides an *eschatological approach to existence and our lives*. By this, I mean that we look at ourselves, our fellow human beings, and creation not from the perspective of the past or the present but as beings entitled to a future. Whosoever deprives any being its future commits a crime, or sin, against that creature. Therefore, Christians are called to be oriented toward the future and to consider the consequences of our actions on our fellow human beings and on creation in general. Paul instructs the church in Galatia to "bear one another's burdens, and so fulfill the law of Christ."[9] This responsibility may entail bearing the consequences of our own actions, instead of deferring burdens to others across time or space.

This approach calls for what may be described as an *ecological asceticism*. It is noteworthy that the great figures of the Christian ascetical tradition were sensitive toward the suffering of all creatures. An equivalent of Saint Francis of Assisi is also abundantly present in the monastic tradition of the East. Accounts of the lives of the desert saints present the ascetic as weeping for the suffering or death of every creature and as leading a peaceful and friendly coexistence even with the beasts.[10] And these accounts do not constitute romanticism. Instead, the actions of the ascetic spring from a loving heart and the conviction that between the natural world and ourselves there is an organic unity and interdependence that makes us share a common fate, just as we have the same Creator.

Asceticism is an unpleasant idea in our present culture, which measures happiness and progress with an increase in capital and consumption and groans unceasingly for "more, more, more." It would be unrealistic to expect our societies to adopt asceticism in the ways Saint Francis and the Desert Fathers of the East experienced it, but the spirit and the ethos of asceticism can and must be adopted if our planet is to survive. The *spirit of asceticism* is the *spirit of restraint*, and restraint is not merely a spiritual virtue or religious recommendation but the necessary response to a finite world of limited resources.[11] Restraint in the consumption of natural resources is a realistic attitude, and ways must be found to put a limit to the immense waste of natural materials. Technology and science must devote their efforts to such a task.

Finally, in and alongside our ecological efforts, we are called to pray, in hope, and to remember our common good. In their Joint Message on the World Day of Prayer for Creation, Pope Francis and Ecumenical Patriarch Bartholomew remind us, "We labor in vain if the Lord is not by our side (cf. *Ps* 126–127), if prayer is not at the centre of our reflection and celebration. Indeed, an objective of our prayer is to change the way we perceive the world in order to change the way we relate to the world . . . to be courageous in embracing greater simplicity and solidarity in our lives."[12]

In *Laudato si'*, Pope Francis offers the following prayer to "all who believe in a God who is the all-powerful Creator":

> *A prayer for our earth*
>
> All-powerful God,
> you are present in the whole universe
> and in the smallest of your creatures.
> You embrace with your tenderness all that exists.
> Pour out upon us the power of your love,
> that we may protect life and beauty.
> Fill us with peace, that we may live
> as brothers and sisters, harming no one.

O God of the poor,
help us to rescue the abandoned
and forgotten of this earth,
so precious in your eyes.
Bring healing to our lives,
that we may protect the world and not prey on it,
that we may sow beauty,
not pollution and destruction.
Touch the hearts
of those who look only for gain
at the expense of the poor and the earth.
Teach us to discover the worth of each thing,
to be filled with awe and contemplation,
to recognize that we are profoundly united
with every creature
as we journey towards your infinite light.
We thank you for being with us each day.
Encourage us, we pray, in our struggle
for justice, love and peace.[13]

At the center of such a prayer is the recognition that we have lost sight of *our common good*. The ecological crisis goes hand in hand with the spread of social injustice, and we cannot successfully face the one without dealing with the other.

Notes

This chapter has, in part, been adapted from remarks made by John Zizioulas at multiple Ethics in Action meetings at Casina Pio IV, in addition to remarks made at the news conference in the New Synod Hall for the presentation of *Laudato si'* in 2015 and at the celebration of its third anniversary in 2018.

1. Bartholomew, "Message for the Day of Prayer for the Protection of the Environment," Greek Orthodox Archdiocese of America, September 1, 2018, https://www.goarch.org/-/message-of-ecumenical-patriarch-bartholomew-for-the-indiction-and-the-day-of-the-protection-of-the-environment-2018-.
2. Bartholomew, "Message for the Day of Prayer."
3. Francis, *Laudato si'* (Vatican City: Libreria Editrice Vaticana, 2015), sec. 66.
4. See Romans 8:19–22 and Colossians 1:15.
5. See, for example, Saint Athanasius of Alexandria (*Letter to Serapion* I, 25); John Chrysostom (*Homilies on the Letter to the Romans*, 14, 5; *Letter to Galatians* 6, 3; Cyril of Alexandria, *On the Epistle to the Romans*, 8).
6. Lynn White, "The Historical Roots of Our Ecological Crisis," *Science* 155, no. 3767 (1967): 1203–7.

7. Genesis 1:28.
8. Romans 8:19.
9. Galatians 6:2.
10. Abba Gerasimos of the Laura near the river Jordan treats the wound of a lion, and a relationship between the two becomes so strong that when the monk dies, the lion sits by his grave until its own death (*Leimon* of John Eucratas, seventh century CE. Migne, *Patrologia Graeca* 87 c, 2985 cy). Saint Macarius of Egypt (fourth century CE) punishes himself for having killed a mosquito without reason by placing himself in a swamp of mosquitos for six months. Saint Silouan the Athonite (+1937) wept for three days and nights for having killed a fly (Archimandrite Sophrony, *St. Silouan the Athonite* [1988], 507f).
11. See Club of Rome, *Limits to Growth: A Report for the Club of Rome's Project on the Predicament of Mankind* (New York: Universe, 1972).
12. Pope Francis and Ecumenical Patriarch Bartholomew, "Joint Message of Pope Francis and Ecumenical Patriarch Bartholomew on the World Day of Prayer for Creation," from the Vatican and from the Phanar, September 1, 2017.
13. Francis, *Laudato si'*, sec. 178–179.

CHAPTER 11

CATHOLICISM AND THE COMMON GOOD

DANIEL G. GROODY

Creation is a gift freely entrusted to humanity by a loving God. The goods of the earth are meant for the benefit of all, not simply a privileged few. In this world human beings are the stewards of the earth's resources, and since when we die we will have to give up everything, only God is the ultimate owner of everything we possess, even our very lives.[1] We will be judged at the end of time by how we have used what we have been given. These beliefs are at the heart and soul of the Christian community and serve as the starting point for Catholic social teaching (CST). In light of the complex problems posed by global poverty, this chapter offers a brief overview of CST and highlights its central focus: fostering union with God and communion with others.

The Disorders of the World and the "Tranquility of Order"

The problems of today present us with many social, political, economic, and even theological challenges, especially as we are confronted by issue of global poverty. Even though creation is richly endowed, the distribution of the world's goods has never been more out of balance. Since 2015, the richest 1 percent own more wealth than the rest of the planet, and the eight richest individuals own the same amount as the poorest half of the world.[2] As the rich get richer, more than 795 million people do not have enough food to lead an active, healthy life.[3] One in five children, adolescents, and youth are not in school.[4] And more than 10 percent of the world live on less than two dollars a day.[5] These alarming statistics not only go against the grain of who we are called to be as creatures, but they are also at odds with the designs of a loving Creator. "The earth provides enough to satisfy every man's need," Mahatma Gandhi noted, "but not every man's greed."[6]

What demands do such social disorders and other issues make on the human conscience? This is the central question of CST. To understand the principles upon on which CST is based, it helps to situate it within the context of the larger frame of the Kingdom of God, which Jesus proclaimed.

The Church always has its sights set on the fullness of the world to come in the next world. But it also believes that, in Jesus, the Kingdom of a new world order has already begun—in this world. Through his death and resurrection, he has reestablished God's reign in the human heart and initiated a new world order. Vatican II describes this kingdom as "eternal and universal, a kingdom of truth and life, of holiness and grace, of justice, love and peace."[7] It believes there is only one salvation history that unites this world and the next, and its social teachings seek to articulate the vision of aligning the order of *this* world with *that* Kingdom in heaven, which Jesus proclaimed.

The Church takes for granted that the effects of original, personal, and social sin have contributed to creating the current social disorders that exist in our world today. But it also believes that grace, good will, and good works are essential to establishing the new world order. Saint Augustine referred to peace as the "tranquility of order" (*tranquillitas ordinis*), and much of CST is directed toward the creation of a peaceful society built on a foundation of justice.[8] The essence of justice from this vantage point is primarily about the work of establishing right relationships with God, others, the environment, and even oneself.[9] When relationships are rightly ordered, true peace is possible. When they are not, disorder results, and people become aliens, even to themselves.

When the Church addresses economic disparities or other social issues, it is concerned with all that affects integral human development and the flourishing of individuals in community. Because war, environmental degradation, oppression, poverty, greed, abuse, drugs, fear, racism, meaninglessness, materialism, and many other perennial problems threaten human beings, the Church seeks a twofold dimension to this mission: to proclaim the God of Life and to build a civilization of love. This means denouncing all that threatens the cultivation of life and announcing the Kingdom of God and its vision of right relationships, which is biblical righteousness. In this light, the Church is concerned not only with what it believes (orthodoxy) but also how it lives out what it believes (orthopraxy), particularly in the exercise of charity and justice.

An Overview of Catholic Social Teaching

In general terms, CST refers to all the principles, concepts, ideas, theories, and doctrines that deal with human life and society as it has evolved since the days of the early Church. Living in the tension between the world as it is and a vision of the world as God intends it to be, CST seeks to challenge those dimensions of society that diminish people's relationships with God, others, the environment, and themselves and to promote those factors that enhance these relationships. While it draws heavily on theology, it also bases its reflections on philosophy, economics, sociology, and other social sciences. By linking theology with the social sciences, CST seeks to better understand the challenges of the current world and provide an ethical foundation for global transformation.

In more specific terms, modern CST refers to contemporary Church and papal documents and encyclicals that address the social problems of today's world. Although the Church has been concerned with social issues since its earliest days, the era of modern CST began with the publication of Pope Leo XIII's *Rerum novarum* in 1891. As the world began to experience greater inequalities in the wake of the Industrial Revolution, the Church made a conscious decision to place itself in solidarity first and foremost with the working class and the poor, not with the rich and powerful. It began to argue that many are poor not because of laziness or neglect but because of a system of structures, policies, and institutions that keep them from developing and flourishing as human beings.

Although modern CST emerges from the Church, it is not directed only to Catholics or Christians. Its message is universal and therefore addressed to all religions, all nations, and all peoples; it is concerned with all aspects of the human being and the full human development of every person.[10] Its purpose is not to organize society but to challenge, guide, and form the conscience of the human community in the development of a new social order.[11] Its primary focus is "action on behalf of justice and participation in the transformation of the world . . . for the redemption of the human race and its liberation from every oppressive situation."[12]

The social mission of the Church, then, expresses itself in terms of liberating the oppressed, calling to conversion the oppressors, and eliminating the structures of oppression. In living out this mission, it comes face to face with life-threatening issues such as immigration, economic inequality, abortion, the death penalty, human cloning, assisted suicide, homelessness, war and peace, human trafficking, just wages, poverty, and the environment.

Drawing upon a rich intellectual tradition, especially that of Saint Thomas Aquinas and later philosophical reflection, CST distinguishes three primary dimensions of social justice: commutative justice, contributive justice, and distributive justice (figure 11.1).[13]

Commutative or contractual justice deals with relationships between individuals, groups, or classes and how agreements are made. It involves the give and take that is part of these relationships and the benefits and responsibilities that go with them. It seeks to ensure that human dignity and social responsibility ground all economic transactions, contracts, and promises. It fosters responsible behavior from both employers and employees: Employers are to

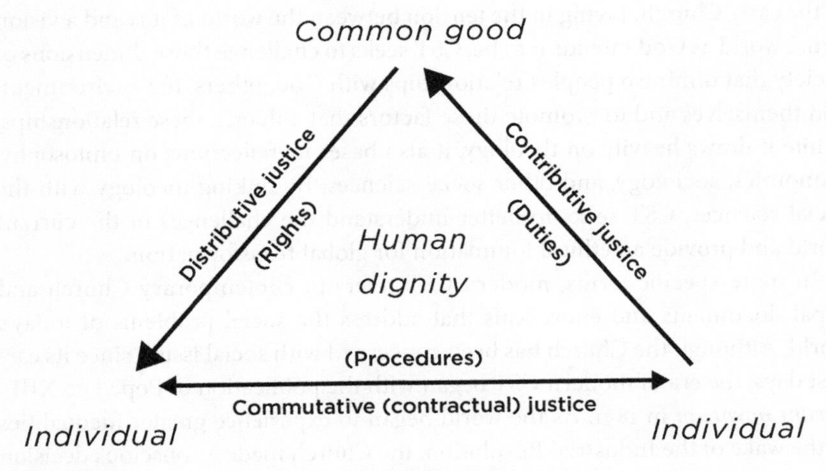

11.1 Social justice as right relationships: a philosophical framework

adequately compensate workers with humane working conditions and fair wages, and employees are to work conscientiously and diligently in exchange for these fair wages.[14]

While commutative justice deals with the relationship of individual units to each other, contributive justice deals with an individual's relationship to society as a whole. It challenges those who seek to take unfair advantage of a system in the name of claiming their rights without any consideration of their responsibilities to a larger, collective body of society. Contributive justice recognizes the responsibility of individuals to give back to the common good, which means that people have a duty to look out for not only their own welfare but also that of others. This duty involves fundamental and basic responsibilities such as voting and paying.

CST also affirms that, while individuals have a responsibility to the common good, the larger society has an obligation to individuals and groups as well. Distributive justice deals with society's duty to the individual, particularly for society's most poor and vulnerable members. Distributive justice seeks the well-being of all members of a community, which means one's basic rights must be safeguarded and protected.

Social justice is a way of bringing the concepts of commutative, contributive, and distributive justice together into a coherent vision. It seeks to speak about the way individuals and institutions are ordered and interact with each other in a positive and life-giving way. At the core of social justice is a respect for each person's human dignity and an overall commitment to the common good. Respect for the human worth of each member of a community is foundational to the flourishing of individuals and to building stable and peaceful communities.

Globalizing Solidarity

While forces like globalization have offered new opportunities to realize the interconnectedness of the human community, CST has also pointed to its shadow side. CST neither naively condemns the process of globalization nor uncritically embraces it but seeks to discern both its positive and negative dimensions, as the Church journeys toward a horizon of hope with faith and trust in a God of Life. It recognizes the positive dimensions of drawing together countries, economies, cultures, and ways of life in new ways and the growing interdependence of the human family and international community. At the same time, it sees that the social problems of today now have a worldwide dimension and have reached a magnitude never before seen in human history.

As CST seeks to steer the human community in a life-giving direction through an indispensable set of ethical coordinates, it bases it moral vision, in brief, on a foundation of human dignity, solidarity, and subsidiarity. It will continue to evaluate social and spiritual progress in terms of how the current structures of society help create a more just social order, how they help the poor, how they contribute to the international common good, and how they foster genuine development. No matter what else changes in society, CST will continue to measure progress as "a globalization without marginalization," or, as John Paul II called it, a "globalization of solidarity."[15]

To advance a "globalization of solidarity" means, in Paul VI's words, "building a civilization of love." When a tsunami hit South Asia on December 26, 2004, and left more than three hundred thousand people dead and millions homeless in its wake, individuals and organizations from around the world stepped in with impressive resources to help those who had suffered from this catastrophe. Such a globalization of institutions for the sake of the human community is one instance of how the human community can come together to build a civilization of love in the face of human suffering around the world.

While such offers of direct assistance are noteworthy, every day more than twenty-five thousand children die of preventable diseases, which adds up to the equivalent of a tsunami every two weeks.[16] The ignorance of and indifference to these lives and other such injustices indicate there is indeed much work still to be done. The disorders of the world and the indifference to those who suffer because of them make pressing demands on the human conscience and call to conversion all individuals and institutions in the world, all of whom share a common responsibility to serve a world in need and to build a peaceful and just social order. Through its social teachings, the Church seeks to be a credible, prophetic voice in a world of injustice to help the human community put ethics in action for the good of others. Even when its own witness to this mission is imperfect, the Church aims to give expression to something greater than itself, namely, the God of Life, who offers us a path to healing and transformation and who calls us to commit ourselves to building a civilization of love.

Notes

1. This chapter is drawn in part from Daniel G. Groody, "A God of Life, A Civilization of Love: Catholic Social Teaching," in *Globalization, Spirituality, and Justice: Navigating the Path to Peace* (Maryknoll: Orbis, 2017), 92–123.
2. For more on these statistics and the methodology used, see "An Economy for the 99 Percent," *Oxfam Briefing Paper*, January 2017, https://www-cdn.oxfam.org/s3fs-public/file_attachments/bp-economy-for-99-percent-160117-en.pdf.
3. "World Hunger Statistics," Food Aid Foundation, accessed April 1, 2021, https://www.foodaidfoundation.org/world-hunger-statistics.html.
4. "One in Every Five Children, Adolescents and Youth Is Out of School Worldwide," UNESCO, February 28, 2018, https://en.unesco.org/news/one-every-five-children-adolescents-and-youth-out-school-worldwide.
5. "Poverty," World Bank Group, accessed April 1, 2021, https://www.worldbank.org/en/topic/poverty.
6. Quoted in E. F. Schumacher, *Small Is Beautiful: Economics as if People Mattered* (New York: Harper & Row, 1975), 33.
7. Paul VI, *Gaudium et Spes* (1965), sec. 39, Vatican website, accessed April 1, 2021, http://www.vatican.va/archive/hist_councils/ii_vatican_council/documents/vat-ii_const_19651207_gaudium-et-spes_en.html.
8. Augustine, "Book 19," *City of God* (New York: Random House, 1993), 690–91.
9. For more on this topic, see Groody, *Globalization, Spirituality, and Justice*.
10. Paul VI, *Populorum progressio* (1967), sec. 42, Vatican website, accessed April 1, 2021, http://w2.vatican.va/content/paul-vi/en/encyclicals/documents/hf_p-vi_enc_26031967_populorum.html.
11. Pontifical Council for Justice and Peace, *Compendium of the Social Doctrine of the Church* (Vatican City: Libreria Editrice Vaticana, 2004), sec. 81.
12. World Synod of Catholic Bishops, *Justice in the World*, November 30, 1971, sec. 6.
13. This figure draws in part from the fine resources available on Catholic Social Teaching from the Office for Social Justice, St. Paul-Minneapolis, available at http://www.cctwincities.org. I am also grateful to Bill O'Neill, S.J., for his insights into developing it further with respect to rights, duties, procedures, and human dignity. See also Groody, *Globalization, Spirituality, and Justice*, 100.
14. National Conference of Catholic Bishops, *Economic Justice for All: Pastoral Letter on Catholic Social Teaching and the U.S. Economy* (1986), sec. 69.
15. John Paul II, *Ecclesia in America* (1999), sec.55, Vatican website, http://www.vatican.va/content/john-paul-ii/en/apost_exhortations/documents/hf_jp-ii_exh_22011999_ecclesia-in-america.html.
16. For more on these statistics, see "UNICEF," UNICEF, accessed April 1, 2021, https://www.unicef.org/factoftheweek/index_53356.html.

Bibliography

DeBerri, Edward, and James E. Hug. *Catholic Social Teaching: Our Best Kept Secret*, 4th ed. Maryknoll, NY: Orbis, 2003.

Himes, Kenneth R., ed. *Modern Catholic Social Teaching: Commentaries and Interpretations*. Washington, DC: Georgetown University Press, 2005.

Himes, Kenneth R. *Responses to 101 Questions on Catholic Social Teaching*. New York: Paulist, 2001.

Holland, Joe, and Peter Henriot. *Social Analysis: Linking Faith and Justice*. Maryknoll, NY: Orbis, 1983.

Kammer, Fred. *Doing Faith Justice: An Introduction to Catholic Social Thought*. New York: Paulist, 1991.

Land, Philip S. *Catholic Social Teaching*. Chicago: Loyola University Press, 1994.

Mich, Marvin L. Krier. *Catholic Social Teaching and Movements*. Mystic, CT: Twenty-Third, 1998.

Pontifical Council for Justice and Peace. *Compendium of the Social Doctrine of the Church*. Vatican City: Libreria Editrice Vaticana, 2004.

Schuck, Michael. *That They Be One: The Social Teaching of the Papal Encyclicals 1740–1849*. Washington, DC: Georgetown University Press, 1991.

Shannon, David J., and Thomas A. O'Brien. *Catholic Social Thought: The Documentary Heritage*. Maryknoll, NY: Orbis, 1992.

CHAPTER 12

ISLAM AND THE COMMON GOOD

HAMZA YUSUF

Religion, to use that word in its broadest sense, comes in many varieties, even within one faith tradition. However, unlike Humpty Dumpty, who claimed that words can mean whatever we choose them to mean, those of us who believe that words have subsisting meanings hold that religions have normative traditions not subject to relativistic interpretations that reduce religion to mean whatever anyone says it means. To the bane of descriptivists, prescriptive religion does indeed exist. Moreover, we believe in revelation—that the Abrahamic God we worship chose to speak to certain individuals among us, Messengers of God, to communicate His intentions for us. And the languages God used have discernable meanings that have stood for centuries despite the claims of those who argue for circularity of language.

The philosopher Leo Strauss argues,

> Man cannot live without light, guidance, knowledge; only through knowledge of the good can he find the good that he needs. The fundamental question, therefore, is whether men can acquire that knowledge of the good without which they cannot guide their lives individually or collectively by the unaided efforts of their natural powers, or whether they are dependent for that knowledge on Divine Revelation. No alternative is more fundamental than this: human guidance or divine guidance.[1]

While I agree essentially with the statement, I would argue that "either/or" statements often entail fallacies that beg the question, Could both be possible or even necessary?

For Muslims, four foundational sources establish our faith: the Qur'an, prophetic practice, consensus (among scholars), and sound reason. Hence, to answer Dr. Strauss, both reason *and* revelation, what the Qur'an calls "light upon light,"[2] revelation upon reason, guide our lives. While the first two, the Qur'an and prophetic practice, comprise revelation, the latter two result from the engagement of reason with revelation to arrive at the given truth being sought.

Islam began in the crucible of revelation, but the Qur'an reminds us that the Prophet, God's peace and blessings upon him, "is not an innovator."[3] According to the understanding of Islam's normative tradition, Muslims embrace a faith that declares a restoration to a primordial religion, the religion of all the prophets sent to all peoples of the world at various times and places. The Qur'an states, "Abraham was neither a Jew nor a Christian, but a monotheist, submitted to God."[4]

Revealed over a twenty-three-year period in seventh-century Arabia, the Qur'an and the Sunnah (the prophetic tradition) spoke to a relatively simple people, almost solely noted for a highly developed oral literary tradition; however, the two central locations of revelation, Mecca and Medina, embodied the two fundamental worldly human endeavors: commerce (in Mecca) and agriculture (in Medina). Hence, from the early period of Islam's beginnings, commercial and agricultural laws involved serious ethical considerations; hence, at the heart of Islam's message to humanity, one finds moral imperatives to the individual (for salvation) and to the community (for social cohesion). These formed the two main bodies of the Islamic moral code: the devotional, which regulate one's relationship with the Creator, and the interpersonal, which governs ethical norms of dealing with our fellow humans.

Aḥmad Zarrūq (d. 899/1493), considered one of the most prominent and accomplished legal, theoretical, and spiritual scholars in Islamic history, argued that three phases comprise the formation of Islamic tradition: revelation, consolidation, and organization. In the third phase, Islamic knowledge becomes organized into a body of doctrine, law, ethics, and spirituality that emerges out of revelation itself—revelation acts as powerful white light that later refracts into divergent colors through the prism of human intellect. In other words, revelation forms into tradition, culture, and empire. Among those formulations, a virtue-ethical tradition emerges relatively early on.

While many verses of the Qur'an directly relate to ethics, and countless traditions of the hadith literature deal directly with ethical matters—perhaps the clearest being the Prophet's statement, "I was sent only to complete noble character"—we cannot understate the impact of al-Farābī's (d. 339/950) commentary on and Isḥāq b. Hunayn's (d. 910) translation of *The Nicomachean Ethics*. At the outset of the fifth century after Hegira, Ibn Miskawayh's (d. 421/1030) *The Refinement of Character* introduced the beginnings of an Islamicized Greek ethics. During this time, philosophical ethics spread among the more cosmopolitan scholars and philosophers, and both Christians and Muslims contributed to the genre of ethical treatises. The Christian ʿAdī b. Yaḥyā (d. 974) and the Muslim Ibn Sīnā (d. 1037), both influenced by the Greeks, wrote books on virtue ethics. Then, in the late fifth century, Rāghib al-Iṣfahānī penned his magisterial *Kitāb al-dharīʿah ilā makārim al-sharīʿah* (The means to the noble virtues of the sacred law) in which he put forward a profound philosophical ethics rooted in the Qur'an and undergirded by an Aristotelean framework. This book heavily influenced Imam al-Ghazālī (d. 505/1111), who then wrote two major ethical treatises that enhanced the philosophical approach begun by his predecessors, Ibn Miskawayh and al-Iṣfahānī. For theory, his book *The Standard of Knowledge* introduced logic as a necessary defense against poor habits of the mind, and for the practical ethical concerns, he wrote *The Standard of Action*, a work that often mirrors the approach Rāghib al-Isfahānī took in his earlier work. These books came to define virtue ethics in the Islamic tradition.

All of the aforementioned ethicists accept the tripartite soul and see its correspondence in the Qur'an with (1) reason (*ʿaql*); (2) zeal for repelling evil (*ḥamiyyah* and *difāʿ*); and (3) appetite for accruing benefit (*hawā* and *shahwah*) featuring prominently in many verses. Moreover, all of them view the cardinal moral virtues of courage, temperance, prudence, and justice as universal. These are the virtues referred to by Imam al-Ghazālī in *The Standard of Action* when he writes, "The matrices of all moral virtues are four."[5] He expounds in great detail the "daughters" of the virtues that spring forth from the matrices and sire other virtues; a daughter of courage, for instance, is generosity, which can help people overcome their fear of loss and privation so they might give to charitable causes. Muslim scholars also recognized Aristotle's intellectual virtues, including love of truth, scientific knowledge, and wisdom. In addition to the moral and intellectual virtues, Imams Isfahānī and al-Ghazālī incorporated the mystical virtues that were introduced from reading the Qur'an by the great mystic scholars of the late second, third, and fourth centuries after Hegira; subsequent scholars incorporated them into any treatment of ethics.

Our scholars clearly distinguished between what they would consider Islamic characteristics derived from the Qur'an and hadith—which they termed *akhlāq maḥmūdah* or *khuluq ḥasan*—and philosophical virtues, termed *faḍā'il*. In addition to these two, they added the mystical virtues, akin to the theological virtues in Christianity. In the Islamic tradition, they also involve a spiritual training (*riyāḍah*), which if practiced culminates in a mystical knowledge of God (*maʿrifah*) that results in true happiness in this world that presages the ultimate bliss of the next world. While the number of these virtues differs among various scholars, nine comprise a core consistent in most presentations. Abū Ṭālib al-Makkī records them in his famous *Qūt al-qulūb* (Nourishment of the hearts) as awe, hope, gratitude, patience, repentance, apatheia, trust, contentment, and love. Love, being the highest, once achieved, results in a state of felicity and contentment in which one reaches an equanimity of soul; and this leaves one prepared to depart the world with a sound heart. The Qur'an says, "On that day, neither wealth nor family will avail them, only the one who brings to God a sound heart (will be saved)."[6]

In essence, the period that each of us has been allotted here on Earth affords us the opportunity to refine our souls. This essentially involves the refinement of character morally and a correct understanding of reality intellectually. The refinement of character involves the ordering of one's understanding that comes through a study of the *sunan* (patterns) of the Divine in creation that are accessed through a deep study of the Qur'an: "Do they not penetrate the Qur'an's meanings, or are their hearts sealed?"[7] For instance, trials in life reside at the heart of the human experience. Embedded in its root structure, the word for "free will," *ikhtiyār,* lies the word for "good." All choices made always attempt for some good (*khayr*); however, that good is either real or apparent. The process of refinement involves discerning the real from the apparent. A key word for "sin" in Arabic includes *khaṭī'ah,* an archery term that means "to miss the mark." The practical application of such a refinement of understanding involves the emulation of prophetic character to approximate perfection. The Qur'an tells us that the Prophet, God's peace and blessings upon him, is on a "vast ethos." The Prophet, God's peace and blessings upon him, represents for Muslims the perfected paragon of virtue. His life from early on involved constant trials and tribulations, and knowing that even God's beloved prophets are not immune to the trials of time provides solace to others. Hence, life on Earth involves constant disruptions that demand ethical and virtuous responses.

Those trials come in two forms: trials of blessings and trials of tribulation. The blessings demand gratitude and the tribulations patience. Hence, patience and gratitude are foregrounded in the Qur'anic hierarchy of virtues: "Surely in that are signs for the grateful, the patient" (31:31). Moreover, privileges and disadvantages are tests from God to reveal who is best in action: "God has made some of you a trial for others: Will you show patience? And your Lord sees everything."[8] Commenting on this verse, Ibn Qayyim said, "The poor are a trial for the rich, and the rich a trial for the poor; the ignorant for the learned, and the learned for the ignorant; the healthy for the sick, and the sick for the healthy."[9] The cultivation of correct responses lies at the heart of spiritual practice until one achieves the state in which the appropriate responses become second nature or habitus (*malakah*). Life on Earth forces upon each of us a journey back to the Divine. The Qur'an gives us a map of that journey.

According to Imam al-Ghazālī, the entire Qur'an has six foundational aims:

(1) To define the God to whom humanity is called
(2) To define the straight path to that God
(3) To define the condition of the one who arrives to God

(4) To define the difference between the virtuous and the vicious
(5) To define the condition of the deniers of God
(6) To define the stages of the path to God and how to walk it[10]

For Imam al-Ghazālī, these can be known only through revelation, but he centers the intellect, in the scholastic sense of the word (*ʿaql* in Arabic, *intellectus* in Latin), at the heart of the pursuit; thus, for him, philosophical ethics aids in the pursuit, and the framework provided by the philosophers is useful wisdom that can be incorporated into our own tradition and work. The Prophet, God's peace and blessings upon him, said, "Wisdom is the lost property of the believer: wherever the believer finds it, he or she has more right to it."[11]

Rāghib al-Iṣfahānī, Imam al-Ghazālī, Qadī Ibn al-ʿArabī, and others, augmented by the Hellenic inheritance the Muslims embraced, systematized the rich virtue-ethics tradition they derived from the Qur'an. Many of the great ethicists of Islam argue that the fundamental virtue one must inculcate in order to begin the journey is humility, which involves what they term *adab* or a deep reverence for hierarchy, from the Greek meaning "sacred order," putting things in their proper place. The Qur'an says, "I will divert My signs from those who display unjustly pride and arrogance."[12] The Prophet said, "I was commanded to humble myself so that no one would boast of being better than another. And whoever humbles himself for the sake of God, God will elevate him." Many scholars have argued that the first obligation of a human being is to reflect on creation in order to arrive at knowledge of the Creator. However, the great scholar and mystic Imam al-Junayd argued that the first obligation of a human being is to humble oneself in order that one's heart might be open to wisdom.

Finally, to return to Strauss's dichotomy, according to Muslims, human reason can determine many truths, such as "knowledge is good and ignorance is bad," but only revelation can truly give us a timeless standard that enables us to ground our understanding of right and wrong, protecting us from emotivism or the dominant zeitgeist morality or lack thereof. Moreover, only through revelation's standard can we discern actions that warrant rewards from those that warrant punishment, and in doing so, revelation grants us the gift of repentance and restitution, when we fall short of that criterion, as well as the hope of recompense for our good deeds in the afterlife. The fundamental morality of the ten commandments, also echoed in the Qur'an, can be achieved through a deliberate practice of self-refinement through the inculcation of moral, intellectual, and mystical virtues to us through both reason and revelation. A man came to the Prophet, God's peace and blessings upon him, and said, "Give me the best advice." He replied, "Say, 'I believe in God,' and then be upright and virtuous."[13]

Notes

1. Leo Strauss, *Natural Right and History* (London: University of Chicago Press, 1953), 64.
2. Qur'an 24:35.
3. Qur'an 46:9.
4. Qur'an 3:67.
5. Abū Ḥmid al-Ghazālī, *Mizān al-ʿAmal* (Cairo: Dār al-Maʿārif, 1964), 264.
6. Qur'an 26:88–89.
7. Qur'an 47:24.
8. Qur'an 25:20.
9. Ibn Qayyim al-Jawziyyah, *Ighāthat al-Lahfān* (Jeddah: Dār ʿĀlam al-Fawā'id, 2011), 894.
10. Muhammad Abul Quasem, *The Jewels of the Quran: Al-Ghazali's Theory* (independently published, Kuala Lumpur, 1977), 21–22.
11. Collection of Imam al-Tirmidhī, no. 2687.
12. Qur'an 7:146.
13. *The Forty Hadith of Imam al-Nawawi*, no. 21.

PART III

AN ETHICAL CONSENSUS ON SUSTAINABLE DEVELOPMENT

Poverty

ETHICS IN ACTION TO END POVERTY

Consensus Statement

The Challenge of Poverty

Although we live in a world of immense wealth—with global economic activity now exceeding US$120 trillion a year—vast numbers of people are still subject to poverty and social exclusion. The Ethics in Action Working Group reviewed the evidence on the nature of poverty in our world today (definitions, prevalence, distribution, causes, and remedies) and deliberated on the practical actions that can be taken by the global community, including the faith communities, to achieve SDG 1, the end of extreme poverty, and SDGs aimed at meeting basic needs and ending various forms of social exclusion.

Analytical Background on Global Poverty

There are two ways to conceptualize poverty: absolute or extreme poverty and relative poverty. Absolute poverty means being below a line that is supposed to measure the ability to meet basic needs. Relative poverty is the idea that even if basic needs are met, an income far below average leads to a life of indignity, hardship, and exclusion. The absolute poverty line is measured globally by the World Bank and is currently set at $1.90 a day (measured in purchasing power–adjusted prices). Relative poverty is typically defined as less than half of household median income.

For most of history, most of the world lived in extreme poverty. Applying the World Bank methodology, we would find that 90 percent of households or more would have been below the poverty line in 1800. Two hundred years ago, life expectancy was around thirty-five years (driven mainly by high child mortality). Today, life expectancy is around seventy years on average and eighty years in the rich countries. The decline in poverty and child mortality is a real accomplishment of the modern economy.

These declines have been especially significant in recent decades. In 1990, according to the World Bank, almost two billion people lived in extreme poverty. By 2015, this had fallen to 702 million. The percent of households living in extreme poverty fell from 37 percent to 9.6 percent over this period. This decline in poverty over the past quarter-century comes mainly from Asia, especially East Asia—and most especially China, which transformed from a village-based, impoverished state to a middle-income country with low poverty in a matter of decades. In recent years, there was also considerable improvement in sub-Saharan Africa, where the poverty rate fell from 57 percent in 2000 to 37 percent in 2015. The Millennium Development Goals deserve a lot of credit for this decline. Today, extreme poverty remains highly concentrated in sub-Saharan Africa and South Asia, with pockets in Central America and Southeast Asia.

Child mortality rates also declined dramatically over this period—this is no surprise, as poverty and death tend to go together. In 1990, 12.5 million children under the age of five years died each year. By 2015, this had fallen to 5.9 million. And almost every one of these remaining deaths—on the scale of the Holocaust every single year—is preventable.

Relative poverty within rich countries is also a profound challenge. For example, while the United States is the richest country in world history, it has not fought poverty since the 1960s. It has the highest poverty rate of any high-income country—about 17 percent. This leads to huge marginalization, most pronounced among the African American community, which experiences entrenched poverty, mass incarceration, terrible social problems, high levels of violence and drug addiction, and low health outcomes and life expectancy. Life expectancy among middle-aged working class white people has also been falling.

Relative poverty is related to inequality. And while inequality is falling *among* countries—thanks to strong economic growth and convergence in places like China and India over the past few decades—it is rising *within* far too many countries.

Societal inequality is typically measured by the Gini coefficient. By this measure, the regions of the world with the most inequality include sub-Saharan Africa and Latin America. Those with the most equality, in contrast, are the social democracies of Northern Europe, especially those of Scandinavia. The Americas experience inequality because they were societies of conquest—the colonists impoverished and enslaved the Indigenous populations, introduced African slaves, and created multiracial societies scarred by extreme inequality. In Africa, inequality is driven by the colonial legacy and a great dependency on primary resources—it is fairly easy for a small group to appropriate the income from a commodity.

The Gini coefficient can be measured in two ways: from market income and from disposable income (which nets out taxes and adds in social benefits). Not surprisingly, the Gini coefficient on market income is greater than the Gini coefficient on disposable income across the economies of the member countries of the Organisation for Economic Co-operation and Development. All governments, at least to some extent, use the tax-transfer system to narrow income inequality. But some countries do this much more than others—this is especially true for the Northern European economies. In the United States, both taxes and social benefits are low, leading to very little redistribution from rich to poor. Unsurprisingly, inequality has soared in the United States—its Gini coefficient is now among the highest in the world (probably the highest since the Civil War period).

Structural Drivers of Poverty

The prevalence of extreme poverty can be traced to a set of interlocking factors:

- The existence of a *poverty trap*, whereby the country is simply too poor to make the basic investments needed to end deprivation—in areas like health, education, water and sanitation, agricultural productivity, and infrastructure.
- *Bad economic policies* that worsen economic growth and income distribution—including closing borders to trade and foreign investment, reliance on central planning, and inflationary finance of government spending.
- The government's *financial insolvency*, given overspending and overborrowing, which pushes the country into a debt trap, diverting money away from needed investments to reduce poverty.
- Impediments of *physical geography*, including being landlocked, mountainous, prone to malaria or other deadly diseases, and highly vulnerable to natural disasters such as droughts and floods (which are made worse by climate change).
- *Weak governance*, whereby the country suffers from corruption, inefficiency, and lack of transparency. This can divert wealth from where it is most needed and impede efforts to fight poverty.
- *Cultural barriers*, especially when women and girls face discrimination and impediments to flourishing. One significant problem is girls being unable to attend school.
- *Geopolitical trends*, especially when some poor countries are treated as pawns in the power games of great powers, the stage for proxy wars, and campaigns of destabilization.

There are two major structural drivers of relative poverty in rich countries. The first is recent technological change, which tends to reward high-skilled over low-skilled workers. The second is globalization, whereby the emergence of global supply chains and competition from developing countries has reduced the wages of low-income workers relative to high-income workers and owners of capital.

Ideological Drivers of Poverty

The main ideological driver of both absolute and relative poverty in the world today is libertarianism, flowing from the idea that self-interest must dominate and that any attempt to frame a common good is a violation of freedom. This ideology insists that all market rewards are fairly and justly earned and that the government has no role in either impeding the natural forces of the market or transferring income from the fortunate to the less fortunate.

The allure of libertarianism thinking is a major reason the global economy tends to be organized largely along corporate lines for the single-minded and aggressive pursuit of profit—which excludes any concern for the poor, the environment, the worker, and even for peace.

A Moral Consensus on Ending Poverty

It is possible to discern a fundamental consensus across the different religious and secular ethical traditions—rooted in mutually shared values and irrevocable standards of conduct—on the obligation to end extreme poverty and social exclusion. At a fundamental level, human beings do not want to inflict suffering on other human beings. Evidence of psychology, evolutionary biology, and neuroscience suggests that human beings are wired for cooperation and endowed with strong tendencies toward altruism and fairness.

The moral codes of the different religious traditions also cohere on the importance of making sure the material needs of all people should be met. Each religion, for example, proposes some version of the Golden Rule.

The Abrahamic religions all agree that human beings are children of God, endowed with a dignity that is possessed rather than earned. They hold that since God is compassionate and merciful, we are likewise called upon be to compassionate and merciful. As a religion born out of the historical experience of marginalization and vulnerability, Judaism contains a specific moral obligation to love the stranger—including the poor and excluded. Christianity believes that human beings are called into existence to become truly themselves only in relation to the other—and this notion finds its highest obligation in the

Beatitudes, which stress that responding to our neighbor in need with compassion benefits not only the recipient but also the benefactor, who becomes the willing recipient of a grace. And in Islam, the Holy Qur'an describes how God placed humanity as his regents on Earth, charged to keep and protect it and it to respect the dignity of all human beings.

This same obligation flows from the Dharmic religions. Buddhism, for example, holds that human beings are trapped in a cycle of suffering as long as they are ruled by self-love and that everything is interrelated, meaning that the self and the other are mutually inseparable. Buddhism stresses the idea of finding one's true self by self-emptying, which encourages self-donation and expresses the true meaning of love and compassion. It too stresses the dignity of the human being, through the teaching of Buddha-nature—the notion that every person has the potential to become Buddha.

A moral obligation to end poverty can also be discerned from some modern streams of thought. Kantian deontology, for example, can be used to argue that poverty violates the autonomy and dignity of the poor, thus presenting a clear obligation to relieve their deprivation. Kantian ethics implies that every person must be given the freedom to become active and dignified agents of their own development, which implies the participation of all. Kant viewed the good as moral perfection of the self directed toward the well-being of others. Likewise, modern consequentialist or utilitarian ethics can also suggest an imperative to end extreme poverty, based on the notion that the greatest happiness of the greatest number is maximized when resources are transferred from the wealthy to the poor. Thus, the "effective altruism" movement, for example, calls upon people to deploy their resources toward doing as much good in the world as possible. This coherence among traditions can give rise to a moral consensus rooted in the reconstituted virtue ethics, which posits a profound reciprocity between unfolding human dignity and advancing the common good. From this perspective, the scandal of poverty goes well beyond material deprivation. It also leads to the diminution or destruction of the capacity to be and to act, to be and to be able to do. The person is no longer able to unfold their dignity by developing their potential across the various dimensions of life. From this perspective, poverty can be seen as a fundamental assault on human dignity.

Likewise, poverty and exclusion undermine the common good—it is simply not possible to build shared well-being, the good of the common social life, when some people are netted out. In one sense, the common good can be likened to a geometric mean rather than an arithmetic mean—when some are deprived and unable to respond to the call to development, society as a whole is wounded.

High levels of inequality can also sever the sense of shared purpose necessary for the realization of the common good. Aristotle himself recognized that the ability of the political community to promote the common good would be impeded by large gaps between rich and poor—because the poor are too poor to embrace civic duty, whereas the rich are more attached to their wealth than to civil obligations. Assured that their rewards are attached to moral desert, the rich grow increasingly segregated from the rest of society, with ever-narrower circles of fraternity and ethical horizons.

Countervailing Forces

Numerous obstacles stand in the way of the moral consensus to end poverty. Again, these forces are both ideological and structural.

Just as the ideology of libertarianism causes poverty, it also hinders efforts to fight it. This is because it rejects values like solidarity and cooperation as a basis for social interaction. The "virtues" it promotes are the false virtues of egoism, competition, and boundless acquisitiveness that reduce all value to financial value, promote an attitude of short-term, self-centered gratification over solidarity and longer-term sustainability, anesthetize moral sentiments, and consign ethics strictly to the private domain. And because it exalts the free market as the best way to improve societal welfare, regarding poverty as a natural by-product of a self-sustaining and virtuous market system, libertarianism admits of no institutional failure and permits no institutional solutions.

The inculcation of libertarian values can also set off of vicious cycle whereby social and moral capital decline, making it even harder to coalesce around cooperative solutions. This is especially the case when the libertarian ethos leads to inequality, corporate impunity, and degradation of the environment.

The main nonideological obstacle to moral action on poverty revolves around problems of commitment. This issue has numerous strands. One strand is distance from the problem, both across space and time. It is difficult to persuade people to make sacrifices that might benefit anonymous people far from home at some point in the future. Relatedly, "ethics" is easier than "action"—it is easier to affirm abstract common values than actually live by them. Accordingly, a "thin" moral discourse needs to be "thickened" to focus on achieving a desired good in concrete terms and in specific situations.

Another strand is moral disengagement, a cognitive process that deactivates the self-regulatory processes that prevent individuals from violating their own moral standards. Moral disengagement is driven by three basic mechanisms: cognitively distorting bad acts so that they appear benign, minimizing personal roles through the diffusion of responsibilities, and holding victims themselves responsible for the harm they are experiencing.

Overcoming these well-known obstacles to action requires some external force, which can come from civil society and the religious communities. This is the real power of Ethics in Action.

Moral Response to Poverty

Before assessing the "moral resources" needed to end poverty, we need to first assess the "economic resources" required. Importantly, is that ending poverty is well within our reach—and capacity prompts moral obligation.

In that context, ending extreme poverty has four key dimensions:

- Most important is meeting basic needs, like health, education, food security, clean water, and sanitation. Here, evidence suggests that education and health for all children are especially important.
- The second dimension is peace. There is a vicious circle here whereby poverty contributes to violence, and violence contributes to poverty. The religious communities are key to breaking such a downward spiral.
- Third, certain pockets of poverty require specific interventions. This is especially true for Indigenous populations, which were historically pushed to the margins and continue to face great deprivation and indignity. Minority communities continue to face hardships everywhere, and girls and women tend to face discrimination all over the world.
- Fourth, global solidarity is needed to help finance social services and infrastructure. While ending poverty starts at home, with the poor themselves, governments often simply lack the needed resources to break the cycle of poverty.

The key point is that the resources needed to end extreme poverty are relatively minor. In a $100 trillion economy, the world poverty gap—what it would cost to meet the basic needs of every poor person—is about $400 billion (less than half a percent of world output). For $40 billion a year—$6 for every person on the planet—we could save five million lives. We could end the AIDS epidemic for $40 billion a year to 2030. We could provide education for every child on the planet through secondary school for another $40 billion. We could solve climate change for about $1 trillion per year (1 percent of world output). We could provide access to clean energy to the bottom three billion for about $200 billion a year.

On the other side of the financial ledger: The resources to combat poverty and social exclusion are well within reach. Only six countries—Sweden, Norway, Luxembourg, Denmark, the Netherlands, and the United Kingdom—deploy the agreed target of 0.7 percent of GDP to official development assistance. Meanwhile, annual military spending in the world is between $1.5 trillion and $2 trillion. After-tax fossil fuel subsidies amount to $540 billion a year (half of what it would cost to solve climate change). The money squirreled away in tax havens is between $20 trillion and $30 trillion. There are 1,800 billionaires in the world, with a combined net worth of $7.1 trillion. And the top ten hedge fund managers in 2015 made $10 billion between them.

Reducing relative poverty also has a number of dimensions, some of which overlap with those of eliminating extreme poverty. The first is ensuring that all people have access to quality education and health care. The second is fiscal redistribution—using the tax and benefit system for purposes of redistribution. This must be underpinned by an abiding sense of social solidarity (the antithesis of the libertarian ethos). The third dimension is an empowered and responsible workers' movement—yet another dimension of social solidarity. An empowered workers' movement can not only boost wages for workers but also act as a counterweight against corporate excess and impunity across the board.

Having identified the practical solutions, we must turn to the related personal and social virtues that need to be cultivated, as well as the renewal of institutions.

Starting with the individual: While efforts to fight poverty rely on policies at the domestic and global levels, they must ultimately be driven by individuals who internalize virtues such as compassion, solidarity, justice, and temperance. Small personal acts can do a lot of good in themselves, but they can also act as models of behavior to others, and they affirm the personal dignity of the donor.

There is a special role for religious believers here, given that religious communities have always been great seedbeds for nurturing virtue. Religious traditions have always endorsed virtuous practices such as moderate and abstemious lifestyles and relieving the suffering of others, including through the deployment of financial resources. Religious communities can mobilize to help the poor directly, both at home and abroad, as well as advocate for political and economic action that prioritizes poverty reduction.

On the social side, the counterpart to individual virtue geared toward helping the poor is the need for institutions oriented by the right values. The main challenge here is to counter the ideology of libertarianism with values based on solidarity, sustainability, and global cooperation. This work begins by recognizing that rights are always attached to duties and that responsibilities have both global and local dimensions. Such value-oriented institutions would contain the modern market economy within a moral framework centered on human flourishing rather than on wealth accumulation detached from ethical concerns. In this context, a key priority is changing the way we look at business—instead of seeing corporations as simple profit maximizers, we must view them as social institutions oriented toward the common good, with responsibilities to a broader array of stakeholders. And we need political institutions that prioritize development and poverty reduction; such priorities would provide a better guarantee of peace than a strong military.

The Sustainable Development Goals Relating to Poverty

The SDGs address poverty in the following ways:

- SDG 1 aims to end extreme poverty.
- SDG 2 aims to end hunger and various forms of malnutrition.
- SDG 3 aims to ensure universal health coverage.
- SDG 4 aims to ensure universal access to education at least through the secondary level.
- SDG 5 aims for gender equality in meeting all basic needs.
- SDG 6 aims for universal access to safe water and sanitation.
- SDG 7 aims for universal access to modern energy services.
- SDG 8 aims for decent work for all and the end of modern forms of slavery.
- SDG 10 aims to reduce income inequality and relative poverty.

The Ethics in Action Working Group considered the following actions needed to achieve these SDGs, on both a global and local scale.

Global Actions to End Extreme Poverty

The Ethics in Action Working Group believes that the most effective way to end both extreme and relative poverty is to support each individual, especially every child, to develop their capabilities to the maximum extent feasible through access to education (SDG 4), health care (SDG 3), and decent nutrition (SDG 3) and to achieve a livelihood with dignity and labor rights (SDG 8). All households should have access to basic infrastructure including water, sanitation, and modern energy services (SDGs 6 and 7). Discrimination and violence against girls and women and against minority groups are important causes of poverty in the world today and should be ended (SDGs 5, 10, and 16). All abusive labor practices, child labor, and all forms of modern slavery should also be ended (SDG 8).

Many of these steps require changes of regulations and perhaps of some social norms but are not financially onerous to implement. Yet other measures (notably universal health coverage, universal access to primary and secondary education, and universal access to basic infrastructure including clean water, sanitation, and modern energy services) require *increased investments* by national and local governments; however, such increased investments are often beyond the budgetary means of governments in the low-income countries. For this reason, global financial support directed toward the poor is a vital component of success in meeting the respective goals and targets. This point has been forcefully made, for example, by the recent International Commission on Financing Global Education Opportunity in 2016.

The magnitude of needed transfers is not large relative to the size of the world economy and the income of the high-income countries. Here is a simple way to see this. Consider a transfer of $1.90 per day (the global line of extreme poverty) directed from the high-income countries toward the eight hundred million poorest people. Such a transfer would amount to around $550 billion per year, or roughly 1 percent of the annual income of the high-income countries (currently estimated by the IMF to be around $51 trillion). In fact, the magnitude of the needed transfers to end extreme poverty is less than $550 billion per year since the low-income countries are already partway toward the poverty line.

A reasonable estimate of financial need is an annual transfer of around 0.7 percent of the GDP of the high-income countries directed toward the low-income and lower-middle-income countries. This is the globally agreed target for official development assistance (ODA). However, actual ODA falls short of this globally agreed target and currently stands at around 0.3 percent of the GDP of the high-income countries. The shortfall per year is therefore around 0.4 percent of high-income GDP, or a shortfall of around $200 billion per year. The needed transfers should generally be targeted toward the support of health (SDG 3), education (SDG 4), basic infrastructure (SDGs 6, 7, and 9), and a social safety net for the disabled, the indigent, and the elderly nonworking poor (SDGs 1 and 2).

General Recommendations

Financial Transfers to the Poor

The Ethics in Action Working Group supports a combination of the following measures to ensure an effective and predictable transfer of resources from the rich countries to the poor countries so that these countries can achieve the SDGs to end poverty:

(1) Establish a Global Fund for Education to mobilize the incremental $40 billion needed each year to provide universal education to secondary level in all low-income countries.

(2) Divert up to 10 percent of current global military spending—approximately $1,700 billion a year—to finance initiatives to reduce poverty and exclusion. This could be called the Isaiah Fund or the Pope Paul VI Fund.

(3) Tax anonymous wealth held in tax havens at a rate of up to 1 percent, raising as much as $200 billion per year.

(4) Establish a billionaires' fund for the SDGs endowed by at least $200 billion.
(5) Urge all wealthy countries to honor the commitment to 0.7 percent of GDP in ODA.

Social Activism for the Poor

In addition to financial support, the poor need global solidarity to defend their basic human rights, including labor rights and environmental rights. The Ethics in Action Working Group supports the following measures to defend the rights of the poor:

(1) Intervene in areas like the Niger Delta to correct the social and environmental devastation brought about by the unjust practices of multinational corporations and governments.
(2) Support grassroots movements and unions so that the poor can become dignified agents of their own development.
(3) Recognize the ownership rights of slum dwellers and smallholders, allowing them access to vital public services.
(4) Change laws and regulations to emphasize a company's duty to a broader array of stakeholders.

Education Regarding Moral Standards and Sustainable Development

The Ethics in Action Working Group believes that global education in basic ethics (human rights, labor rights, support for the poor, social solidarity, environmental sustainability) and sustainable development can support a more peaceful, equitable, and sustainable world and promote the achievement of the SDGs. In this regard, the Ethics in Action Working Group supports the following measures:

(1) Reform educational curricula in schools to teach the ethics of sustainable development (a core component of all virtue traditions).
(2) Reform business and economics curricula to better incorporate ethics and prosocial values.
(3) Establish a youth movement for peer educational efforts on ethics and sustainable development.

Actions by the Faith Communities, Ethicists, and Religious Leaders

Global religious and secular ethical leaders should organize special and sustained efforts to promote the SDGs and the shared ethical underpinnings of the world's religious traditions to achieve sustainable and integral development. The Ethics in Action Working Group endorses the following steps:

(1) Enlist high-level religious and other ethical leaders to play a leading role in restoring social and moral capital, including through participation in official forums such as the UN High-Level Political Forum and the G20.
(2) Empower and equip religious communities and congregations around the world for multireligious action centered on the values of hospitality and shared well-being and on partnerships for the SDGs.

Ethics in Action Working Group Action Item

The Working Group will begin the preparation of an Ethics in Action handbook on sustainable development, building on *Laudato si'* and related texts, that emphasizes the shared ethical underpinnings of sustainable and integral human development in the major faiths and the importance of a new "virtue ethics" to promote the fulfillment of the SDGs and the Paris Agreement on climate change.

CHAPTER 13

THE CHALLENGE OF GLOBAL POVERTY

JEFFREY D. SACHS

The first Sustainable Development Goal—indeed, the top priority of the sustainability agenda—is the end of poverty.

We are a rich world. As of 2019, the average net national income per person on this planet is about US$8,800. However, when we examine the national net income of individual countries, we see stark differences between the wealthy and the poor countries. The adjusted net income per person in the United States is about $51,000 per year, whereas in other countries such as India or Malawi, the adjusted net income is about $1,600 or $250, respectively.[1] There is vast inequality between and within countries, as well. Ours is a world of great aggregate income and wealth where there are still hundreds of millions of people suffering from extreme deprivation, even dying from their poverty.

What is extreme poverty? Extreme poverty in basic usage means that people do not have the essential means to ensure their survival and basic human dignity. They do not have enough nutritious food to eat, safe drinking water or access to sanitation, access to health care when they need it, or places in school for their children.

How many people in the world live in extreme poverty? The estimates vary depending on the precise definition, but according to the World Bank, which is the official scorekeeper of extreme poverty, the number is around 735 million people (as of 2015), calculated based upon the international poverty line of US$1.90 per day (in terms of 2011 purchasing power parity).[2] This number is roughly 10 percent of the world's population. The good news is that the proportion of people living in extreme poverty has been coming down.

In the centuries leading up to the Industrial Revolution, global GDP per capita rose from about $400 per person globally to about $1,200 per person in Western Europe by 1800 and to between $400 and $700 per person in Eastern Europe, China, Japan, the United States, India, and Africa.[3] Once the Industrial Revolution went into full swing at the beginning of the 1800s, standards of living began to improve in the industrialized countries of Western Europe and the United States. Then, after World War II and the end of the European Imperial era, economic improvement spread far more broadly.[4] Because of technological advance around the world, and the accompanying economic progress, the proportion of people facing extreme deprivation has diminished to around one billion of the world's eight billion people at the start of the 2020s.

In 1990, the proportion of people living below the international poverty line was around 36 percent, according to the World Bank's estimates.[5] In the year 2000, all governments of the world adopted the Millennium Development Goals, focused on fighting extreme income poverty and other forms of extreme deprivation (hunger, lack of access to health care and education, lack of access to safe water and sanitation, and others). MDG 1 called for the world, and for each country, to cut the rate of extreme poverty by at least half as of 2015 compared with the 1990 rate. On an aggregate global level, that goal was actually achieved five years ahead of deadline, in 2010. And by the end of the fifteen-year period of the MDGs in 2015, more than one billion people had been lifted out of poverty; worldwide, poverty had fallen from an estimated 36 percent to around 10 percent.[6],[7] On a country level, China achieved the most remarkable gains in eradicating extreme poverty. By the time of the founding of the People's Republic of China in 1949, China had become an exceedingly poor country, following more than a century of foreign invasions, civil war, and other turmoil. Since the beginning of the 1980s, with the dramatic economic reforms initiated by Deng Xiaoping, more than one billion people have escaped from extreme poverty, which was finally ended in 2020.

The remaining concentrations of extreme poverty are in sub-Saharan Africa and South Asia. In 2015, more than 85 percent of the global poor lived in these regions, with more than half in sub-Saharan Africa. In sub-Saharan Africa in 2015, about 41 percent lived in extreme poverty, and three-fourths of the sub-Saharan countries had poverty rates above 18 percent. Yet those rates of extreme poverty were falling, in some countries quite rapidly in the years preceding the COVID-19 pandemic. In South Asia, the rate of extreme poverty in 2015 was about 12 percent. While on the global level, rates of extreme poverty are declining, regionally and within countries, these numbers are stagnating or even increasing in certain cases, such as in the Middle East, where poverty rose from 2.6 percent in 2013 to 5.0 percent in 2015.[8]

It is shocking, however, that even with marked progress in poverty reduction, hundreds of millions of people still live below the basic needs line estimated by the World Bank. About eight hundred million people worldwide are without access to an improved water source, and more than 2.5 billion are without access to improved sanitation.[9] The World Health Organization estimates that roughly half of the world's population is without access to essential health services.[10] Roughly one billion people do not have access to basic electrification. The "bottom billion," as it has been called, are without access to reliable, safe water and sanitation.[11] The World Health Organization estimates that in 2018, about 5.3 million children under the age of five died[12] of conditions that are utterly preventable or completely treatable with enough means, enough finance, and enough access to basic services to address those conditions. For example, no child has to die of malaria, which is a 100 percent treatable disease with a dose of medicine that costs less than one dollar. Nonetheless, hundreds of thousands of children still die of malaria each year, because they lack access to affordable and timely life-saving health services.

A world as rich as ours can readily end extreme poverty. SDG 1 in fact calls on the world to end extreme poverty by the year 2030. If each government carries out its responsibilities with decent and competent policies directed at health (SDG 3), education (SDG 4), and basic infrastructure including safe water and sanitation (SDG 6), sustainable energy (SDG 7), and digital connectivity (SDG 9), and if these goals are accomplished with gender equity (SDG 5), reduced inequalities (SDG 10), and access to justice for all (SDG 16), the end of poverty is achievable. However, to achieve these goals, the rich countries must live up to their long-standing but long unfulfilled promise of providing official development assistance of at least 0.7 percent of gross national income as agreed upon since 1970.[13] That additional financial help could then be directed toward enabling the world's poorest people to meet their basic needs (health care, nutrition, water, sanitation, electrification, and skills development), climb the ladder of development, and rise above the threshold of extreme poverty.

In addition to SDG 1 to end extreme poverty, we also have the challenge, the means, and the opportunity to end *relative* poverty. Relative poverty is inequality so extreme that it diminishes the dignity of those low in income distribution, thereby creating tremendous stresses on poorer individuals and households because they can't participate adequately in the mainstream of society. Relative poverty is defined in relation to the economic status of others in a given society.[14] One convenient and widely used definition of relative poverty is an income less than 50 percent of the national median income of the population.[15] If a household is below half of the median, it's going to have a harder time being part of normal social life, participating in good citizenship, having the human dignity and the means to ensure that children have access to good neighborhoods, quality education, and the skill development they will need to grow up as healthy and productive citizens.

We know that relative poverty, just like extreme poverty, is extremely debilitating for well-being because it puts tremendous stresses on individuals, even in rich countries. Those at the bottom of the income distribution, those in relative poverty, though they have incomes well above an absolute global poverty line, have life expectancy many years less than those at the top of the income distribution. SDG 10 calls on countries not only to end extreme poverty, as in the first of the SDGs, but to dramatically reduce relative poverty by narrowing the income inequalities within countries. When countries have high levels of relative poverty and high degrees of inequality, such as the United States, it turns out, not surprisingly, that children born into poor families have much worse life prospects than children born into wealthier families. Following this, intergenerational relative poverty tends to persist, and social mobility in societies with high relative poverty tends to be low.

In the United States, we find that strong predictors of a child's likelihood of growing up to have a life of prosperity are the education and income levels of their parents. Rather than enjoying a society of high social mobility in which each child has the chance to make their own way, poor children face tremendous obstacles from birth, from the stresses of their environment, the strains of impoverished neighborhoods, the difficulties of schooling, and the physical difficulties of cognitive development and decent nutrition. Rates of social mobility in the United States turn out to be very low or quite stagnant,[16] especially compared to those of countries such as those of Scandinavia, which have low rates of relative poverty and high degrees of intergenerational social mobility.[17] By narrowing the income gaps among households in society, we also enable all children to have the chance to achieve their potential for the coming generation.

Relative poverty can be reduced substantially through fiscal and labor market policies. In 2021, using the OECD's measure of relative poverty (less than 50 percent of median income), the rate of relative poverty among the high-income OECD countries ranged from lows of 4.9 percent in Iceland and 6.1 percent in Denmark and the Czech Republic to a high of 17.8 percent in the United States.[18] The countries with low rates of relative poverty generally achieved that through extensive fiscal and labor market measures, including universal access to publicly funded health care and education, generous transfer payments to families in need, universal access to child care and family leave, and active labor market policies, including publicly ensured job placement services and job training. The United States, by contrast, has no nationwide commitments to such basic policies, and the high rate of relative poverty is the adverse consequence.

The SDGs provide a road map for ending extreme poverty, calling on all nations to meet basic needs like health care, education, water, sanitation, electrification, and addressing the challenges of relative poverty, which afflict even the richest countries of the world. The SDGs calls on us to pay careful attention to ensuring that households, especially of minority groups and the disadvantaged, do not fall below the levels of subsistence and human dignity.

The end of poverty has been dreamt of throughout human history. That goal is now within reach. It's a major step that the 193 UN member states have taken on that very goal in SDG 1, with a target date of 2030. Yet we are far off track from achieving this goal, buffeted by the COVID-19 pandemic and by the neglect or indifference of too many governments, including those of many of the richest countries in the world. Our sacred duty is to intensify our efforts to 2030, recognizing our vast power and potential to reduce suffering and promote well-being.

Notes

1. "Adjusted Net National Income per Capita," World Bank, accessed October 7, 2021, https://data.worldbank.org/indicator/NY.ADJ.NNTY.PC.CD?view=map.
2. "Poverty and Equity Data Portal," World Bank, accessed October 7, 2021, http://povertydata.worldbank.org/Poverty/Home.
3. Derek Thompson, "The Economic History of the Last 2000 Years: Part II," *The Atlantic*, June 12, 2012, https://www.theatlantic.com/business/archive/2012/06/the-economic-history-of-the-last-2000-years-part-ii/258762/.
4. Clark Nardinelli, " Industrial Revolution and the Standard of Living," Library of Economics and Liberty, accessed October 7, 2021, https://www.econlib.org/library/Enc/IndustrialRevolutionandtheStandardofLiving.html.
5. "Poverty and Equity Data Portal."
6. "We Can End Poverty: Millennium Development Goals and Beyond 2015," United Nations, accessed October 7, 2021, https://www.un.org/millenniumgoals/poverty.shtml.
7. "Decline of Global Extreme Poverty Continues but Has Slowed: World Bank," World Bank, September 19, 2018, https://www.worldbank.org/en/news/press-release/2018/09/19/decline-of-global-extreme-poverty-continues-but-has-slowed-world-bank.
8. World Bank, *Poverty and Shared Prosperity 2018: Piecing Together the Poverty Puzzle* (Washington, DC: World Bank, 2018), https://openknowledge.worldbank.org/bitstream/handle/10986/30418/9781464813306.pdf.
9. "Global Water, Sanitation, and Hygiene Fast Facts," Centers for Disease Control and Prevention, last modified April 1, 2021, https://www.cdc.gov/healthywater/global/wash_statistics.html#two.
10. "World Bank and WHO: Half the World Lacks Access to Essential Health Services, 100 Million Still Pushed Into Extreme Poverty Because of Health Expenses," World Health Organization, December 13, 2017, https://www.who.int/news-room/detail/13-12-2017-world-bank-and-who-half-the-world-lacks-access-to-essential-health-services-100-million-still-pushed-into-extreme-poverty-because-of-health-expenses.
11. Paul Collier, *The Bottom Billion: Why the Poorest Countries Are Failing and What Can Be Done About It* (New York: Oxford University Press, 2007).
12. "Child Mortality and Causes of Death," World Health Organization, accessed October 7, 2021, https://www.who.int/gho/child_health/mortality/mortality_under_five_text/en/.

13. "Financing for Sustainable Development," OECD, accessed October 7, 2021, https://www.oecd.org/dac/stats/the07odagnitarget-ahistory.htm.
14. "Migration and Inclusive Societies," UNESCO, accessed October 7, 2021, http://www.unesco.org/new/en/social-and-human-sciences/themes/international-migration/glossary/poverty/.
15. "Indicator 65: Percentage of Households with Incomes Below 50 Percent of Median Income ('Relative Poverty')," UN Sustainable Development Solutions Network, accessed October 7, 2021, https://indicators.report/indicators/i-65/.
16. Richard V. Reeves and Eleanor Krause, "Raj Chetty in 14 Charts: Big Findings on Opportunity and Mobility We Should All Know," Brookings Institution, January 11, 2018, https://www.brookings.edu/blog/social-mobility-memos/2018/01/11/raj-chetty-in-14-charts-big-findings-on-opportunity-and-mobility-we-should-know/.
17. Rasmus Landersø and James J. Heckman, "The Scandinavian Fantasy: The Sources of Intergenerational Mobility in Denmark and the US," *Scandinavian Journal of Economics* 119, no. 1 (2017): 178–230.
18. "Inequality—Poverty Rate—Data," OECD, accessed October 7, 2021, https://data.oecd.org/inequality/poverty-rate.htm.

CHAPTER 14

ETHICAL ACTIONS TO END POVERTY

ANTHONY ANNETT

In this chapter, I will discuss some of the ethical actions to end extreme poverty, reflecting on the deliberations of the Ethics in Action initiative. A starting point is the recognition that the moral codes of all of the world's religions emphasize relieving suffering, affirming human dignity, and meeting basic needs. Most traditions have some variant of the Golden Rule, calling on people to be compassionate, empathic, loving, and merciful to one other. Across the secular ethical traditions, we can discern a similar imperative to end extreme poverty.

Many virtue ethicists in the tradition of Aristotle, for example, would argue that poverty is more than mere material deprivation, because it diminishes or destroys the very capacity of the human being to achieve the fullest development across all dimensions of life. In other words, people are blocked from being or becoming who they are meant to be or become when they are mired in poverty; when they lack access to health care, education, nutrition, water, sanitation, social protection, basic finance, and clean energy; when they are impeded from achieving a livelihood with dignity and respect, including through unjust wages and labor practices; and when they face forms of discrimination and violence, including because they are women or girls. In this sense, poverty is an assault on both human dignity and the common good, two important features of shared well-being identified in all the religions and secular ethical traditions.

There are a number of exceptions to this great ethical consensus—libertarianism in particular, which essentially argues that a person should be free to do whatever they want and to accept obligations that are not freely chosen, including from the government. The second exception to this broad ethical imperative is neoclassical economics, which arises out of the utilitarian tradition. Neoclassical economics is based on the notion that a person maximizes their own preferences subject to a budget constraint. Such preferences are maximized under a condition known as the Pareto efficiency, whereby it is impossible to make somebody better off without making somebody else worse off.

These ways of looking at the economy influence the current state of the world. We are immensely wealthy. The global economy is around US$128 trillion in international prices. According to the International Monetary Fund (IMF), we spend about $540 billion a year on post-tax fossil fuel subsidies. And about $20 trillion to $30 trillion is stashed away in tax havens. There is a clear ethical misallocation of resources in the world today. The global economy is not acting out a shared moral imperative to end suffering, poverty, and social exclusion.

How should things change on a practical level? The cost of financing the Sustainable Development Goals as a whole is probably around $3 trillion to $4 trillion a year—around 3 percent to 4 percent of global GDP. That is not trivial, but it also is not huge from a global perspective.

Recently, the IMF attempted to estimate what it would cost to achieve key SDGs in sectors such as health, education, roads, electricity, water, and sanitation, across forty-nine low-income developing countries, some of the poorest in the world. The number they came up with was around $0.5 trillion a year. For the countries themselves, that's a very large number—about 15 percent of GDP on average.

Is this achievable? The first point to make is that low-income developing countries should take ownership of their own development. That is subsidiarity in action. Doing this, they could possibly raise 5 percentage points of GDP in extra domestic revenue through raising taxes and improving revenue efficiency. But that is an ambitious goal. They could also try to make spending more efficient. There's also scope there. But even if countries do all of this, there will still be a gap—which the IMF estimates at $350 billion a year. Again, a large number, but this additional spending amounts to only 0.3 percent of global GDP—a small cost for meeting a major ethical imperative. How can this be met in practice? The first place to start is with official development assistance—official donor funds. If all countries meet the target of contributing 0.7 percent of gross national income, up from 0.3 percent today, it would be possible to raise another $230 billion a year. That would go a long way to filling that gap.

Another option discussed by Ethics In Action, through the Move Humanity campaign, is to ask the world's approximately 2,200 billionaires—with a combined net worth of $9.1 trillion—to donate a mere 1 percent of this net worth every year. That would raise $91 billion a year. Again, this would go a long way toward meeting the basic needs of some of the poorest people in the world.

The private sector also has a role to play. A basic tenet of the Ethics in Action initiative is that we are all co-responsible for the common good—person and society, nation-state and international community, public and private sector. Private business and finance have a particularly important role to play in building infrastructure and financing the development of renewable energy. However, doing so will require a strong partnership between governments and the private sector. We need investors and corporations to take a risk on sustainable development, to focus more on building long-term value and less on short-term profits. And governments must meet the private sector in partnership, creating an enabling environment that is transparent and free from partiality and corruption.

Against this backdrop, Ethics in Action has come up with specific mechanisms to transfer resources from rich to poor. One is to establish a global fund for education to mobilize $40 billion a year to provide universal education to the secondary level to all children in all areas of the world. In the grand scheme of things, this is a tiny investment. And yet it would deliver incalculable returns. How can the children of the world meet the grave challenges that face them today—the challenges of technology, climate change, the future of work—without at least a secondary education? It is just not possible.

Global funds are a tried and tested strategy, especially in the health space. They have been instrumental in improving health outcomes in the world's poorest countries, including by significantly reducing deaths from AIDS, malaria, and tuberculosis. They pool resources in a way that avoids overlap and duplication. They are based on clear metrics, clear evaluation, independent reviews, and good governance structures. They have worked before, and they can work again.

Another suggestion of Ethics in Action is to divert 10 percent of current global military spending to finance initiatives to fight poverty, social exclusion, and climate change. Such an initiative could be called either the Isaiah Fund (given the prophecy in Isaiah 2:4 that the many people of the nations will beat their swords into ploughshares) or the Pope Paul VI Fund (because Pope Paul VI made a similar recommendation in his 1967 encyclical, *Populorum progressio*).

A further suggestion of Ethics In Action is to tax the wealth in tax havens at a rate of around 1 percent a year. This is a huge pool of money that should not be divorced from the common good.

Another dimension of addressing poverty pertains to social activism for the poor. Key here is the prioritization of decent work as a central source of human dignity and meaning, which must take preference over the quest for profits, especially short-term profits. This effort would entail support for grassroots movements, unions, and worker cooperatives to promote the interests of the poor and workers—enabling them to become dignified agents of their own development and counteracting the influence of business and finance.

Relatedly, we need to change laws and regulations at the company level to emphasize that companies have a duty toward a broader array of stakeholders, not just shareholders—and this duty encompasses an acknowledgment of and efforts to address their social and environmental impact.

We also need the full implementation of the Paris Agreement on climate change by all nations without delay. In addition to addressing environmental concerns, this is a pro-solidarity, pro-poor policy.

Finally, ethical education must play a key role in fighting poverty and social exclusion. A reformed education system would seek to inculcate the values and virtues of sustainable development rather than those that support egoism and materialism. Achieving this goal entails reforming educational curricula in schools, economics programs, business schools, and within businesses more generally to incorporate ethical responsibility and prosocial values.

Change takes place not only at the level of government policy or intergovernmental initiatives but also at the level of the person and in the institutions in which that person operates. This notion is also a vital dimension of the actions we need to take to end extreme poverty, social deprivation, and marginalization.

CHAPTER 15

COMMUNITY-BASED POVERTY REDUCTION

JENNIFER GROSS

Our communities are built on an outdated construct of a market economy that is a breeding ground for indifference, selfishness, jealousy, and distrust. We have no vision of virtue. The entity that has taken over, the infamous "invisible hand," is not one hand but many billions of hands, each pulling in its own direction, guided by self-interest and egoism. We need to reintroduce the idea of the common good and cultivate virtue through integral and sustainable development. Such an ethical economy respects the human dignity and moral rights of all people. There is a profound reciprocity between unfolding human dignity and advancing the common good. If we value sustainable development, we need to cultivate related virtues to make this value habitual.

The practice of virtue is the path to unfolding human dignity, potential, and creativity. We need to build on the idea of Aristotle that one can exercise the process of virtue ethics and integral human development only within a community. We must engage in the practice of humanizing others in our community. We must abandon this age of indifference.[1] We need to stop living as if we should be concerned only with self-interest. We must humanize others in a global community across the world, including in sub-Saharan Africa. The current global market economy undermines the integrity of those living in extreme poverty.

The most effective way to end both extreme and relative poverty is to support all individuals, especially all children, to develop their capabilities to the maximum extent possible. The Millennium Villages Project was an interdisciplinary research project demonstrating that the Millennium Development Goals could be achieved with an integrated package of scientifically proven development interventions. Many of the interventions created social institutions in which virtue was cultivated and practiced to lift people from poverty.

As an example, a case study in Sauri, Kenya, at Nyamninia Primary School, focused on virtue development and community-based poverty reduction through a simple meal program. The Nyamninia school meals program started in 2005 with support from the Millennium Villages Project. Each class community (consisting of a class's teachers, students, and parents) involved in the program voted on and agreed to the value of school meals and agreed that an adequate meal consisted of maize and beans.[2] Once agreed, the required amount of funding was contributed for each student's meals, supplementing nutrition through school feeding programs with farm and animal produce.

To engage the community, teachers asked students and their parents to donate planting materials like maize, vegetable seeds, and legumes, as well as chickens. A sudden change of attitude toward farming and agriculture in students and parents began to occur at the school, in the household, and in the community. Contributing gave people a sense of community ownership and responsibility. If challenges arose in harvesting, everyone faced them together. When harvests were successful, communities engaged in entrepreneurship by selling produce to their schools and in their communities.

Income generation kept the program sustainable. This income was used to buy seeds for the next planting season and to expand into new projects with dairy cows, pigs, and goats. Produce marketing provided an opportunity for students to develop positive relationships with adults (their teachers and adults in the community) through interactions of buying and selling. Students developed marketing, bookkeeping, money management, and entrepreneurial skills.

Within this market setting, students learned to work not only for self-interest but also for the good of the entire community. Successful programs were able to contribute funding to other school feeding programs and to an HIV program. Each class community was also asked to identify the vulnerable children within that community, and parents with the available means agreed to contribute more to meet the needs of those children. The students thus developed deep care for and respect for the human dignity of one another. The program was not only financially successful but also rich in the creation of ethical and moral fiber.

The Nyamninia school meals program nurtured leadership skills and good citizenship among the children involved. Self-confidence and respect came through respecting one another. The program "mobilized resources for the realization of good health, through improved food and nutrition security, improved family income, environmental protection, and enhanced literacy levels in an equitable, just, sustainable and gender sensitive manner."[3]

Owing to engagement in a participatory learning setting, academic performance also rose. The students created a market of their own in which the needy were cared for and actively instituted a moral structure in their community. An outward blossoming of ethics and moral character radiated through their school and households, changing attitudes of parents and educators, and into the village. Students developed virtues through learning and through practice: As Aristotle says, "For the things we have to learn before we can do them, we learn by doing them."[4]

The experience of the Nyamninia school meals program gives rise to an important lesson:

> At the heart of the virtue approach to ethics is the idea of "community." A person's character traits are not developed in isolation, but within and by the communities to which he or she belongs. As people grow and mature, their personalities are deeply affected by the values that their communities prize. The virtue approach urges us to pay attention to the contours of our communities and the habits of character they encourage and instill.[5]

Working within a framework of social sustainability underpinned by ethics, the students were able to build economic sustainability in their community and lift others from poverty—two key pillars of sustainable development. The third pillar of sustainable development was exercised through care for the land and agriculture. The fourth pillar, good governance, was actively practiced by giving the students the power and agency to be leaders in their community.

The Nyamninia school meals program excelled in teaching school-going children the fundamental aspects of the Sustainable Development Goals: the importance of food and nutrition security through farming; good leadership; good citizenship; and an innovative and entrepreneurial approach to society and security. By actively engaging in ethical conduct, the students developed compassion, inner peace, and happiness. As Pope Francis says, "Inner peace is closely related to care for ecology and for the common good. Because, lived out authentically, it is reflected in a balanced lifestyle together with a capacity for wonder which takes us to a deeper understanding of life."[6]

In a broader context, there is no way to develop inner peace and unfold people's capabilities in a global society that perpetuates atrocities affecting the dignity of those living in extreme poverty. In sub-Saharan Africa, around 10 percent of children die before reaching the age of five. Maternal death rates are scandalously high. Many people suffer unnecessarily from preventable and treatable diseases such as malaria, diarrhea, tuberculosis, and HIV/AIDS. Those afflicted by extreme poverty have little or no access to the most fundamental aspects of modern medicine. Many countries in the developing world struggle to make progress toward the health-related SDGs because millions live in rural areas beyond the reach of modern health care.

The solution to this problem lies in within the community itself. Across Ghana and other countries across Sub-Saharan Africa, governments have deployed community health workers (CHWs) to rural villages. Community health workers are nonclinical care providers who live in the communities they serve. They volunteer to be trained through a government program that aims at meeting the basic health needs of the community.

These workers move from home to home in their villages. Each is given a smartphone that enables them to provide instant diagnoses through SMS, allowing them to provide care for all members of their community. In addition to diagnosing malaria and prescribing treatment, the community health workers are trained to give first aid, provide nutritional advice, and help women give birth. They keep track of disease outbreaks and overall public health concerns and offer a vital link between their community and the broader health care system of doctors, nurses, clinics, and hospitals.

A key factor in the success of this program is that the community health workers are members of, and chosen by, the community they serve. They are accepted as leaders by the their community and have natural cultural awareness.

This factor is so important because, in sub-Saharan Africa, 53 percent of the poorest households are disengaged from the health system[7] because of poor road infrastructure, clinic fees, and community beliefs. By overcoming these barriers to receiving health care, the community health worker program has transformed the health of rural communities. Expanding such programs could thus improve the health of a nation.

A health system that involves community health workers providing care in rural areas has the potential to unite all communities in a region and a nation. It is a system that works and that saves millions of lives. As Aristotle states, "For even if this is the same thing for an individual and a city, to secure and preserve the good of the city appears to be something greater and more complete: the good of the individual by himself is certainly desirable enough, but that of a nation and of cities is nobler and more divine."[8]

OECD countries have a moral obligation to provide technology and means to help poor communities. With digital technology, Africa could make revolutionary advances in health care. If this olive branch were extended, teams across all countries in Africa would be empowered in a common goal to uplift the human dignity of all by providing them with health care. The OECD countries can learn an important lesson from the children of Nyamninia Primary School and the community health workers of Ghana in terms of developing poverty-sensitive institutions and creating systems that foster the dignity of all.

Social sustainability is one of the least understood, discussed, and practiced pillars of sustainable development. However, society can learn from the examples of Nyamninia and Ghana, which are models for community-based poverty reduction. Strengthening social interactions based on trust, ethics, and social support networks to reduce inequality within communities works in reciprocity with the other complex pillars of sustainable development. The four pillars of sustainable development—economic, social, environmental, and governance—mirror one another and, when implemented together, allow us to glimpse the profound beauty of peace.

We are dependent on one another. We are called to live in solidarity with one another now and in the future. This profound reciprocity calls for the patient and resolute cultivation of personal and social virtues. Aristotle puts it this way: "Our happiness depends on the cultivation of virtue. Since happiness is an activity of soul in accordance with perfect virtue, we must consider the nature of virtue; for perhaps we shall thus see better the nature of happiness."[9]

Fostering and practicing virtue and happiness within a community creates effective social institutions oriented by shared values. Just as each is called to contribute to the common good, each in turn is to be supported by it. To truly have inner peace, there must be active engagement and practice of virtue, mutual mirroring of human dignity in one another, and a constant unfolding of what it is to be human. Then, the common good, or "the human good, turns out to be the soul's activity that expresses virtue."[10]

By respecting human dignity and practicing virtue, the children of Nyamninia Primary School and the community health workers of Ghana are lifting themselves and their communities from poverty. Virtue ethics is really about integral human development—the good of each person and all people and how we develop to become more human. Being fully human is not narrow self-interest but relational: relation to one another, to creation, and to God. Sustainable development fosters these relationships and brings us to a place of happiness and peace. Pope John Paul II declared that "development was the new name for Peace."[11] In our shared development, we must practice and share in this peace.

Notes

1. Francis, *Laudato si'* (Vatican City: Libreria Editrice Vaticana, 2015), http://www.vatican.va/content/francesco/en/encyclicals/documents/papa-francesco_20150524_enciclica-laudato-si.html.
2. Jessica Masira, "The Nyamninia Primary School Meals: A Sustainable Success Story."
3. Mary Njeri Karanu and Ruth Khasaya Oniang'o, "Impacting Nutrition Through Schools: Rural Outreach Program (ROP) Africa 4H Program Profile," *African Journal of Food, Agriculture, Nutrition and Development* 17, no. 3 (2017), https://www.ajfand.net/Volume17/No3/ROP-4H%20Program%20Profile%20-%20Impacting%20nutrition%20trhough%20schools.pdf.

4. Aristotle, *The Nicomachean Ethics*, trans. W. D. Ross, ed. Lesley Brown (Oxford: Oxford University Press, 2009),1103a33.
5. Manuel Velasquez, Claire Andre, Thomas Shanks, and Michael J. Meyer, "Ethics and Virtue," *Issues in Ethics* 1, no. 3 (1988), https://www.scu.edu/ethics/ethics-resources/ethical-decision-making/ethics-and-virtue/.
6. Francis, *Laudato si'*, sec. 225.
7. Sarah Hall, *People First: African Solutions to the Health Worker Crisis* (Nairobi: African Medical and Research Foundation, 2007), https://chwcentral.org/wp-content/uploads/2013/07/People-First-African-solutions-to-the-health-worker-crisis.pdf.
8. Aristotle, *Nicomachean Ethics*, 1094b7–10.
9. Aristotle, *Nicomachean Ethics*, 1102a5–7.
10. Aristotle, *Nicomachean Ethics*, 1098a15–17.
11. Paul VI, *Populorum progressio* (1967), sec. 76.

CHAPTER 16

JUDAISM AND POVERTY

DAVID ROSEN

While Jewish teaching on the value of modesty extends to consumption,[1] poverty is not seen as a virtue. On the contrary, the rabbis of the Talmudic period declared that "he who is crushed by poverty is as one subject to all the afflictions of the world,"[2] and even charitable giving is limited in Jewish law lest the person becomes a burden on society.[3] Nevertheless, while poverty is seen as a challenge to be combatted, it is also viewed as an opportunity for those with the means to assist in its alleviation.

The Hebrew word that is translated as charity is *tzdakah*, meaning "righteousness," which is precisely "the way of the Lord" that Abraham's descendants are instructed to follow,[4] manifesting the quality of compassion that is described as the essential characteristic of those who are truly of the seed of Abraham.[5]

And on the holiest day of the Jewish calendar, the Day of Atonement, Isaiah chapter 58 is recited in synagogue. It declares that the fast that God desires is that which leads

> You [to] share your bread with the hungry, and to take the wretched poor into your home; When you see the naked, you clothe him, and not to ignore your own flesh. Then shall your light burst through like the dawn and your [spiritual] healing shall spring forth quickly; your righteousness shall go before you, the presence of the Lord shall be your rear guard. Then, when you call, the Lord will answer; When you cry, He will say: here I am. If you banish the yoke from your midst, send away the accusing finger and evil speech; And you offer your sustenance to the hungry and satisfy the famished soul—then shall your light shine in the darkness, and your gloominess shall become (bright) like noonday.[6]

In extolling the virtue of *tzdakah*, the Talmud[7] quotes the verse from Proverbs that states, "He who pursues charity and mercy [or lovingkindness] finds life, righteousness and honor."[8] The text goes on to tell of a conversation between the second-century Roman governor in the land of Israel, Turnus Rufus, and Rabbi Akiva, in which the former expressed a challenge: "If your God loves the poor, for what reason does He not support them Himself? Rabbi Akiva said to him: He commands us to sustain the poor, so that through them and the *tzdakah* we give them, we will be delivered from the judgment of Gehenna."

Rabbi Judah Loew ben Bezalel (16c.) explains that "the Torah (Pentateuch) obliged us concerning *tzdakah* and tithes in order to improve one's very essence as a human being, so that one should give (to the needy) for the good of one's own soul."[9] And in the words of Rabbi Abraham Isaac HaCohen Kuk (20c.), "the goal of *tzdakah* is twofold; firstly in order to benefit the poor . . . ; the second (reason) is that he activates the potential good virtue into actual substance by performing goodness and loving kindness and thus perfects his soul."[10] Judaism affirms that good actions have religious primacy in forming the character of the human soul (and thus its ultimate destiny), as opposed to faith in itself. In the language of the author of *Sefer Hahinukh*, "a person is the product of his/her deeds."[11]

The idea that material blessings bring with them responsibilities toward the less fortunate is predicated on the idea that none of us are the actual owners of our material possessions. They are but gifts from God of His bounty of which we are purely custodians,[12] with the responsibility to use them for the common good, in particular for the needy. Thus, the Hebrew Bible is replete with injunctions regarding responsibilities toward the poor and vulnerable, in particular the widow, orphan, and stranger.[13] These include leaving for them the corners of harvested fields, as well as the fallen sheaves and gleanings from fields, vineyards, olive groves, and their threshing floors and presses, in addition to an annual tithe for the poor that must be given every two of three years.[14] Notable is the prohibition in Exodus against exploiting the poor, for whom God is presented as having special care because of His compassion.[15]

The explicit precept in the Torah that demands donations to the needy appears in the fifteenth chapter of Deuteronomy:

> If there is a poor person among you, do not harden your heart and shut your hand. Rather, you must open your hand and lend him sufficient for whatever he needs. Beware lest you harbor the base thought in your heart . . . so that you are miserly towards your needy bother and give him nothing and he cries out to the Lord against you and you will have sinned. Give to him readily and have no regrets in your heart when you do so, for because of such action the Lord your God will bless you in all your efforts and in all your undertakings. For there will never cease to be needy people in your land, which is why I command you: open your hand to your brother, to your poor and your needy in your land.[16]

Maimonides codifies this obligation as follows:

> We are obliged to provide for the poor according to their needs. If he has no clothes, we clothe him; if he has no household goods, we purchase them. If he has no wife, we marry him off . . . even if this poor person used to ride on a horse and have an attendant, we purchase such for him.[17]

Moreover, Jewish law requires us to give the alms seeker the benefit of the doubt: "Whoever comes and says 'feed me' we do not check him out to see whether he is a deceiver, but immediately feed him."[18] The Torah extends these responsibilities to include non-Jewish people: "If your brother becomes poor and cannot maintain himself with you, you shall support him, similarly the stranger and sojourner, he shall live with you.[19]

On the basis of the guiding statement in the Talmud that "the whole of Torah [Judaism] is for the sake of Peace," the sages also demand charity for the poor of idolaters, as well as care for their sick and burial of their dead.[20] Maimonides codifies these responsibilities and places them in their scriptural context:

> We are obliged to maintain even the poor of idolaters, attend to their sick and bury their dead, as we do with those of our own community, for the sake of the Ways of Peace. Behold it is said: "Her ways are pleasant ways and all her paths are Peace" (Proverbs 3:17); and it is written, "God is good to all and His mercy extends to all His creatures" (Psalm 145:9).[21]

In other words, not only are these "ways of peace" the essence of Torah but they also express the essential Divine attributes that we are called to emulate.

The extension of these obligation beyond the confines of the covenanted community under the rubric "the ways of peace" affirms the bonds of common humanity necessary for promoting peace in society. Indeed, the Hebrew root of the word *shalom*, "peace," is *shalem*, meaning "complete and united." It is through practical response to the needs of the other that we promote that unity and harmony that are the essence of true peace and redemption for society.[22]

Because our responsibility to care for the poor and eradicate poverty is meant to be the expression of human righteousness and compassion, the manner in which this duty is performed is significant, both when we provide materially and even when we are unable to.

Maimonides summarizes this notion as follows[23]:

> If a poor man requests money from you and you have nothing to give him, speak to him consolingly. It is forbidden to upbraid a poor person or to shout

at him, because his heart is broken and contrite, as it is said, "A broken and contrite heart, O God, You will not despise."[24] And it is written, "[God works] to revive the spirit of the humble, and to revive the heart of the contrite."[25]

Beyond our responsibility to respond to the immediate needs of the poor, we are obliged to work for the eradication of poverty. However, the most virtuous and praiseworthy response is that which facilitates the fullest expression of the dignity of the needy by enabling them to become self-sufficient. Maimonides describes eight levels of charitable response, the highest being to enable the poor to exit from poverty through gift, loan, or partnership or by finding employment.[26]

The responsibility to care for the poor is not just the duty of the individual but of the community as a whole. There is thus an obligation upon communities to establish a fund for the poor to be maintained by trustees who must be accountable and transparent. Maimonides states that these trustees must be "well known and trusted persons who engage the citizens from one week to the next taking from each one what it is appropriate to give and that which he is allocated [to give]." He adds that the sages (of the Talmud) taught that one should not live in a town that does not have a fund for the poor.[27]

At the heart of the challenge of poverty is not only human responsibility to ensure the wise and moral distribution of resources but, above all, the imperative of human solidarity itself. And even this imperative, from a religious perspective, serves a higher goal. One of the most powerful Rabbinic homilies in this regard is based on the verse in the Book of Psalms[28]: "Let the Lord arise and scatter his enemies and let those that hate Him flee from before him."

Regarding this verse, the sages say,

> Five times [in the Book of Psalms, we find that] David calls on God to arise and scatter his enemies. Yet we do not find [in the Book of Psalms any mention] that the Lord rises up [in response]. When does He arise [in the book of Psalms]? "For the oppression of the poor and the cry of the needy, then will I arise," says the Lord (Psalm 12:6).[29]

In this dramatic midrash, the sages teach that even if one is King David himself, one may not assume that God is on one's side. In a similar vein, Abraham Lincoln is reputed to have replied to a question as to whether he thought that God was on his side in the Civil War that what is important is whether we are on God's side![30]

We are truly on God's side, declares the midrash, when we care for the needy. We are only a Godly society when we address the needs of the poor in our society as a whole.

Notes

1. E.g., Avot 2:8, 4:6.
2. Midrash Rabbah, Exodus, 31:12.
3. TB Ketubot 50a; Maimonides, *Mishneh Torah*, Gifts for the Poor, 7:1,5 and 10:18.
4. Genesis 18:19.
5. Babylonian Talmud, Betzah 32b; Yalkut Shimoni, Deuteronomy 13:889; cf. Babylonian Talmud Yevamot 79a.
6. Isaiah 58:7–10.
7. Babylonian Talmud, Bava Bathra 10a.
8. Proverbs 21:21.
9. *Gur Aryeh* commentary, Numbers 21.
10. *Nitzanei Eretz* sec. 4, pp. 33–34.
11. *Sefer Hahinukh*, Frankfurt edition, commandment 16.
12. Leviticus 25:23.
13. Deuteronomy 14:29, 26:12.
14. Mishnah, Pe'ah 4:2, 8:5; Tosefta, Peah 4:2; Babylonian Talmud, Hullin 131b.
15. Exodus 22:24–26.
16. Deuteronomy 15:7–11.
17. Maimonides, *Mishneh Torah*, Gifts for the Poor, chapter 1.
18. Code of Jewish Law, *Shulhan Arukh*, Yoreh De'ah 251:10.
19. Leviticus 25:35.
20. Babylonian Talmud, Gittin 59b.
21. Maimonides, *Mishneh Torah*, Melakhim 10:12.
22. See Isaiah 32:17, 54:14; *Mishneh Torah*, Gifts for the Poor, 10:1.
23. Maimonides, *Mishneh Torah*, Gifts for the Poor, 10:4–5.
24. Psalms 51:19.
25. Isaiah 57:15.
26. Maimonides, *Mishneh Torah*, Gifts for the Poor, 10:10.
27. Maimonides, *Mishneh Torah*, Gifts for the Poor, 9:1–3.
28. Psalms 68:2.
29. Bereishit Rabbah 75:1.
30. Francis B. Carpenter, *Six Months at the White House with Abraham Lincoln* (New York: Hurd and Houghton, 1866).

PART IV

AN ETHICAL CONSENSUS ON SUSTAINABLE DEVELOPMENT

Peace

ETHICS IN ACTION FOR SUSTAINABLE AND INTEGRAL DEVELOPMENT ON PEACE

Consensus Statement

This statement was issued by Ethics in Action after deliberations held at Casina Pio IV in the Vatican on February 2 and 3, 2017. Ethics in Action is a partnership cohosted by the chancellor of the Pontifical Academies, headquartered at Casina Pio IV, the UN Sustainable Development Solutions Network, Religions for Peace, and the University of Notre Dame. The founders of Ethics in Action include the Blue Chip Foundation, the Fetzer Institute, Christina Lee Brown, and Jacqueline Corbelli. Representatives of several religious bodies participated in preparing this statement, along with scholars and representatives of nongovernmental organizations specialized in questions of peace and conflict.

Positive Peace and Its Pillars

The world's religions are based in peace, call for peace, and promote peace. Religious leaders since the time of the prophets have been urging that we "beat swords into ploughshares." "Blessed are the peacemakers," declares Jesus in the Beatitudes. Muslims, Hindus, Buddhists, Jains, Sikhs, Indigenous peoples, and other religions' believers understand—each in their own ways—that peace is the true "name" of their religion. Across the diverse religions, the injunction not to kill and to respect life is deeply shared. The obligation to advance peace is a foundational moral and spiritual imperative across religious traditions. Accordingly, interpretations of religion that go against peace are self-contradictory.

For the world's religions, peace goes beyond the mere absence of war. These traditions increasingly share a vision of "positive peace" rooted in the dignity of each and the unity of all, grounded in each religion's experience of the transcendent. This vision of positive peace is built on four essential pillars, stated explicitly in Pope John XXIII's encyclical *Pacem in terris* (1963) and acknowledged by the other religions in their own terms as reflecting the deepest desires of the human spirit: truth, justice, charity, and liberty. So understood, positive peace calls for the unfolding of human dignity in a way that is linked directly to honoring rights and executing reciprocal responsibilities. Positive peace is also realized through our common obligation to seek the good of the other and avoid evil by advancing shared well-being, which includes living in harmony with nature.

The diverse religions affirm that this positive vision of peace calls for the patient and resolute cultivation of personal virtues and value-oriented institutions. Indeed, the pillars of positive peace must be supported by virtues like mutual respect, trust, and nonviolence. Tolerance, while necessary, must be strengthened with genuine solidarity. Likewise, justice must always be paired with mercy, as otherwise human justice would be an imperfect grounding for peace. Also, the cultivation of personal virtues is not enough—these virtues must also find institutional expression in a way that challenges and transforms structures of violence and injustice into those that nurture peace. As Pope Francis noted, peace is an "active virtue" that calls for the engagement and cooperation of each individual and society as a whole. Peace is our true human destiny, which makes it our responsibility to pursue and our right to attain.

Challenges to Peace

Peace is jeopardized whenever truth, justice, charity, and liberty—the pillars of positive peace—are undermined; whenever their related virtues of mutual respect, trust, solidarity, and mercy are denied; and whenever institutions offend human dignity and fail to serve the common good.

There are many proximate causes of war. Some wars are rooted in fear, desperation, perceived threats, and real deprivations and injustices. A particular problem today is the risk of conflict exacerbated by extreme poverty and inequality, persistent marginalization and social exclusion, and the alarming pace of environmental degradation. In this light, climate change can be seen as a silent war on the planet and the Paris Agreement on climate change as a treaty of peace. Other wars have more ignoble underpinnings—motivated by a quest for profits, land, resources, glory, revenge, revenue, or geopolitical advantage. Whatever the underlying causes, wars violate human dignity and rend the fabric of the common good. They provide fertile ground for demagogues to spread fear and hate. Wars represent the failure of politics and are typically based on lies rather than truth. It must also be acknowledged that this is a particularly perilous time for peace, with tensions brewing across the globe. As of January 2021, the Bulletin of Atomic Scientists' Doomsday Clock stands at 100 seconds to midnight, signaling the greatest risk of apocalypse since 1953.

Many traditions have recognized that some wars can be considered in a very limited sense "just," because they are defensive, a last resort, legal, spare noncombatants, and advance strictly limited military violence that is "proportional" to the offending cause. Today, in an era when indiscriminate weapons are used in areas co-inhabited by civilians and the military, the bar for a technically "just" war must be set extremely high. And even when a war can be deemed "just," it remains a profound failure in terms of the ideal of nonviolence.

While wars often occur despite the ardent efforts of many religious leaders and communities to prevent them, religions, tragically, are too often implicated in wars. Demagogic leaders may proclaim that the local or national community is endangered in its core religious identity and seek to mobilize religion to justify violence or to gain political power. Promoters of violence may appeal selectively to religious texts and traditions in the defense of violence. Working together in mutual respect, the world's religions have courageously begun to reject violence in the name of religion—and they must do so ever more firmly.

The Role of Religions

Working together, religious communities advance peace in three main ways. First, they provide a shared ethical basis for peace, rooted in their understandings of human dignity, the common good, and the Golden Rule. Second, they are uniquely well positioned and equipped for "strategic peacebuilding"—coordinating local, national, and transnational resources for ending violent conflict, striving for social justice, and building bonds of cooperation and solidarity. Third, they can deploy their deepest experiences of unlimited mercy, compassion, and capacity for loving self-giving to absorb the grave sufferings caused by human cruelty, thereby advancing healing, reconciliation, and forgiveness of one's enemies.

Many interreligious initiatives have made decisive contributions to peace; for example, different configurations of Indigenous, Hindu, Buddhist, Jain, Sikh, Yazidi, Jewish, Christian, Muslim, and other religious representatives (including women and youth) have helped to mediate between conflicted groups and have advanced reconciliation, justice, and mercy in their communities. As a result, their conflict-ridden communities were helped to reimagine their futures, reclaim hope, address their burdened pasts, heal, and move forward together.

There are also numerous examples of profound, world-changing interventions by religious leaders in nudging political leaders to pursue peace. *Pacem in terris* played a historic role following the Cuban Missile Crisis to help the United States and Soviet Union find a path toward arms control, notably through the Partial Nuclear Test Ban Treaty of 1963 and the Treaty on the Non-proliferation of Nuclear Weapons of 1968. The 1989 Taif Agreement ended the Lebanese Civil War, disarmed militias, and created political power sharing among various national communities. The Vatican has also played a pivotal role in ending conflict between Argentina and Chile and between Azerbaijan and Armenia, as well as in the civil war in the Central African Republic. Most recently, the Catholic Church played a direct and enormously fruitful role in bringing conflicting parties in Colombia to the negotiating table and reaching a peace agreement, ending the longest-running military conflict on the planet.

The Role of Ethics in Action

The challenge facing Ethics in Action is to find practical steps for the unique wisdom, beauty, and shared moral convictions of the world's great religious traditions to help guide the world back from the brink and reality of war—and toward a vision of positive peace rooted in the unbreakable link between unfolding human dignity and advancing shared well-being.

Ethics in Action is therefore resolved to do whatever it can to help the world's religious leaders promote peace. We aim to mobilize the scientific, academic, and international communities to spread the message that religion should not be instrumentalized and manipulated in the name of political agendas. Rather, religious leaders are committed to reducing the fear in their communities, combating the lies that accompany the drumbeats of war, and actively promoting the virtues of positive peace and the institutions in their communities that embody and enact these virtues. This work must include tackling the spread of hatred and violence via the internet and social media.

In this regard, we propose the following measures, both in terms of advocacy and our own engagement.

General Recommendations: Advocacy

(1) Devise and implement a media-savvy, cross-cultural, inclusive strategy to change the narrative of Islam in the United States and Europe and in minority communities in Islamic-majority countries.
(2) Push for the establishment of a fund to reduce military spending and divert resources to finance sustainable development (to be called the Isaiah Fund or the Pope Paul VI Fund)

(3) Push for a world free of nuclear weapons (following the call of many religious leaders, including in Pope Paul VI's plea for multilateral disarmament at the United Nations in 1965 and in Pope Francis's moral condemnation of nuclear weapons in his 2017 World Day of Peace message).
(4) Push for full implementation of the Paris Agreement, and raise awareness of the links between climate change and conflict.

General Recommendations: Engagement

(1) Produce a joint public statement by religious leaders that represents a collective call to action for reawakening morality and ethics to underpin the promotion of positive peace.
(2) Widely promote the virtue of nonviolent conflict resolution.
(3) Ensure wide participation by religious leaders and communities in the worldwide climate march on April 29, 2017.
(4) Work with foundations to support grassroots interreligious initiatives in conflict-ridden multireligious communities.
(5) Reach out to leaders of nuclear disarmament groups to offer the support of religious leaders and communities.
(6) Use and deepen the channels of communications of peace, in pulpits, in congregations, and through social media.

Ethics in Action Working Group Action Items

(1) Request the UN secretary-general to put the issue of religion and peace on the UN Security Council agenda for 2017 in support of global peace and sustainable development.
(2) Recommend the creation of an interreligious contact group for the UN secretary-general and UN Security Council.
(3) Organize a joint campaign by Ethics in Action and Religions for Peace for healing and reconciliation in Syria.
(4) Develop and disseminate through religious networks an Ethics in Action education curriculum to promote the culture of peace.

CHAPTER 17

ON PEACE AND A MORAL FRAMEWORK FOR STATECRAFT

JEFFREY D. SACHS

The Message of *Pacem in terris*

In the opening lines of his April 1963 encyclical, *Pacem in terris*, Pope John XXIII implored the world to achieve a peace that had been desired for centuries yet had seemed all but impossible to achieve. He wrote, "Peace on Earth—which man throughout the ages has so longed for and sought after—can never be established, never guaranteed, except by the diligent observance of the divinely established order."[1]

Pacem in terris represents the powerful argument that international statecraft must be guided by a moral framework, one grounded in human dignity. The encyclical recognizes people's natural rights to respect, education, culture, decent work, and legal protection, and their right to choose freely to emigrate and to immigrate. Importantly, the encyclical also highlights people's responsibility toward their fellow humans. Statecraft must not be Machiavellian in its approach if we are to achieve global peace. It cannot be about the purported advantages of war but about a moral framework for the global good. No country can justly or morally dominate or take advantage of other countries.

The encyclical was written in the immediate aftermath of the October 1962 Cuban Missile Crisis, when the Soviet Union and the United States came to the very precipice of nuclear war.

If President Kennedy had listened to his advisors in how to respond to the Soviet placement of nuclear missiles in Cuba, the world would likely have ended in nuclear war. Most of Kennedy's advisors preached an immediate attack against the Soviet Union, in particular air strikes against the Soviet nuclear missiles installed in Cuba. We know in retrospect that such an attack would have almost inevitably led to a full-scale thermonuclear war. Even as the crisis was being resolved by diplomacy between Kennedy and his Soviet counterpart, Nikita Khrushchev, a Soviet submarine commander came within a breath of launching a nuclear-tipped torpedo that might have triggered global nuclear annihilation. The submarine commander did so because of a misunderstanding, incorrectly believing that his submarine was under direct attack by U.S. forces. This misunderstanding nearly destroyed the world.

By the grace of God, one can say, President Kennedy had the wisdom to hold back and play for time; to listen to the sage advice of his UN ambassador, Adlai Stevenson, who argued vigorously for a peaceful and diplomatic resolution of the crisis; to send his brother Robert to secret meetings with the Soviet ambassador in Washington; to reopen channels of diplomacy; to ignore hostile messages from the Kremlin and accept the more peaceful messages; and ultimately, to find the way toward a peaceful resolution of the crisis.[2]

Pope John XXIII's message of peace was conveyed not only to Kennedy but also to Khrushchev. Both Kennedy and Khrushchev were deeply inspired by *Pacem in terris*. By the end of the Cuban Missile Crisis, Khrushchev and Kennedy recognized each other not as combatants but as leaders who needed each other to avoid a future global disaster.

President Kennedy was fundamentally and profoundly affected by the Cuban Missile Crisis, and in the last year of his life, he devoted his leadership to making a lasting peace with the Soviet Union. On June 10, 1963, Kennedy made a historic speech, deeply influenced by *Pacem in terris*. Kennedy's "Peace Speech," given as a commencement speech at American University, was unique.[3] Unlike perhaps any other major foreign policy speech by an American president, the peace speech was directed not toward the Soviet Union with threats and warnings but to the U.S. citizenry. Kennedy called on Americans to change their attitudes regarding peace with the Soviet Union. He urged Americans to reject the idea that "war is inevitable, that mankind is doomed, gripped by forces we cannot control."

President Kennedy famously said that day,

> So, let us not be blind to our differences—but let us also direct attention to our common interests and to the means by which those differences can be resolved. And if we cannot end now our differences, at least we can help make the world safe for diversity. For, in the final analysis, our most basic common link is that we all inhabit this small planet. We all breathe the same air. We all cherish our children's future. And we are all mortal.[4]

Peace Today: SDG 16

In our current moment, the message of peace is central to the global community's drive toward sustainable development. SDG 16 calls for the promotion of peaceful and inclusive societies, as well as access to justice for all. It holds central the issue of human dignity and equitable access to institutions of law and justice for all to achieve peaceful societies. The targets of SDG 16 lay out the broad strokes for how this should be achieved, including the following:

- Target 16.1: Significantly reduce all forms of violence and related death rates everywhere.
- Target 16.3: Promote the rule of law at the national and international levels and ensure equal access to justice for all.
- Target 16.7: Ensure responsive, inclusive, participatory and representative decision-making at all levels.
- Target 16.10: Ensure public access to information and protect fundamental freedoms, in accordance with national legislation and international agreements.[5]

These targets are ambitious and idealistic, and the challenge of securing peace remains immense. The *Global Peace Index* report, published annually, reported in 2019 that the average level of global peacefulness had increased slightly over the previous year. Yet the world is still less peaceful than it was a decade ago, and the trends of the last decade indicate that global peacefulness is deteriorating, especially in the domains of ongoing conflict and safety and security. This important and useful report also points out that climate change can indirectly increase the likelihood of violent conflict through its impacts on resource availability, livelihood, security, and migration. The world's nations must cooperate to address these environmental crises and to secure peace more generally.[6]

The *Global Peace Index* report also usefully tries to quantify the cost of violence and militarism and the economic benefits of peace, as difficult as that quantification is in practice. In 2018, according to the metrics cited in the report, the economic impact of violence, military spending, security spending, and deaths on the global economy amounted to some US$14.1 trillion in terms of purchasing power parity. The costs of violence on our planet were equivalent to 11.2 percent of global GDP and averaged a staggering $1,853 per person across the globe.

In countries such as Syria, the economic cost of violence as a percentage of GDP was measured at 67 percent.[7] The costs of the physical destruction of infrastructure and buildings, homicide, terrorism, militarism, and other violence consumed vast resources. Every basic need in Syria could be met many times over using the money squandered on conflict. Global expenditures on violence, if redirected properly, could easily end extreme poverty, create a place in school for every child, and ensure decent livelihoods for everyone on this planet. (For information on the cost of ensuring basic education for all, see the part 7 opener.)

Some countries are more peaceful than others. The Peace Index ranks every country for peacefulness based on twenty-three indicators including military expenditure as a percentage of GDP, incarceration rate, terrorism impact, and the number of refugees and internally displaced peoples. In 2019, Iceland was found to be the most peaceful country on Earth, followed by New Zealand, Portugal, Austria, and Denmark. Syria and Afghanistan were the two least peaceful countries. Regionally, the Middle East and North Africa remained the least peaceful region of the world, although the region had become more peaceful in the previous year. The region of Central America and the Caribbean had the largest deterioration of peacefulness in the previous year.[8]

A Brief Case Study: Syria

Let us look more closely at the case of Syria, the second least peaceful country in the world in 2019 and the site of one of the world's major conflicts in recent years. The example of the Syrian crisis can be examined to understand more deeply the cause of the crisis and what we can do to achieve lasting peace in Syria.

Syria was a relatively stable and peaceful country for the first decade of the twenty-first century. It had many pressing economic, social, and political challenges, including an authoritarian government, minorities suffering under the political system, and tremendous economic stresses. Yet despite these enormous challenges, there was little major conflict.

At the end of the first decade of the twenty-first century, Syria was hit by a number of shocks, most remarkably by an extreme ecological shock: drought. This kind of environmental shock is becoming more and more prevalent owing to human-induced climate change. The Syrian drought lasted from the winter of 2006 and 2007 until 2010 and, by accounts of climate scientists, was the worst three-year drought ever experienced in the country. Scientists have also determined that anthropogenic climate forcing had made the occurrence of such a severe drought three times more likely.[9] It is not only the severity and the length of the drought that are noteworthy, however; it is also important to consider how strongly this drought affected the livelihoods and situations of people living in both rural and urban areas.

The drought caused agricultural productivity to decline. Before the drought, about 25 percent of Syria's GDP came from agriculture; in 2008, this fell to just

17 percent. It is certain that the drought also displaced large numbers of people. Desperately poor farmers who could not earn a livelihood and could not grow their crops ended up on the margins of cities. In addition to internal displacement, an estimated 1.2 million to 1.5 million Iraqi refugees entered Syria between 2002 and 2007. By 2010, Iraqi refugees and internally displaced Syrians made up about 20 percent of the population, which contributed to the smoldering discontent growing in the country owing to the decline in living standards and the crisis of hunger that were gripping much of the country because of the drought. Syrian cities had experienced a population growth of more than 50 percent in less than a decade, for a total urban population of 13.8 million in 2010.[10]

The so-called Arab Spring, which spread through the region through early 2011, lit a fuse in Syria. After the first spark of revolution in Tunisia in December 2010, Syria saw its first protests in March 2011. The protest and unrest in Syria can be attributed in part to hunger and displacement, as well as to deep divisions within Syrian society. The government of President Bashar al-Assad quickly repressed the protests, shot into crowds, and created a worsening political crisis of sides for and against the regime. This was a deeply stressed and increasingly unstable environment, but not yet one of war.

One thing that tipped Syria into war was a decision made in the United States, where the U.S. government, namely President Obama and Secretary of State Hillary Clinton, decided that the Syrian regime had to go. Concerned about the "murderous regime in a combustible region," the United States asked the international community to take action to prevent civil war and unrest that could spread to neighboring countries.[11] Ironically, U.S. support for Assad's ouster provided the context for the civil war that the United States was ostensibly trying to avoid.

In late 2012 or early 2013, Obama signed a presidential order directing the Central Intelligence Agency to work with Saudi Arabia and other forces in the region to train and arm fighters to overthrow the Assad regime. Violence surged as weapons flowed into Syria. As a result, from 2012 onward, a beautiful country with unique historical and archaeological sites that had been preserved for millennia was bombarded and plundered. A widening war brought in fighters from afar, and the Syrian civil war became a proxy war, pitting the United States and its allies against Russia, Iran, Syria, and others.

The UN led a number of attempts to find a diplomatic solution but to no avail. The Obama administration's insistence that Assad must leave as a condition for a peace deal led to a stalemate at the negotiating table and continued bloodshed throughout Syria. Numerous iterations of UN-mediated diplomacy have failed to bring about peace in the region.[12]

It's estimated that at least five hundred thousand people have died, including civilians and combatants on all sides, that more than 6.6 million have been displaced internally, and that more than 5.6 million have fled to Lebanon, Jordan,

Turkey, and some parts of Western Europe. Much of the country's infrastructure, housing, and commercial buildings have been destroyed. It is estimated that reconstruction will cost upward of $250 billion, but this reconstruction will not take place until after a political transition.[13]

So, what are the lessons from the Syrian civil war? First, we are reminded that conflicts erupt because of underlying pressures. They erupt from underlying conditions of extreme poverty. They erupt because of hardships within a society, inequalities, a lack of access to basic needs, and ecological shocks that render parts of a country uninhabitable, unsuitable for farming, or unable to support life, causing displacements of the population. We also learn that in a world filled with armaments and competing powers ready to take sides in a local conflict, what starts as a local skirmish can escalate dramatically.

We learn how important it is to put diplomacy first, and to say to every superpower—the United States, Russia, or any other—that it is essential that local conflicts be addressed through the principles of international law, through the UN Charter, and through the work of the UN Security Council, rather than through the individual actions of any country. When Secretary of State Clinton declared on behalf of the United States in 2012, "Assad must go," she sowed the seeds for a regional proxy war that ended in the deaths of hundreds of thousands of people. Had the UN Security Council properly done its job; had the United States directly but diplomatically worked with the permanent members of the Security Council (Russia, China, the United Kingdom, and France) and the whole fifteen-member body; and had the UN special envoy for peacemaking at the time, the former secretary-general, Kofi Annan, been supported fully to accomplish his task of finding a resolution of this conflict back in 2012, how many lives could have been saved? How many tens of billions of dollars of could have been diverted from destruction?

Peace and Sustainable Development

Conflicts occur because of underlying conditions of inequality, injustice, ecological shocks, and extreme poverty. They must be resolved by addressing those underlying causes, not by pouring money into armaments for a conflagration. In our world today, no country, no matter how powerful, should demand the following of another: "Your government must fall because we want a different one." That kind of regime change operation is invariably a precursor of more violence and war.

We should be using diplomacy rather than force to address the difficult but solvable challenges of border disputes and contested lands. The UN, and especially the UN Security Council, were established precisely to find a path of rationality, decency, and transparency to avoid conflict and to avoid the chance of a local conflict becoming a global disaster, which is an ever-present possibility in our time. The UN was formed in the aftermath of World War II for the primary purpose of securing peace, and then, on that foundation of peace, securing the economic, social, and political development of societies around the world.

The route to peace can be found through promoting decent, fulfilling livelihoods for every person by meeting the basic needs of sustainable development everywhere on the planet. We should be investing in solving the underlying conditions leading to conflict, ending extreme poverty, and ensuring that all people have access to basic resources, including energy, for example, in a way that doesn't require or lead to wars over oil fields or control of other natural resources. If we address the underlying challenges of poverty and social justice, if we address the underlying worsening conditions of the physical environment that are displacing millions around the world, and if we promote international diplomatic approaches to peacemaking, we can find a way to lasting peace. As Jesus said in the Sermon on the Mount, "Blessed are the peacemakers, for they are the children of God."

Notes

1. John Paul XXIII, *Pacem in terris*, Vatican website, April 11, 1963, http://www.vatican.va/content/john-xxiii/en/encyclicals/documents/hf_j-xxiii_enc_11041963_pacem.html.
2. The story has never been better told than in the recent book *Gambling with Armageddon: Nuclear Roulette from Hiroshima to the Cuban Missile Crisis*, by the historian Martin J. Sherwin (New York: Knopf, 2020).
3. The speech is both inspiring and remarkable in its eloquence. I wrote at length about it, and its historical context, in Jeffrey D. Sachs, *To Move the World: JFK's Quest for Peace* (New York: Random House, 2013).
4. John F. Kennedy, "Commencement Address at American University, Washington, D.C., June 10, 1963," John F. Kennedy Presidential Library and Museum, https://www.jfklibrary.org/archives/other-resources/john-f-kennedy-speeches/american-university-19630610.
5. "Goal 16," United Nations, accessed October 8, 2021, https://sustainabledevelopment.un.org/sdg16.
6. Institute for Economics and Peace, "Executive Summary," in *Global Peace Index 2019: Measuring Peace in a Complex World* (Sydney: Institute for Economics and Peace, June 2019), https://www.visionofhumanity.org/wp-content/uploads/2020/10/GPI-2019web.pdf.
7. Institute for Economics and Peace, "Chapter 3: Economic Impacts of Violence," in *Global Peace Index 2019*.
8. Institute for Economics and Peace, "Chapter 1: Results," in *Global Peace Index 2019*.
9. Colin P. Kelley, Shahrzad Mohtadi, Mark A. Cane, Richard Seager, and Yochanan Kushnir, "Climate Change in the Fertile Crescent and Implications of the Recent Syrian Drought," *Proceedings of the National Academy of Sciences* 112, no. 11 (March 2015): 3241–46.

10. Kelly, Mohtadi, Cane, Seager, and Kushnir, "Climate Change in the Fertile Crescent."
11. Kate Cavanaugh, "Hillary Clinton on Syria: Not Exactly Consistent," *GlobalPost*, July 18, 2012, https://www.pri.org/stories/2012-07-18/hillary-clinton-syria-not-exactly-consistent.
12. Katy Collin, "7 Years Into the Syrian War, Is There a Way Out?" Brookings Institution, March 16, 2018, https://www.brookings.edu/blog/order-from-chaos/2018/03/16/7-years-into-the-syrian-war-is-there-a-way-out/.
13. "Syria: Events of 2018," Human Rights Watch, accessed October 8, 2021, https://www.hrw.org/world-report/2019/country-chapters/syria.

CHAPTER 18

ADVANCING SHARED WELL-BEING AS A MULTIRELIGIOUS VISION OF POSITIVE PEACE

WILLIAM F. VENDLEY

Each religion has a vision of peace that is sincerely held by its believers. While one ought to avoid a "syncretistic" blending of the beliefs of different religions, it remains evident that diverse religious visions of peace provide foundations for carefully discerning the elements of a shared vision of peace. This discernment expresses a consensus in shared values, notwithstanding the differing beliefs and doctrines that are unique to each religious tradition. Consensus in terms of shared values provides a foundation for Religions for Peace's expression of principled commitment: *Different Faiths—Common Action.*

"Peace is *more* than the absence of war." This widespread insight points to the *positive* dimension of peace. Each religious tradition has a positive vision of peace rooted in its respective experience of the sacred. Each positive vision of peace is a fecund notion of flourishing that summons people to unfold their human dignity and "welcome" the dignity of the other. Each calls people to advance communal flourishing with just institutions and enjoins people to live in harmony with the natural world. Each calls people to live in harmony, love, and compassion and directs them toward an ultimate state of positive fulfillment. Each religious community's positive vision of peace also helps its believers to bring into the light the profound gaps, contradictions, and personal and social failures that mark human experience and threaten peace.

Groups such as Religions for Peace have long labored to discern both shared elements of positive peace and the major threats to peace. Ultimately, advancing positive peace and addressing the grave threats to peace are inextricably related. This chapter sets forth the notion of shared well-being[1] as an anticipatory and heuristic notion of positive peace.

The Challenge of Our Time

A World in Pieces

> "We are a world in pieces. We need to be a world at peace."[2]
>
> UN Secretary-General António Guterres

In this quote from the United Nations Secretary-General, the first sentence refers to our "brokenness," and the second calls for a coherent state of peace, one that in terms of this analysis necessarily includes a tacit notion of positive peace.

The statement that "we are a world in pieces" refers to a disturbing catalog of threats to peace: There was a 408 percent increase in battle deaths and a 247 percent increase in deaths by terrorism between 2007 and 2016;[3] the number of refugees doubled during that time period;[4] conflicts continue in many places, including Ukraine, Syria, Nigeria, the Democratic Republic of the Congo, South Sudan, the Central African Republic, Yemen, the Holy Land, Myanmar, on the Korean Peninsula, and within the Central American states, which are beset by gangs; there is ominous growth in the sophistication of military technologies in space, new energy weapons, and artificial intelligence coupled with robotization;[5] 70 percent of the global population faces high restrictions on religious freedom;[6] 767 million people (more than 10 percent of the population) live on less than US$1.90 per day;[7] and virtually all states are behind in their commitments to the Paris Agreement on climate change.[8] Addressing these threats to peace is an urgent responsibility.[9]

Importantly, the secretary-general's declarative statement, "We are a world in pieces," hints at the broader challenge of establishing peace. The threats to peace just mentioned have arisen within what could be termed a "meta-crisis" of the Modern Order: Modern democratic tenets—including guarantees of free and fair elections, the rights of minorities, freedom of the press, and the rule of law—have come under attack around the world,[10] while on the economic front, the nine richest people today have more wealth than the bottom four billion.[11] Such economic distortions prompted Mark Carney, the governor of the Bank of England from 2013 to 2020, to note, "Just as any revolution eats its children, unchecked market fundamentalism can devour the social capital essential for the long-term dynamism of capitalism itself."[12]

Exacerbating the political and economic dimensions of the meta-crisis of the Modern Order, there is today a meta-crisis of "truth" within which the very notion of truth is contested[13] and so-called fake news is tailored for selected audiences for either commercial or political gain.[14]

It follows that we need to evaluate the Modern Order in relationship to a robust, multireligious notion of positive peace. What are the genuine enduring strengths of the Modern Order regarding the establishment of a holistic vision of positive peace? What are its weaknesses? How can a multireligious vision of positive peace preserve the genuine achievements and strengths of the Modern Order? How can a multireligious notion of positive peace fill in any gaps or strengthen any weaknesses in the Modern Order? These are highly demanding "second-order" questions that address the context of today's threats to peace. While prosecuting these second-order questions is demanding, and while related answers can seem abstract, the practical impact of preserving the strengths and addressing the weaknesses of the Modern Order in relationship to a multireligious vision of positive peace is—in the long run—great. Patience is encouraged.

The use of the term "Modern Order" requires a comment: There is a diversity of "orders" unique to different cultures and states, each potentially fostering particular dimensions of true positive peace. Examining diverse orders in detail requires discernment by competent religious and moral people living within those orders. However, the Modern Order impacts all countries in varying ways. For example, the entire UN system within which all states work is largely expressed within the framework of the Modern Order. Therefore, our examination of *positive peace* in relation to the Modern Order can be relevant to all states, even those largely or partially organized by a different, even in some ways competing, order.

Strengths and Limits of the Modern Order

Among the great strengths of the Modern Order are its commitments to human freedom, universal human rights, and the tolerance of others. These powerful principles, along with notions of free trade, constitute an essential part of the foundation of the Modern Order that lies behind its extraordinary achievements such as the establishment of the United Nations, the production of the Universal Declaration of Human Rights, the adoption of remarkable UN conventions, the recent adoptions of the Paris Agreement on climate change, and the Sustainable Development Goals.

At present, the strengths and weaknesses of the foundational organizing principles of the Modern Order—such as freedom, human rights, and tolerance—are becoming increasingly evident. Additionally, the Modern Order has an ambiguous relationship to the notion of the common good. It is necessary to examine these core principles to discern the ways religions can together affirm and complement them. Doing so can help us to identify key areas of a shared multireligious vision of positive peace.

Strengths of the Modern Order's Notion of Freedom

Freedom is a profound mystery at the heart of human dignity. Freedom allows people to commit themselves to what they hold to be true and valuable. Through their free choices, people engage in self-determination and self-actualization. The Modern Order's notion of freedom includes perhaps the greatest freedom, religious freedom, through which people commit themselves to their experience of the sacred as the source of ultimate meaning and value. Across religious traditions, many religious scholars note that freedom is the great ally of religion, as forced religious belief is self-contradictory in its disregard for personal conscience. Respect for freedom also allows a sincere nonreligious believer to declare their worldview and take actions based on it. In short, the Modern Order is premised on a radical commitment to freedom.

By protecting personal freedom, the Modern Order has empowered many to assume unprecedented degrees of autonomy, allowing them to shape their lives in ways they value. Moreover, with the freedom of the Modern Order, the human family communicates, travels, and trades more extensively than ever before. Global information and encounters are now a matter of course to a constantly increasing number of world citizens. Today, lifestyles from distant corners of the planet interact, bringing a rich and diverse set of options before the eyes of world citizens.

Limitations of the Modern Order's Understanding of Freedom

No one needs to explain the value of freedom to the oppressed. But opposing a lack of freedom is perhaps easier than determining how achieved liberties should operate in a free society. When the harsh, black shadow of oppression is swept aside, the seemingly black-and-white struggle between freedom and oppression is replaced by the permissiveness of open societies. This begs the question, Whose freedom is to be upheld when the freedoms of some collide with the freedoms of others?[15]

Does the freedom of the environmental campaigner have priority over economic freedom? Does the freedom of the mother override the potential freedom of her developing fetus? Ought we to prioritize the freedom of those living today over the freedom of coming generations? How is economic freedom related to political freedom? Do they strengthen each another, or does one undermine the other?

In short, do we adequately grasp the idea of freedom when equating it with a decrease of limitations and an increase in options? Do voluntarily borne obligations constitute denials or true manifestations of freedom?

The Modern Order is challenged by these questions—the idea of freedom is in a struggle with itself. Having eaten from the tree of knowledge and having learned the bitter lesson that the freedom of a few can undercut the freedom of others, the Modern Order's notion of freedom has lost its innocence. It seems that, from now on, proponents of the Modern Order's notion of freedom make their home in a world both built and endangered by freedom itself.

Religions and the Recovery of the Foundations of Freedom: The Growth of the Modern Soul

If the Modern Order's most perverse and lethal expression of freedom is genocide, its most tragic expression nihilism, and its most banal expression the rampant consumerism that dominates so much of modern culture, the question to religions today is how they can affirm and deepen the Modern Order's championing of freedom by clarifying its meaning.

The challenge is profound, and if a "common-sense" answer is that the deepest meaning of freedom is the capacity to choose the good, it is the concrete examples of men and women willing to struggle deeply with the ambiguity of the Modern Order's notion of freedom—people like Mahatma Gandhi, Fyodor Dostoevsky, the young Dutch Jewish woman Etty Hillesum, Aleksandr Solzhenitsyn, Dietrich Bonhoeffer, and Mother Teresa—that offer profound and usable insight.

Each of these men and women struggled profoundly with the deepest meaning of freedom—many in jails or, in the case of Etty Hillesum, in a concentration camp living in solidarity with fellow Jewish prisoners. Each ultimately found that the ground of freedom is mysteriously spiritual and that freedom is anything but arbitrary. Each found that the deeper they yielded to the innermost "pull" of freedom, the more it sustained them in their commitment to truth, care for others, and resistance to a distorted order that injured human dignity.[16]

If the Modern Order is understandably quiet about freedom's foundation and goal owing to its respect for diversity, its commitment to freedom is an especially profound strength in that it is extended to all people; however, it is simultaneously vulnerable to degraded, capricious, or otherwise distorted notions of freedom. Religions, through the examples of their remarkable members, can make clear that freedom is grounded in sacred mystery, that it is radically spiritual, that the exercise of freedom opposes any dis-order that humiliates the meaning of being a person.

Today, religious communities are called to affirm the Modern Order's recognition of the foundational importance of freedom. At the same time, they are called to show by example the sacred grounding of freedom that leads through the despair of nihilism, that rejects the narcissism of mindless consumerism, and that expresses itself as radical care—care for all and care for the order that would help each person to actualize their human dignity.

Strengths of the Modern Order's Notion of Human Rights

Another pillar of the Modern Order is the notion of human rights. While the antecedents of human rights can be found in a wide variety of historical religious and other cultural streams, their most salient global manifestation is linked to the United Nations.

Human rights are rights inherent to all human beings, regardless of race, sex, nationality, language, religion, or any other status. Human rights include the right to life and liberty, freedom from slavery and torture, freedom of opinion and expression, the right to work, and education. Everyone is entitled to these rights without discrimination.

Neither religions, nor cultures, nor states, nor social groups, nor families are the true sources of rights. Rather, all social entities are called to respect the human rights that inhere in people because of their intrinsic dignity. Indeed, the world's religions' respective experience of the sacred is understood by them—each in its own way—as the ground for the human dignity from which human rights flow.

The Modern Order is a champion of human rights, and it is hard to overestimate the profoundly *positive* impact of the human rights regime on human well-being.

Limits of the Modern Order's Notion of Human Rights

While human rights can be conflictual, with the rights of one clashing with the rights of another, there is a more profound limitation to the Modern Order's foundation in human rights.

While it is true that "rights impose responsibilities," it remains that rights do not explicitly summon people to become "good." For example, if the catalog of basic rights makes clear that a person's dignity should be honored and protected, this catalog does not make clear a person's inner obligation to develop their many potentials. The right to education is surely a basic entitlement essential for entrance into society, but that right does not make clear the obligation a student has to patiently unfold—step-by-step—their potential to learn.

Thus, even as religious communities affirm the foundational importance of rights, decry their violation, and labor for their recognition, a basic question to religious communities is, How can they also complement the Modern doctrine of rights?

Virtue as a Complement to Rights

It can be argued that virtues are an essential complement to rights. Virtues are habits, and they differ from skills in that they intend habitual orientations to value. Virtues and skills are both similar and profoundly different. They are similar in that both are habits that take patience and repetition to master. Their profound difference lies in the classes of objects they intend. Skills are related to tasks, from the rudimentary task of tying one's shoelaces to the myriad tasks related to advancing the standard of living. From the simple to the most complex, skills relate to the effective, efficient, and repetitive achievement of tasks. Virtues, on the other hand, relate to values, to decisions on what is worthwhile, what is valuable. A virtue is a habitual orientation to a value. The patient and resolute acquisition of personal virtues linked to all dimensions of the human being brings degrees of perfection to these dimensions, allowing people to unfold the myriad potentials of their human dignity in accord with their respective value.[17]

Virtues include habits linked to unfolding our personal potentials, as well as those linked to our just and caring relationships with others. The former include habits related to health, honesty, intellectual curiosity, love of learning, prudence, temperance, and fortitude, while the latter add kindness, justice, tolerance, solidarity, harmony with nature, love, compassion, and mercy.

Although the catalog of virtues varies across religions, it remains that religions have historically regarded the cultivation of virtue as the royal road for unfolding and realizing human potential, achieving just relations with one other, and arriving at religiously sublime states of harmony, love, and compassion.

When widely shared, virtues help to knit the fabric of social cohesion; they generate social trust, which is even more fundamental than a social contract such as citizenship. Without virtuous people, even materially prosperous societies dig their own graves. People in these societies work at cross-purposes; they mutilate social trust; they distort their respective scales of value—pursuing selfish gains at the expense of the sacrifice essential to building society as the agent of holistic development.[18]

Virtues equip people for community, and community is essential for societies to be effective agents of their own development. The way virtues can complement rights can be seen in the relationship among virtue, community, and development.

It can be argued that the basis for holistic development is community. Community, in this sense, means fellow feelings, shared meanings and values, and a shared commitment to the common good. People are born into and abide in community, and community is the basic underpinning for all institutions that serve society—from families, schools, civil society organizations, economic institutions, and governments to religious and other holistic life-stance communities. These institutions, including the UN, cannot function, or even survive, without a large measure of community.

Take governments: Even if there are good leaders, good policies, and enough resources, without a healthy society rooted in community, there is little likelihood of a government serving effectively. Resources and skills can be scarce, adequate, or lavish, but without a healthy measure of community they cannot be easily and efficiently deployed for the well-being of the members of society.

So, we must inquire about what undermines community. It has been argued that the main "enemies" of community are (1) ignorance, (2) individual egoism, and (3) group egoism.[19]

Understandably, these enemies of community were not on either of the lists developed in the two tracks at the United Nations (the member-state track and the secretary-general track) that advanced the formulation of the SDGs. These respective UN processes rightly developed their own lists of categories that dealt explicitly with the substance of the seventeen SDGs. Nevertheless, ignorance, egoism, and group egoism—can be understood as a complementary list of development challenges with significance across the SDGs and other areas of needed development.

Ignorance is a threat to community when groups of people are denied the education they need to function in their societies. In this situation, people are effectively barred from entering the mainstream. Importantly, ignorance is also a threat whenever people and societies unknowingly or semi-knowingly refuse to analyze what thwarts or facilitates their development.

Individual egoism is a threat to community because egoists find loopholes in the social setup, which they exploit to enlarge their share of particular goods. To the extent that people have figured out and used loopholes to avoid their debt to community, they exploit it. Worse, it's not just individuals who are selfish. Groups, too, learn how to work the system. And they find ways to justify to themselves their taking of unfair advantages. They can develop an ideological facade that justifies their ways before the bar of public opinion. If they succeed in their deception, the social process is distorted, and community is eroded. Group egoism calls forth resentment. So, the body-social begins to seethe with hostility. Trust is lost. Cynicism sets in.[20]

Ignorance, individual egoism, and group egoism threaten community and therefore human development. Without a constant renewal of community, the measure of community currently enjoyed is easily squandered. If community is the genuine basis for society, and a healthy society is the basis for development, the practical question becomes, What can be done to build up the community that underpins a healthy society as the subject of development? Surely the answer involves many factors. However, the cultivation of virtue is a key response to the threats to community discussed here.[21] Becoming virtuous is hardly a solitary act; rather, it is an act of solidarity.

Today's question to the diverse religious and other schools of virtue is, What gifts of virtue cultivation do you have to both sculpt the characters of people working to become good and to help build the sense of community essential for development on all levels—local, national, regional, and global?

Far from diminishing the foundational importance of individual human rights, the cultivation of virtue can complement their importance and undergird their recognition. Indeed, widely embraced virtues could nurture coherent communities committed to defending human rights as an integral part of true development.

The Modern Order's Notion of Tolerance

A fundamental premise of the Modern Order is a commitment to the tolerance of others. The European Council of Religious Leaders (an interreligious council within the Religions for Peace global network) have defined tolerance as "an active recognition of diversity and means of respecting the otherness of the other with whom we differ religiously, culturally or otherwise, with compassion and benevolence."[22]

Fifteen years earlier, in 1995, the United Nations Educational, Scientific and Cultural Organization declared, "Tolerance is not only a cherished principle, but also a necessity for peace and for the economic and social advancement of all peoples."[23] The full implication of the true nature of tolerance is perhaps best understood by its opposites: the ugly faces of intolerance, prejudice, discrimination, marginalization, and deprivation that shape the daily life of hundreds of millions. Importantly, promoting tolerance must not be confused as a proxy for lack of conviction, indifference, or neglecting one's values. Tolerance does not constitute "concession, condescension or indulgence . . . . The practice of tolerance does not mean toleration of social injustice or the abandonment or weakening of one's convictions."[24]

On the contrary, true tolerance is threatened by a culture of indifference, in which truth claims remain uncontested or, worse, are no longer even seriously made. To dissent from and dispute alternative viewpoints—respectfully and based on reasonable arguments—honors their defenders as worthy of intellectual engagement in a shared pursuit of the truth. Dialectical engagements, far from detracting from tolerance, defend and strengthen a culture of tolerance through the theoretical acknowledgment they generate of the respective worldview of others:

> Conflicting interests and views are not in themselves a threat to peace. They present a challenge to creatively harmonize different interests. In a culture of peace, everyone should strive to transform situations of conflicting interest so that their power and dynamism are channeled into creative development which promotes peace and harmony.[25]

In short, the Modern Order's advancement of tolerance continues to have incalculably *positive* impacts, not least in the area of religious freedom. Moreover, tolerance is virtuous when it becomes a habitual orientation to respecting the dignity and freedom of the other. Religious communities around the world are beneficiaries of tolerance and need to be its champions.

Solidarity as the Complement to Tolerance

Even while appreciating the vital and perdurable importance of tolerance as advanced by the Modern Order, two limitations can be noted.

First, there is the dilemma of tolerating views that one feels to be seriously morally wrong. A standard retort is that "people are free to do whatever they wish so long as they do not infringe on the rights of others." While a rights-based answer is significant, it is not fully satisfactory, because rights do not express an inner summons to become good, as discussed previously. It would seem that principled dialogue and teaching by example are a necessary response to this dilemma, while just how one is to negotiate moral differences in truly serious cases is typically not adequately addressed by the proponents of tolerance.

Second, and more importantly, religious moral imperatives both include and go beyond tolerance. Religions enjoin their believers to be in solidarity with others. Solidarity expresses an existential identification with and commitment to the well-being of the other. Thus, even as the world's religions can affirm the vital importance of tolerance and work to advance it, they can also position tolerance within the wider concern of active solidarity. The two virtues—tolerance and solidarity—can co-abide, with tolerance functioning as a necessary inner moment within the more profound religious commitment to solidarity. Solidarity calls us to make the other person's well-being our own vital concern. Solidarity suffers and rejoices with the burdens and beauties of the other. In the solidarity of love and compassion, the relational self experiences its own well-being as connected to the well-being of others.

The Modern Order's Notion of the Common Good

The Modern Order has an ambiguous relationship to the notion of a common good (a good for the whole of a society; e.g., a good educational system for all) in contrast to a personal good (e.g., a good education for a particular person). Importantly, the common good is not adequately understood as simply the aggregate of personal goods. Rather, the common good includes among other factors the shared meanings and values of a society, the personal commitments of civic virtue, and the value-based institutions that serve and support all in society to unfold the many dimensions of their dignity.

While a wide variety of views on the common good exist within the Modern Order, a typical position focuses quite heavily on personal goods. The philosopher John Rawls expresses this tendency toward the priority of personal goods over the common good in his pithy phrase "rights over goods," by which he expresses the Modern Order's wariness that one person's notion of the common good may be experienced as an oppressive imposition by another. Rawls, therefore, places the emphasis on the rights and freedoms of individuals, although he acknowledges that some form of the common good can and must be pursued.

Within the Modern Order itself, there are serious counterreactions to the notion that a healthy social order can be built largely on autonomous individuals pursuing their private goods to the greatest degree possible. Such an extreme approach, it is being argued, "undermines the notion of a good society, leaving its participants ever more isolated, asocial, selfish, calculating and spiritually barren."[26] In short, the Modern Order manifests a range of views regarding the common good, including new so-called communitarians[27] who identify deeply with the Modern Order yet argue that community and a common good that are more than an aggregate of personal goods are essential for the good society.

Emergent Multireligious Notions of the Common Good

If the Modern Order has an ambiguous relationship with the notion of the common good, is there an emergent multireligious notion of it that can be put in the service of the human family? This is not a small question, as the reality referred to by the term "common good" varies within religious traditions. Indeed, the term does not exist in most traditions, although there are corresponding notions in those traditions. Is there, then, a notion of the common good that is acknowledged across traditions?

The Ethics in Action initiative led to a discernment of substantial and still-emergent areas of consensus on a shared notion of the common good.

From the religious point of view, the most original common good is the sacred—God if you are a theist, the Eternal Buddha if you are Mahayana Buddhist, or nature infused with the Divine if you are a nondualist. In the realm of worldly life, the common good includes the earth with its air, soil, water, and web of biodiversity that supports all forms of life. So, too, it is the store of cultural wisdom and all institutions that support human dignity, ranging from manners and mores to the complex social, economic, and political institutions that are integral to societies. Even personal monetary wealth is considered a common good in some traditions. People can be caretakers of it, but its ultimate "universal destination" is to build up the common good. Moreover, today, there is an ever-growing notion of a global common good that is essential for a shared life on Earth. In short, highly significant—if formal—areas of consensus on the common good were discerned by Religions for Peace representatives.[28]

Importantly, a multireligious notion of the common good can both affirm some of the notions of the common good associated with the Modern Order and serve as a powerful catalyst to build a more robust notion where needed.

Shared Well-Being: An Emergent Multireligious Notion of Positive Peace

The following section sets forth the outline of an emergent multireligious consensus on positive peace as shared well-being. The core of shared well-being addresses three basic questions:

(1) How does an individual become a good person?
(2) How do we build a good society?
(3) What is the integral and reciprocal relationship between becoming a good person and building a good society?

Answers to these questions point toward an integral whole that offers a normative heuristic notion of human flourishing, or *positive peace*.

Shared well-being builds directly on the previous section's analysis of the strengths of the Modern Order and the complementary strengths of religions. Shared well-being affirms the profound value of freedom, the great importance of human rights, and the perduring significance of tolerance. Shared well-being also affirms the recovery of the spiritual ground of freedom and the role of virtues as complementary to the importance of rights. The coupling of the virtues of solidarity and tolerance is emphasized. Importantly, shared well-being expressly links the protecting and unfolding of rights-protected human dignity by the cultivation of virtues with the development of the common good, which includes institutions that honor and support human dignity.

Affirming an emergent multireligious consensus on shared well-being could be easily misunderstood as a naive or reckless trivialization of the foundational differences that mark the ways diverse religious communities understand themselves. Therefore, a series of precisions are offered as qualifiers of this emergent notion. To facilitate the flow of the text, these important precisions are elaborated for interested readers in the annex at the end of this chapter.

A key starting point for the notion of shared well-being resides in the fact that across diverse religious traditions, people are considered to be intrinsically relational. As a result, becoming a truly realized person is organically linked to all reality to which people are related: the sacred, other people, and the common good, which includes the environment. It follows that if all people are intrinsically relational, in a profound sense, each one's well-being is necessarily shared. Shared well-being follows from religions' understandings of people as radically relational.

The heart of shared well-being is the unbreakable link between actualizing human dignity and building up the common good that serves and supports it. If people are called to actualize all dimensions of their being, including the vital, affective, aesthetic, intellectual, moral, and religious, then they are called to cultivate the virtues essential to unfolding those dimensions of their being. Simultaneously, the common good, with its value-based institutions, is to be developed to assist all people in these same dimensions of virtue cultivation.

The relationship between unfolding human dignity and advancing the common good is to be mutually beneficial: What is good for the person is also good for society, and vice versa. In practical terms, the common good is to be evaluated in terms of its adequacy in supporting people to virtuously unfold their rights-protected dignity. In turn, each person is to actualize their relational being by also contributing in their own way to building up the common good. Importantly, institutions are for society what virtues are for people. Like virtues, institutions (notwithstanding their related skill sets) are oriented by social values. Institutions are both informal and formal, ranging from social manners and etiquette, families, schools, and civic groups, to companies and related economic institutions, governments, intergovernmental organizations, and religious bodies. All are oriented to values. As with virtues for people, social institutions seek to provide efficient, repetitive results in accord with particular values. Institutions with values oriented to shared well-being are essential for the realization of those values.

In a highly general fashion, one that needs to be complemented by an appropriate dialectical analysis, shared well-being provides a remote criterion for the evaluation of institutions. Do they foster freedom? Do they honor human rights and nurture virtue? Do they, in short, thwart or support the unfolding of human dignity? Do they drain or build up the common good? Do they honor and foster the nexus between unfolding dignity and building up the common good? Do they honor a transgenerational common good, so that today's actions do not undercut future possibilities for flourishing? Answers to these questions are likely to be mixed, calling for a dialectical analysis as the basis for requisite social critique and creative reform.

For example, while business exchanges of goods and services are inherently good, utterly essential for development, and have contributed dramatically to human well-being, it would be naive to ignore the short-term, profit-driven character of many of today's largest companies. It would be naive to ignore the great power of some of these institutions on governments or the power they exert over the media by their advertising or their power over the market itself.

When these institutions work against human dignity and the common good, we are invited to recall how group egoism distorts the community essential for development.

Bucking vested interests and realigning institutions around values that support shared well-being will call for large reserves of civic virtue. Working together, religious communities can help to nurture and animate the reserves of civic virtue necessary for *positive* social change. Importantly, if shared well-being calls for an economics[29] and politics of the common good, it also calls for the cultivation of virtuous consumers and committed citizens.

The public notion of the common good is negotiated and emergent from divergent notions of the common good held by diverse communities and people. In pluralistic societies, it is typically negotiated in terms of public language using public warrants. Religious communities—both individually and in a collaborative alliance such as Religions for Peace—can and should be vital partners in these social negotiations. Such a negotiated, consensual notion of the common good is necessarily emergent, and we should expect development in the notion of the common good.[30]

While each person is to be supported by the common good in their efforts to unfold rights-protected human dignity by relevant virtues, each person is simultaneously called to help build up the common good that can support others. In short, each is to receive support from the store of the common good in their efforts to become virtuous, and each is to contribute to the common good to help all others unfold their dignity. This root relationship is diachronic: We are heirs of a common good built significantly in the past, and we are to advance a common good for the future. Shared well-being provides a framework for engaging the common good with a summons to unfold the human dignity that is protected by rights, cultivated by virtues, and supported by value-based institutions that are vital components of the common good. This reciprocity of unfolding human dignity and advancing the common good is to be mediated by two principles: solidarity and subsidiarity.[31]

Solidarity

Building on what was noted earlier, solidarity acknowledges that all are to be concerned for all,[32] and this concern is diachronic and must extend to future unborn generations. Solidarity calls for the concrete action of care accessible to human agents in their particular circumstances. The opposite of solidarity has been termed the "globalization of indifference" and refers to the widespread apathy regarding others' well-being.[33]

Subsidiarity

All people and all institutions on every level of social organization—from the simplest and most local to the most complex and global—are to be agents of development. Yet, successively higher levels of agency should not arrogate to themselves the legitimate agency of more basic levels of social organization. This principle requires constant reinterpretation given new ways of social organization facilitated by the World Wide Web.

It has long been noted that people perceive reality in conformity with the received notions they use to examine it. Insofar as shared well-being honors but also expands the Modern Order's dominant ways of perceiving reality, we can anticipate attention to vital dynamics of human flourishing "under-noticed" by a more restricted horizon and related analytical tools. Importantly, the notion of shared well-being sketched here can welcome the moral insights of deontological and utilitarian moral approaches, as well as the virtue approach already noted. Religious communities, with their profound traditions of spirituality, billions of members, thousands upon thousands of far-flung places of worship, and global spokespeople, will need to play their roles to advance shared well-being as a religious and moral responsibility. Today it is also a pragmatic necessity, for even hard-headed materialistic empiricists are coming to the pragmatic realization that we are all no safer than the least secure among us. From a pragmatic point of view, our personal well-being depends on the well-being of the other; personal well-being is necessarily a species of shared well-being.

Empirical Findings

The earlier sketched multireligious notion of shared well-being is partially corroborated by groundbreaking research undertaken by the Institute for Economics and Peace. Besides its celebrated national rankings of peacefulness, it seeks to discern by empirical analysis the drivers of positive peace. For that purpose, the Institute for Economics and Peace has developed a framework based on rigorous analysis that isolates the factors statistically associated with highly peaceful societies. These factors are grouped into eight pillars that interact as a system:

(1) Well-functioning government—delivering high-quality public and civil services, engendering trust and participation, and generating political stability by the rule of law
(2) Sound economic regulations—leading to competitive businesses and industrial productivity
(3) Equitable distribution—ensuring fairness in access to resources such as education and health, as well as crucial private and public goods
(4) Assuring the rights of others—safeguarding tolerance among different ethnic, linguistic, religious and socioeconomic groups within the country, as well as among genders and age groups
(5) Good relations with neighbors—being conducive to regional integration, foreign direct investment, tourism, and human capital inflows
(6) Free flow of information—ensuring, through free and independent media, that citizens are well informed and thus better prepared for participatory decision-making and more resilient in times of crisis
(7) High levels of human capital—ensured by a broad and deep system of education that helps people in the process of lifelong learning and adaption to change
(8) Low levels of corruption—improving the efficiency of resource allocation and the running of essential public services, which in turn improves confidence and trust in institutions

It is critically important to note that these eight pillars, while empirically validated as drivers of peace, express *qualitative* (value-based) social choices of the societies marked by them. In the language of shared well-being, these societies honor freedom, human rights, and tolerance. Furthermore, they have chosen value-based institutions that build up a common good that serves citizens broadly.

Further research could be undertaken to assess the *positive* impacts of religious and other communities showing by example the spiritual depth of freedom, the complementarity of human rights protection and virtue cultivation, and the coupling of tolerance with solidarity.

Combining the multireligious insights of shared well-being with the empirical findings of the Institute for Economics and Peace could lead to a more robust understanding of how societies flourish and thereby provide an enriched base for public policy decisions designed to enhance the drivers of positive peace. The eight pillars give an "empiric" of positive peace that can assist policy makers. They begin with data and work "from below" to develop a notion of positive peace, while the Religions for Peace approach to positive peace as shared well-being works "from above." We can anticipate that dynamic interaction between both approaches will—over time—greatly clarify the qualitative or value-based drivers of positive peace.

Religions' Virtues to Heal the Damage

Earlier we saw that the core of shared well-being builds on the answers to three basic questions:

(1) How does an individual become a good person?
(2) How do we build a good society?
(3) What is the integral and reciprocal relationship between becoming a good person and building a good society?

We saw that answers to these questions combine in an integral whole that is a normative—if heuristic—dynamic notion of positive peace understood as shared well-being.

Reality, however, is more complex. In addition to these three questions, experience painfully demands a fourth: How do we collaborate with the sacred to heal our personal and social faults? It is a question that arises in every person who has the courage to recognize that the line between good and evil runs through their own heart. It is a question that acknowledges that we, people and societies, contradict and fail our deepest potentials for goodness; we conspire to inflict damage—sometimes lethal damage—upon others, often perversely calling it "good"; as relational beings, in hurting the other, we lacerate ourselves. We use self-screening rationalizations; we invert the scales of value, prizing selfish personal or group gain over the well-being of all. We are victims and victimizers. Our social body is infected; our social "facts" are a tangled amalgam of the authentic drive to develop and distortions that arise out of ignorance, egoism, and group egoism. The infection extends to our institutions, to the very order by which we organize our collective lives, and, of course, to the succession of orders we call history.

Although UN Secretary-General António Guterres offered a pithy expression of our contemporary situation—"We are a world in pieces"—human fault is not a new phenomenon. Brilliant research on what can be called an "archeology of fault" makes clear that symbols of evil traverse the entire religious history of the human family and that common ones are variants of "defilement, sin and guilt."[34]

If evil is a pervasive problem, the deepest question for religious people is, What is the sacred doing about it? Does the sacred reach only to the surface of our reality? Or does it call upon and enter into our hearts, so that—humbled, healed, and newly empowered—we might engage in collaboration to help transform evil?

The question is utterly practical: Just what do the religions counsel when the victims of unjust suffering cry out from the depths of their hearts, when the innocent are slaughtered and the cloth of connection shredded? Just what do the religions counsel when the institutions that are to be built to help us hurt us? Just what do the religions counsel when people face situations that appear hopeless, when they must bear the unbearable? On a more personal level, just what do our religions counsel when, in moments of unvarnished honesty, we realize that we collude in fault and hide it from ourselves?

Religions are sometimes accused of being unrealistic, and they can indeed be interpreted in a naive fashion. But, religions' ruthless prosecutions of human fault in relation to the sacred disclose profound possibilities of collaboration to overcome evil.

Each tradition counsels its own practices to heal the ravages expressed in their symbols of evil. These include a commitment to repair injustice based upon unflinching honesty, repentance, restitution, and reconciliation; calls for the transformation of social structures that hurt us into ones that nourish us; sober calls for self-sacrifice for others and the common good; calls for the voluntary bearing of innocent suffering; calls for returning good for evil; and calls for forgiveness, unrestricted compassion, and love.

If the path of collaboration for shared well-being offers the taste of joy that gladdens the human heart, it will, nevertheless, be long and difficult, marked by personal and collective failures. In the blaze of sunshine, the religions counsel the virtues that build up the human heart. In the dark night, they speak of hope anchored beyond human vicissitudes.

Annex: Precisions Regarding Shared Well-Being—A Heuristic, Multireligious Notion of Positive Peace

The following five precisions are useful when considering the account of the emergent, heuristic notion of shared well-being:

- *Homologous relations*: The emergent model recognizes that each religious tradition has nested foundational terms and fundamental presuppositions; thus, the notion of positive peace of a particular tradition can be fully understood only in relation to the terms and presuppositions of that tradition. The notion of homology can, however, provide a modest but helpful way of comparing elements of positive peace across diverse schools. Homologies, in my use of the term, compare functional equivalencies across diverse systems. Homologous elements of positive peace are those that have a similar function in diverse religious systems. The model of a modest emergent consensus acknowledges the homologous character of its assessments of similarity and thereby retains the dissimilarity of the respective religious traditions.

- *From compactness to differentiation*: If religious communities have originating experiences that serve as their respective grounds, the dynamism of these communities manifests in their responses to new historical circumstances through a process of reexpressing themselves called "traditioning." In the process of traditioning, a community encounters new challenges and labors to faithfully reexpress itself in relation to these challenges. The relatively compact and highly fecund originating experiences within a given tradition are thereby further differentiated. It follows that an emergent multireligious notion of positive peace may be based on each community further differentiating its own relatively compact tradition in relationship to positive peace.
- *Bilingualism and the need for public language*: Religious communities today are challenged to become bilingual.[35] Each religious community communicates among its follower in what can be termed its primary language, which includes all the carriers of meaning of that tradition. The employment of a religious primary language is the typical way communities communicate with their members. However, sectarian religious language is not as useful when trying to communicate beyond the boundaries of a religious community. Today, religious communities are challenged to be doubly creative: to engage contemporary challenges with the creative use of their respective primary languages for their internal use and to transpose their expressions of religious care into a species of "public" language. The great strength of the latter is that it provides a medium for diverse communities to find consensus on shared values and thereby establish a common framework for action, even as each religious community continues to hold and develop its respective primary language.
- *Local and global*: Religious communities exist on local, national, regional, and often global levels. Each context is distinct; therefore, the highly generic emergent consensus noted in this chapter needs to be adapted and filled out in every discreet context.
- *Qualitative and quantitative*: The emergent consensus discussed in this chapter is expressed in qualitative language. It needs an appropriate mediation into relevant forms of quantitative analysis.

Notes

This chapter was adapted from a paper written by the author for the seventh World Assembly of Religions for Peace held in August 2019 in Lindau, Germany.

1. Religions for Peace first began to speak of shared well-being more than a decade ago as a multireligious cipher for positive peace. The strategic plan of the World Council (the

global leadership body of Religions for Peace), issued in 2013, called for the development of shared well-being as an expression of positive peace.

2. António Guterres, "Address to the General Assembly," United Nations, September 19, 2017, https://www.un.org/sg/en/content/sg/speeches/2017-09-19/sgs-ga-address.
3. Institute for Economics and Peace, *Global Peace Index 2017: Measuring Peace in a Complex World* (Sydney: Institute for Economics and Peace, June 2017), https://www.visionofhumanity.org/wp-content/uploads/2020/10/GPI17-Report-1.pdf.
4. *Global Peace Index 2017*.
5. *Global Peace Index 2017*.
6. "Global Uptick in Government Restrictions on Religion in 2016," Pew Research Center, June 21, 2018, https://www.pewforum.org/2018/06/21/global-uptick-in-government-restrictions-on-religion-in-2016/.
7. "Poverty," World Bank Group, accessed April 3, 2019, https://www.worldbank.org/en/topic/poverty/overview#1.
8. "Climate Action Tracker," Climate Action Tracker website, http://climateactiontracker.org.
9. This catalog of threats to peace should not blind us to the highly promising developments of our time: The nature of the problems humanity faces requires systems thinking on a planetary scale and corresponding modes of moral consciousness that call people to act both locally and collaboratively toward the entire world. This cosmopolitan shift in our conceptions of challenges and related moral responsibilities is already being nurtured by web-based media that reinforce interdependence among world citizens and their moral awareness. Moreover, this emergent global mindset is complemented by technological advances that—if morally guided—can further liberate the human family to work together for the positive peace I term shared well-being as set forth later in this chapter.
10. Freedom House, *Freedom in the World 2019: Democracy in Retreat* (Washington, DC: Freedom House, 2019), https://freedomhouse.org/sites/default/files/Feb2019_FH_FITW_2019_Report_ForWeb-compressed.pdf.
11. Oxfam International, *Reward Work, Not Wealth* (Oxford: Oxfam International, January 2018), https://www-cdn.oxfam.org/s3fs-public/file_attachments/bp-reward-work-not-wealth-220118-en.pdf.
12. Mark Carney, "Inclusive Capitalism: Creating a Sense of the Systemic" (speech, Conference on Inclusive Capitalism, May 28, 2014), https://www.bis.org/review/r140528b.pdf.
13. Benedict XVI, "Address of His Holiness Benedict XVI," Vatican website, September 8, 2007, http://www.vatican.va/content/benedict-xvi/en/speeches/2007/september/documents/hf_ben-xvi_spe_20070908_vespri-mariazell.html.
14. Alvin Chang, "The Facebook and Cambridge Analytica Scandal, Explained with a Simple Diagram," *Vox*, May 2, 2018, https://www.vox.com/policy-and-politics/2018/3/23/17151916/facebook-cambridge-analytica-trump-diagram.
15. Claus Dierksmeier, *Qualitative Freedom—Autonomy in Cosmopolitan Responsibility* (New York: Springer, 2019). This writer benefited from fruitful exchanges with Dr. Dierksmeier.
16. David Walsh, *The Growth of the Liberal Soul* (Columbia: University of Missouri Press, 1997). This author is indebted to Walsh's brilliant work.
17. Advancing positive peace requires the dynamic interplay of developing skills, virtues, and value-based institutions in a virtuous cycle in which each makes its contribution but also fosters and reinforces the development of the other two. Weakening one inevitably places distorting stress and overemphasis on the other two, whereas strengthening one can invite a concomitant development of the other two. Animated by freedom and protected by rights, this dynamic interaction of virtues, skills, and institutions is the

true engine of authentic sustainable integral development. The living matrix of this interactive dynamic cycle is community.

18. In this paragraph and the following section on virtues and community, I draw heavily on and gratefully acknowledges the work and "wording" of Bernard Lonergan's *Method In Theology* (London: Darton, Longman & Todd, 1972), 360–61.
19. Lonergan, *Method in Theology*.
20. Lonergan, *Method in Theology*.
21. As we shall see, the development of just value-based institutions that honor human dignity and build up the common good is another essential response.
22. "Our Commitment to Justice, Equality and Sharing (Istanbul Declaration): Istanbul Declaration of Tolerance," European Council of Religious Leaders/Religions for Peace, February 15, 2010, https://ecrl.eu/our-commitment-to-justice-equality-and-sharing/.
23. "Declaration of Principles on Tolerance," UNESCO, November 16, 1995, http://portal.unesco.org/en/ev.php-URL_ID=13175&URL_DO=DO_TOPIC&URL_SECTION=201.html.
24. "Declaration of Principles on Tolerance," articles 1.2, 1.4.
25. European Council of Religious Leaders of Religions for Peace, *Lille Declaration on a Culture of Peace* (New York: Religions for Peace, May 27, 2009), https://ecrl.eu/wp-content/uploads/Lille-Declaration-EN.pdf.
26. As summarized by Linda Raeder in "Liberalism and the Common Good: A Hayekian Perspective on the Common Good," *Independent Review* 11 (1998): 519.
27. Anglo representatives of the communitarian stream of the Modern Order include Charles Taylor, Alasdair MacIntyre, Michael Sandel, and Michael Walzer, among others. A major German representative is Hans Joas. In the field of economics, Christian Felber has started the Economy for the Common Good movement.
28. Even with general areas of agreement across religious traditions, two key points are that a consensual common good is to be negotiated and that the consensual common good of one country can be expected to differ from that of another. Thus, the common good is necessarily an analogous notion, differing from place to place and across time, even as significant areas of consensus continue to emerge, not least on the global level.
29. An economics of the common good can harness the remarkable power of the market as long as the market functions within the moral envelope of the common good.
30. When the public notion of the common good is less robust in some areas than in a particular religious community's notion, that community is invited to serve the public based on its notion of the common good, with respect for the freedom and human dignity of those who are not members of their community and have differing notions of the common good.
31. The precisions on homologies and from compact to differentiated noted in the annex apply to these terms, which are borrowed from a particular stream of moral religious reflection.
32. Concrete responsibility is related to capacity and in part to proximity to others. Thus, one can have a more general responsibility for all and a more concrete responsibility for those close by. However, this notion only modifies and does not annul the thesis that, in principle, all are for all.
33. Pope Francis uses this term, and it is widely resonant across traditions.
34. Paul Ricœur, *The Symbolism of Evil*, trans. Emerson Buchanan (New York: Harper & Row, 1967). Importantly, without symbols of evil, we do not even know that we are going astray. Thus, the symbols of evil and the symbols of salvation are directly linked.
35. William Vendley, "Religious Differences and Shared Care: The Need for Primary and Secondary Language," *Church and Society* (1992): 16–22. (Published by the Social Justice and Peacemaking Unit of the General Assembly Council, Presbyterian Church, United States.)

CHAPTER 19

BUILDING PEACE

Strategies, Resources, and Religions

R. SCOTT APPLEBY

Building peace is not accomplished in the short term or through the application of a simple process or a single method. Depending on how long the various parties to a conflict have been fighting, and how violent the fight has been, simply opening the space for a mutually secure cessation of violence may involve repeated failed efforts before a cease-fire holds. Negotiating terms for a peace agreement is, in turn, a complex process involving many local and external actors.

Yet *conflict management* and *conflict resolution*, which address the presenting symptoms of a conflict, rather than the underlying disease, are only the first steps in building a just peace—the only kind of peace that is sustainable in the long run. Treating the underlying disease is known as *conflict transformation*, which requires still other kinds of processes, expertise, and timelines. It includes the initial healing of the trauma endured by the parties, the routinization of constructive relations, and the postwar (re)creation of schools, courts, and other crucial institutions of governance and civil society. Last but not least, forgiveness and reconciliation among erstwhile combatants are achieved piecemeal, if ever, over years if not decades.

The term "peacebuilding" encompasses all of these practices, and more. *Conflict prevention*, for example, is an ongoing process necessary at every stage along the arc from the initial outbreak of violence to the sustaining of peaceful relations; it may entail security measures such as early warning of violence among antagonistic parties and social policies such as peace education from early childhood through adulthood.[1]

In many settings peacebuilders are confronted by a vexing landscape and a long history of simmering or raging conflict sparked and sustained by a variety of bad actors. To take one example of many, consider the deadly regional violence between the Hutu and the Tutsi that unfolded in Great Lakes Africa over thirty-five years following the Social Revolution of 1959 to 1961, culminating in the Rwandan genocide of 1994. For decades preceding that revolution, many actors, including German, Belgian, and French colonizers, Catholic missionaries, internal African political parties, and extremist militias, set the stage for the mass violence that erupted in Burundi and Rwanda. While there were intermittent years of low-level violence following the Social Revolution, no sustained "ceasefire" was achieved, arguably, until the aftermath of the 1994 genocide and the establishment of the International Criminal Tribunal for Rwanda that year. Only then did attempts at conflict prevention, management, and resolution gain traction; it remains to be seen whether the tense relations between the two major ethnic groups can be pacified and transformed.[2]

Strategic and Just

In the simplest terms, peacebuilding nurtures constructive human relationships at every level of society and across the potentially polarizing lines of ethnicity, class, religion, and race. Peacebuilding that is *strategic*, however, sees these rebuilt or new relationships as making possible successful initiatives for sustained, constructive, structural change over time. Strategic peacebuilding, that is, addresses the roots of the conflict as well as its branches; accordingly, it focuses not only on transforming inhumane social patterns but also flawed structural conditions that weaken the conditions necessary for a flourishing human community (e.g., social and economic discrimination against a minority or marginalized ethnic group, as in the case of Rwanda during the period preceding the 1994 genocide).

Ideally, peacebuilding that is strategic addresses every stage of the conflict cycle and involves all members of a society in the nonviolent transformation of conflict, the pursuit of social justice, and the creation of cultures of sustainable peace. Peace*building* is an ongoing, recurring, yet also dynamic, set of practices, quite distinct from peace*making*, a term once used to refer primarily to the diplomatic efforts of high-level officials who attempted to impose peace from the top down. At least since the end of the Cold War, however, the making or building of sustainable peace has been understood to proceed on the local as well as the regional, national, and international levels, with players at each level managing conflict and providing early warning of possible violence, resolving disputes, accompanying victims, healing wounds, and attempting to foster or rebuild broken relationships. The end goal of these disparate but interrelated phases of conflict transformation is peace based on justice, or *justpeace*—a dynamic state of affairs in which the reduction and management of violence and the striving for social and economic justice are mutually reinforcing dimensions of constructive change.[3]

Not least among the hallmarks of strategic peacebuilding is the recognition and cultivation of *interdependence* among former antagonists. As Fanie du Toit argues in his reflections on the aftermath of the struggle against apartheid in South Africa, the pragmatic recognition that *my* side needs *your* side if either of us is to prosper is the first step away from the mutual self-destruction ensured by repeated cycles of violence and policies of political and economic exclusion. Indeed, du Toit argues, the acceptance of interdependence is a necessary social and political condition for good governance, human rights, prosperity, and peace. The practices of interdependence include the promotion of transparent communication across sectors and levels of society in order to engage as many voices and actors as possible in the reform of institutions and creation of partnerships for the common good. Equally desirable is the coordination and (where possible) convergence of resources, programs, and practices of the institutions, agencies, and movements that influence the causes, expressions, and outcomes of conflict.[4]

The initial collaboration between erstwhile foes may be—in some cases, *must* be—stimulated and supported by outsiders. But sustaining the peace over time must eventually become the ordinary practice of local citizens and institutions. Building on this awareness, scholars have begun to identify indicators of this kind of sustained, routinized, or "ordinary" peace and to develop metrics to track its growth (or decline) in local, regional, and national communities. This attempt to define peace in a more nuanced and complex way—that is, beyond the narrow metrics of "negative peace" (the absence of outright war), leads them to speak of "categories" or "levels" or "qualities" of peace, with each country positioned at some point along a continuum. The countries rated most highly on the peace spectrum are those that score well on qualities such as an independent judiciary, observance of the rule of law, and a robust civil society with freedoms of speech and assembly guaranteed to all, regardless of ethnicity, race, class, or religion.

To return to our example of Rwanda, by the narrow metrics of negative peace, the country is no longer experiencing war or high levels of political violence and so might be said to have enjoyed a sustained peace for almost a quarter-century after the genocide. Yet the *quality* of the peace leaves much to be desired. Despite a series of elections, for example, Rwandan politics remains dominated by a small group of ethnic Tutsis, and society remains bifurcated and unequal.[5]

Given the range of settings in which one strives for a quality peace—that is, for an everyday existence marked by just and constructive relationships—there is no standard approach to building peace, no "one-size-fits-all" set of tools and resources appropriate for every situation. Accordingly, the strategic peacebuilder must become something of a diagnostician. What particular ills beset this particular community? Who are the local allies in building peace? What assets do they bring to the effort?

Religions, to take a misunderstood but important example, are potential contributors to building a just peace. Although mainstream media often depict religion exclusively as a source of intolerance and discrimination, in many and perhaps most settings, religious actors bring crucial resources to the multilayered, culturally nuanced art of peacebuilding.

Religions as Resources for Strategic Peacebuilding

The three distinct transformative processes at the heart of peacebuilding—ending violent conflict, striving for social justice, and building healthy cooperative relationships in conflict-ridden societies—are interrelated most fundamentally at the local level; even when violence originates and occurs at the national or regional level, its impact is felt most keenly and directly in neighborhoods, towns, villages, and cities—in local communities. There, religious communities are custodians of the symbolic, ritual, and practical resources that articulate meaning and purpose and reach beyond the surface of a conflict to its very depths—to what is really at stake.[6] And, unlike international relief, humanitarian and peacebuilding organizations, religious communities do not come and go depending on levels of violence or the availability of funding; they *are* the people at risk. That is why, despite the failings of some religious leaders and institutions, religious communities remain among the most trusted actors on the local scene. They are poised to perform several functions essential to conflict transformation. For example, the morally prophetic character of religious communities gives them a heightened sensitivity to rifts in the social fabric—to subtle as well as open forms of social discrimination, political oppression, and other forms of injustice. Their close reading of the society and its fault lines enable religious actors to diagnose social ills and identify early warning signs of approaching conflict.[7]

Owing to their pivotal position in society, local religious leaders can also become critical players in countrywide and regional peace processes. Bringing representatives of warring sides to peace talks typically requires concerted effort by those wielding high levels of political and social authority. But they cannot replace religious and cultural agents who, operating on the local level, interpret agreements and prepare the society for their implementation.[8]

Religions are also trans-local actors; they operate beyond as well as within national borders. Religious officials such as bishops, imams, rabbis, and abbots have been called upon to mediate international conflicts, lead truth and reconciliation commissions, accompany victims, and defend human rights in their own nation or in foreign settings.[9] As part of the proliferation of transnational social movements for global–local justice, faith-based organizations such as Catholic Relief Services, World Vision, and Islamic Relief have brought some of the distinctive symbolic and human resources of religion to the work of development, humanitarian relief, and peacebuilding. Whether staffed by secular or lay religious professionals, faith-based organizations aspire to embody the virtues and charisma of the host religious community.[10]

Peacebuilding, as we have seen, requires more than management of the conflict, reduction of violence, or agreement on political issues. It must alleviate the suffering of people scarred by the lived experience of sustained violence. The wounded survivors of deadly violence and their broken communities need the kind of healing that restores the soul, the psyche, and the moral imagination. Its preferred modalities are therefore symbolic, cultural, and religious—dimensions of existence that touch the deepest personal and social spheres, which directly and indirectly shape the national and political spheres.

The spiritual and moral resources of religious communities are also available to the healers. The occupational hazards of the multiphased, recurring, relentlessly demanding cycles of conflict and conflict transformation that compose strategic peacebuilding include burnout and depression. Ministers of many kinds—religious, spiritual, and psychological—specialize in the restoration of mind and body, the recommitment of energy, and the renewal of hope. Builders of peace, no less than victims of war, benefit from such ministrations.[11]

Conclusion: SDG 16

SDG 16 calls for peaceful and inclusive societies that provide "access to justice for all" and "effective, accountable and inclusive institutions at all levels." The notion of *just peace* provides an analytical lens through which to understand and evaluate progress toward the goal. Similarly, the targets specified for this goal correspond roughly to the escalating "degrees of peace" found along the "quality peace" continuum, reflecting the strategic approach to building a peace that becomes more inclusive and durable at every stage. The first target to be reached in achieving SDG 16, for example, is to "significantly reduce all forms of violence and death rates everywhere." While this could be considered an (ambitious) expression of negative peace (the absence of war or deadly violence), the target indicators suggest that the target is being reached only to the degree that everyday life takes on certain positive qualities, such as a taken-for-granted sense of security, measured by "the proportion of population that feels safe walking alone around the area they live."[12]

In short, I argue that strategic peacebuilding—peacebuilding that is holistic, just, and sustainable, operating on all levels and enacted by many agents, while rooted in the local communities affected most severely by deadly conflict—is a necessary means to the successful achievement of SDG 16. It should be endorsed and supported by governments and intergovernmental agencies that seek to ensure the success of the 2030 Agenda.

Bringing religious communities more fully into the circle of actors dedicated to achieving SDG 16 is a recommendation warranted by the definition and conceptualization of strategic peacebuilding presented in this chapter. The sustainable transformation of conflict requires the redress of legitimate grievances and the establishment of new relations characterized by equality and fairness according to the dictates of human dignity and the common good. Experts on communal violence and ethnic and civil wars increasingly regard transitional justice and the healing of individuals as well as communities as both a form of post-conflict therapy and a precondition for the prevention of renewed conflict and the transformation of destructive social and structural patterns. In that sense, the healing that leads to reconciliation is the sine qua non of peace. How might religious communities renew their mission and self-understanding to answer these needs of peacebuilding? How might they bring spiritual, ritual, symbolic, and moral resources to bear on the challenges to peace in our complex twenty-first-century world? How might we support the education and formation of religious peacebuilders? The key to building peaceful and inclusive societies that provide "access to justice for all" may well be the evocation of the tender mercies and prophetic indignation of religions drawing on their deepest, most humane sensibilities and commitments.

Notes

1. See Tobi P. Dress and Gay Rosenblum-Kumar, "Deconstructing Prevention: A Systems Approach to Mitigating Violent Conflict," in *From Reaction to Conflict Prevention*, ed. Fen Osler Hampson and David M. Malone (Boulder, CO: Lynne Rienner, 2002), 229–49.
2. See Gérard Prunier, *Africa's World War: Congo, the Rwandan Genocide, and the Making of a Continental Catastrophe* (New York: Oxford University Press, 2009), 1–36.
3. John Paul Lederach and R. Scott Appleby, "Strategic Peacebuilding: An Overview," in *Strategies of Peace: Transforming Conflict in a Violent World*, ed. Daniel Philpott and Gerard F. Powers (New York: Oxford University Press, 2010), 34–35.
4. Fanie du Toit, *When Political Transitions Work: Reconciliation as Interdependence* (New York: Oxford University Press, 2018).
5. See Christian Davenport, Erik Melander, and Patrick M. Regan, *The Peace Continuum: What It Is and How to Study It* (New York: Oxford University Press, 2018), 1, 12–13. The concept of quality peace is developed in the context of post-conflict settlements in Peter Wallensteen, *Quality Peace: Peacebuilding, Victory and World Order* (New York: Oxford University Press, 2015).

6. See Lisa Schirch, *Rituals and Symbols in Peacebuilding* (Bloomfield, CT: Kumarian, 2005).
7. David Little and Scott Appleby, "A Moment of Opportunity? The Promise of Religious Peacebuilding in an Era of Religious and Ethnic Conflict," in *Religion and Peacebuilding*, ed. Harold Coward and Gordon S. Smith (Albany: State University of New York Press, 2004), 6.
8. For case studies of the achievements of religious peacebuilders, see David Little, ed., *Peacemakers in Action: Profiles of Religion in Conflict Resolution* (New York: Cambridge University Press, 2007); and Joyce Dubensky, ed., *Peacemakers in Action*, vol. 2, *Profiles in Religious Peacebuilding* (New York: Cambridge University Press, 2016).
9. R. Scott Appleby, *The Ambivalence of the Sacred: Religion, Violence and Reconciliation* (Lanham, MD: Rowman & Littlefield, 2000), 207–44.
10. Loramy Conradi Gerstbauer, "The Whole Story of NGO Mandate Change: The Peacebuilding Work of World Vision, Catholic Relief Services, and Mennonite Central Committee," *Nonprofit and Voluntary Sector Quarterly* 38, no. 6 (2009).
11. John Paul Lederach and Angela Jill Lederach, *When Blood and Bones Cry Out: Journeys Through the Soundscapes of Healing and Reconciliation* (New York: Oxford University Press, 2010), 3, 11, 53, 68–72, 100.
12. "Indicator 16.1.4: Proportion of Population that Feel Safe Walking Alone Around the Area They Live (Percent)," United Nations Department of Economic and Social Affairs, accessed April 9, 2021, https://www.sdg.org/datasets/93cea5544d6a41859f14ee6ed7a4cc95?geometry=125.858%2C-35.024%2C-93.166%2C64.702.

PART V

AN ETHICAL CONSENSUS ON SUSTAINABLE DEVELOPMENT

Migration

ETHICS IN ACTION

Statement on Migration

Migration is a shared condition of all humanity. We have all been strangers in a strange land. All people live today as a result of migration, by themselves or their ancestors. Migration is a matter sometimes of choice, often of need, and always an inalienable right.

All helpless people deserve to be helped. Offering such help is a commandment and a blessing shared among all religions. Accordingly, as Pope Francis reminds us, our duties to migrants include "to welcome," "to protect," "to promote," and "to integrate."

Most people want to reside and prosper in the land of their birth. This is natural. Yet to do so, they require safety, food security, economic opportunity, freedom from environmental distress, and prospects for their children's future. Forced migration is the result of wars, poverty, environmental degradation, and climate change that compel people to leave their homelands. Because of these factors, we are currently facing the largest humanitarian crisis since the Second World War. And the face of the migrant is increasingly a youthful face—for the first time in history, half of all refugees are children and youths, and one in every two hundred children in the world today is a refugee. Preventing the mass forced displacement of peoples has become one of the great ethical challenges of the twenty-first century.

The great religious traditions all emphasize the dignity of each person and the unity and common destiny of the entire human race in our common home. Accordingly, each faith calls upon individuals and communities to welcome, assist, and protect the refugees, migrants, and displaced in our midst.

Each religion provides a foundation for building a world of respect, solidarity, and safety for migrants.

Responding to the fundamental causes, the deep solutions to forced migration are peace, prosperity, and sustainability. Pope Paul VI declared that "development is the new name of peace." We embrace that wisdom, with the restatement that today, "sustainable development is the new name of peace." We recognized that frequent human-caused ecological disasters constitute a new and growing threat in our own time and a spur to mass migration. We affirmed a moral obligation to welcome refugees and that such an obligation extends in particular to the countries responsible for causing the wars and environmental disasters that force people to move in the first place. We recognized that children in particular need a home, a safe haven, a decent education, and an appropriate response to any physical and mental health challenges. We also affirmed the need to prevent the emergence of new technological refugees given the potential effects of technology on employment and work in the near future.

Overall, we called for an approach based on sustainable and integral human development—the fullest development of each person and all people, allowing them to become active agents of their own development. This notion includes the full integration of migrants into the economic, social, political, and cultural life of the nation in which they settle or the choice of a speedy and safe return to their homelands as circumstances permit.

General Recommendations

(1) Push for a political response to migrants guided by three levels of responsibility:
  a. The most basic principle that "in case of need all things are common" because "every man is my brother"—these issues are related to existence and subsistence and condition other related issues (such as accommodation, food, housing, and security).[1]
  b. As a component of the fundamental rights of peoples, legal guarantees of primary rights that foster an "organic participation" in the economic and social life of the nation in which migrants settle. Access to these economic and social goods, including education and employment, will allow people to develop their abilities.
  c. A deeper sense of integration, reflecting responsibilities related to protecting, examining, and developing the values that underpin the deep, stable unity of a society—and, more fundamentally, create a horizon of public peace, understood as Saint Augustine's "tranquility in order."

  Especially in this latter context, policies on migration should be guided by prudence, but prudence can never mean exclusion. On the contrary, governments should evaluate "with wisdom and foresight, the extent to which their country is in a position, without prejudice to the common good of citizens, to offer a decent life to migrants, especially those truly in need of protection."[2]

(2) Encourage and promote humanitarian corridors, targeting the most vulnerable asylum seekers and refugees, especially in nations that have space. One such positive example is the program created by the governor of San Luis in Argentina that gives humanitarian visas to Syrian refugees.

(3) Promote moral education—among children and adults alike—to inculcate the norms, virtues, and values of empathy, compassion, solidarity, and care for our common humanity and common home, which incorporates obligations toward migrants and refugees.

(4) Encourage businesses to provide decent work and employment opportunities for newcomers so that they may in turn support their families in dignity and security and contribute to their communities. This includes finding creative and effective solutions in using technological advances in service of the common good.

(5) Encourage governments to increase budgets for peace and support for sustainable development, not for arms and military spending—the best response to the migrant crisis is to cultivate solidarity for people and care for the planet, not wars.

(6) Call upon all stakeholders—religious communities, civil society, business, and government—to take concrete steps to end wars, stop the arms trade, overcome poverty, and halt human-caused environmental degradation and climate change, as guided by *Laudato si'*, the encyclical of the Holy and Great Council of the Orthodox Church, the SDGs, and the Paris Agreement on climate change.

(7) Mobilize key stakeholders—the UN, development leaders, businesses, individuals with a high net worth, and religious communities—to direct billions of dollars of new financial resources to achieve the SDGs, including the protection of migrants and the end of human trafficking and modern slavery.

Ethics in Action Working Group Action Items

(1) Forge partnerships with the UN Sustainable Development Solutions Network, academicians of the Pontifical Academy of Sciences, and other experts to provide scientific evidence of the roots of today's forced migrations in wars, extreme poverty, social exclusion, climate change, and environmental degradation. We pledge to be guided by best practices on mental health and trauma, legal protections, and education, as well as the well-being of asylum seekers, refugees, and irregular migrants in varied destinations.

(2) Help religious leaders of all major faiths to proclaim the common value of all religions in extending generous solidarity toward migrants and refugees, including by acting together. For example, the "Faith Over Fear" campaign of Religions for Peace and UNICEF seeks to mobilize religious communities to welcome and support refugees. The UN Alliance of Civilizations is also called upon to prioritize this concern.

(3) Support the Holy See's new Dicastery for Promoting Integral Human Development in its mission to help the Church its institutions to protect and support refugees and migrants in their daily struggles.

(4) Work with the United Nations to support the intergovernmental negotiations on the new UN Global Compact for Migration to address large movements of refugees and migrants. International migration policy should be based on a co-responsibility among origin, transit, and destination countries—including distribution mechanisms for refugees based on international agreements (ideally within the United Nations) and compensation by the rich countries for the first-entry countries that carry the large drivers of migration, human trafficking, and modern slavery.

Notes

1. Pope Francis addressed these issues in Lampedusa in 2013 and in Lesbos in 2016. See "Homily of Holy Father Francis," Vatican website, July 8, 2013, http://w2.vatican.va/content/francesco/en/homilies/2013/documents/papa-francesco_20130708_omelia-lampedusa.html; and "Visit to Refugees: Speeches of His Beatitude Ieronymos, Archbishop of Athens and All Greece, of His Holiness Bartholomew, Ecumenical Patriarch of Constantinople and of Pope Francis," Vatican website, April 16, 2016, http://w2.vatican.va/content/francesco/en/speeches/2016/april/documents/papa-francesco_20160416_lesvos-rifugiati.html.
2. "Address of His Holiness Pope Francis to the Members of the Diplomatic Corps Accredited to the Holy See for the Traditional Exchange of New Year Greetings," Vatican website, January 9, 2017, http://w2.vatican.va/content/francesco/en/speeches/2017/january/documents/papa-francesco_20170109_corpo-diplomatico.html.

CHAPTER 20

THE DRIVERS OF MIGRATION

JEFFREY D. SACHS

Since the beginning of humanity, people have been on the move.[1] We are a species of migration. Our species, *Homo sapiens*, evolved in Africa some two hundred thousand to three hundred thousand years ago and began to migrate out of Africa in small bands perhaps 125,000 years ago. A major dispersion from Africa occurred around seventy thousand years ago, and the evidence suggests that it was in this period that the main migration to the rest of the world occurred.

Migration is a normal process, but at certain times and under certain circumstances, forced migration and refugees can be considered as a measure of crisis, not merely a measure of the normal tendencies of people and their families over time to choose places to live different from their birth places. A global refugee is someone who has been forced to flee their home owing to "fear of persecution for reasons of race, religion, nationality, political opinion, or membership of a social group."[2] Refugees may also flee from violence, war, conflict, or environmental disaster. The number of worldwide migrants and the number of refugees in the world have been rising sharply in recent years.[3] This issue has become central to our humanitarian concerns, and central to contested politics all over the world.

According to the UN's most recent estimates in 2019, about 272 million people around the world are living in countries different from their country of birth.[4] This is about 3.4 percent of the human population. The number of international migrants has increased by about 50 percent since 2000,[5] but that it still is a relatively small part of the global population of about 7.7 billion people. This figure indicates something very important, which is that most people in most places of the world would like to be able to live in their country of birth. Many people would like to raise their children in their homelands with their local languages and cultures. It is true that by no means did every one of these international migrants move out of desperation. But many did, either because they were forced to leave because of macro-factors, such as violence and environmental destruction, or they were induced to leave because they could not earn a basic livelihood and there were limited opportunities for their children's futures, or they lacked the hope that their children would be able to grow up and have a decent future in their homeland.[6] We should look at this large number of people and understand the roots of the migration, as well as how we might best address these human needs and the economic realities that they reflect.

The estimates of refugees—people forced to flee from their homes—is also a complex number to comprehend. As of 2019, it's estimated that around 70.8 million people had been forced to flee from their homes. Of these, about 25.9 million were refugees forced to cross national borders and 41.3 million were internally displaced, meaning that they have been forced from their homes but found refuge within their home country.[7] It's also estimated that about 18.8 million people had been internally displaced within their countries owing to natural disasters including droughts, cyclones, and floods in 2017. These people were not forced to flee because of conflict or violence but because of environmental hardship resulting from natural disasters or from places on the planet becoming uninhabitable because of global warming and other climate changes that cause tremendous hardship. The number of environmental refugees or "climate refugees," as they are called in the media, will continue to rise in the twenty-first century.[8]

Whatever the exact numbers, vast numbers of migrants are moving because of economic desperation to seek material improvements in their livelihoods in other countries. Vast numbers of refugees have fled their homes and crossed international borders because of violence. And vast numbers of people who have been displaced from their homelands continue to cross international borders because of environmental crises. Consider the migrant crisis in the United States right now. Since the 1980s, displacement, economic instability, and insecurity in countries such as El Salvador, Guatemala, and Nicaragua have led to a population of about 3.5 million Central American refugees relocating to the United States. Civil war and internal violence cause many people from Central America who are desperately trying to find a means of survival and livelihoods to cross into the United States.[9] People are fleeing because they cannot survive in their homelands. We know that people are fleeing not only because of poverty but also because of violence. Central America is wracked by conflict. This is partly conflict related to the desperation of poverty and partly conflict related to the narcotics trafficking in which cocaine and other illegal drugs are being smuggled to the United States from South America through transit points in Central America. Violence is also emanating from an incredible proliferation of small arms that come into the hands of local gangs, typically flooding in from the United States. It's a perfect storm of poverty, ecological disaster, gang warfare, a flood of weapons, and narcotics trafficking. These factors lead desperate people to make a desperate attempt to find a way to survive.

What we have right now is fear. We have a politics of fear, a politics of extreme backlash against the most desperate people in the world, those who have nothing and are trying to save their families. What we need is a politics of knowledge, awareness, reason, and compassion and a politics that follows the basic strictures of international law. This will save lives. Therefore, the basic principle is that if conditions are decent, the vast majority of humanity would like to stay, live, and raise their children in their homelands. They would prefer to speak their native languages and live among people of their own cultures and traditions. Most important of all, we must help to ensure that every part of the world is ecologically, economically, socially, and politically viable so that people can live and thrive in their homelands. Mobility is important, and we should visit one other as tourists, in our professional capacities and to promote art and cultural exchange. But we must create a world in which it is not necessary for tens of millions of people to flee to save their lives or to secure a decent future.

What should be our responses to forced migration? Well, at the core, we should understand the causes of this desperate migration and address those underlying conditions. At the same time, we should identify and maintain a compassionate and law-based way to help those who must leave their homes. Three fundamental forces cause people to migrate, forcing tens of millions of people into internal displacement and cross-border migration: conflicts, environmental shocks, and extreme poverty. When we think about the political and social responses to forced migration, we should therefore start by analyzing the sources of those conflicts, ways to address environmental devastation, and ways to address extreme poverty.

When it comes to conflict, the most important solution to forced migration is to end the conflicts. In recent years, the Syrian refugee crisis has seen more than 5.6 million Syrians forced to flee their homes since 2011.[10] Over three million have resettled in Turkey, as well as in Lebanon, Jordan, and Western Europe. In addition to the Syrian crisis, Europe has experienced floods of migration from Afghanistan, Libya, and other conflict areas. In all of the upheaval, there was a tremendous amount of debate about what to do with the incoming populations, but there was not enough discussion about stopping the conflicts that were leading to that desperate forced migration. When it comes to conflict, we must address the source of the conflict and find a peaceful rather than militarized approach to ending it. In Iraq, Syria, Afghanistan, and

Libya, the international community has often responded to conflict through military operations. This is true too in Central America and Mexico, where violence, often related to narcotics trafficking to bring drugs to the U.S. market, has been met as through a heavily militarized "War on Drugs." This response has not solved any problems but instead has caused violence so severe as to massively displace populations.

Climate shocks, such as the extreme droughts that Syria experienced in the first decade of the twenty-first century and the one ongoing in Central America, lead to massive displacement. Two kinds of vital political responses are needed to address climate-caused migration. First, there must be international support to address the urgent challenges posed by the shocks, which these countries did not create. Many of these countries are on the receiving end of global climate change processes that have been caused by the pollution and emissions of major economies like the United States, not by themselves. What kind of help do these countries get when there's a three- or four-year mega-drought? Under our current system, they receive almost no help at all. UN climate treaties call this category "losses and damages from climate change" but there is no associated budget attached to it to help those hit by these shocks. We need a short-term emergency response so that people don't have to flee desperately from their homes when such shocks occur. But we also need long-term responses to end the human-induced climate change, which is already on the brink of being out of control. Environmental forced migration will expand tremendously, even disastrously, unless we get the planet onto a safer trajectory, in line with the commitments that we've made in the SDGs and in the Paris Agreement on climate change.

When it comes to the third cause of forced migration, extreme poverty, the answer is economic development in impoverished places. Migration owing to poverty can be seen in the flow of desperate people from sub-Saharan Africa to Europe and from Central America and Southern Mexico to the United States. In both cases, people are trying to find a place where they can guarantee livelihoods for themselves and their families because their home region does not provide the jobs or the incomes that make for a viable life. Once again, by focusing only on symptoms—for example, the violence and dislocation of the United States' War on Drugs—we don't get to the underlying challenge of extreme poverty. The president of Mexico has stressed in recent times that instead of the billions of dollars that Mexico receives from the United States to fight the War on Drugs, money should be channeled to economic development in the poor regions of Central America. The president of Mexico is saying, "*Abrazos, no balazos*": "Hugs, not gunfire."[11] The Mexican government is urging a financial reprogramming of billions of dollars to be directed to economic development in places like Guatemala and Honduras, so that they become viable economies, rather than sources of forced migration. Sadly, and ironically, the Trump administration, in its desperate but very confused idea of stopping this kind of poverty-related forced migration, cut aid to Central America rather than increasing development assistance. In other words, it was trying to punish populations, as if forced migration were a matter of casual choice rather than what it is: a reflection of desperation. In this case, the United States government had it exactly backward. Rather than cutting aid, redirecting militarized financing toward development, as Mexico suggests, would get to the crux of the matter.

We also need appropriate social and political responses at the receiving borders of the United States, Europe, and other countries receiving refugees and migrants. For that, we need compassion and a sense of our shared humanity to understand that people are fleeing desperate conditions. We must stop ripping children out of the arms of their mothers and fathers, as has been occurring at the U.S. border with Mexico in a kind of brutality that is shocking to witness from a country that is supposed to be governed by law and guided by decency. We need to understand that refugees have rights, as they are human beings. Refugees have the rights of the Geneva Conventions, and even if they will not become long-term residents or citizens of the receiving countries, those countries should provide them decent and hospitable conditions of dignity and temporary support until they can return home—when those homelands can offer opportunities for economic development, for jobs, and for a future that is safe for themselves and their families.

Notes

1. A version of this chapter was originally delivered as a lecture in May 2017, which is reflected in some of the statistics mentioned later on.
2. European Commission, "Refugee," Migration and Home Affairs, accessed June 14, 2021, https://ec.europa.eu/home-affairs/what-we-do/networks/european_migration_network/glossary_search/refugee_en.
3. See United Nations Department of Economic and Social Affairs, Population Division, *International Migration Report 2017: Highlights* (New York: United Nations, 2017), https://www.un.org/en/development/desa/population/migration/publications/migrationreport/docs/MigrationReport2017_Highlights.pdf.
4. "The Number of International Migrants Reaches 272 Million, Continuing an Upward Trend in All World Regions, Says UN," United Nations Department of Economic and Social Affairs, September 17, 2019, https://www.un.org/development/desa/en/news/population/international-migrant-stock-2019.html.
5. United Nations, *International Migration Report 2017*.
6. Francesco Castelli, "Drivers of Migration: Why Do People Move?" *Journal of Travel Medicine* 25, no. 1 (2018), https://doi.org/10.1093/jtm/tay040, https://doi.org/10.1093/jtm/tay040.
7. "Figures at a Glance," UNHCR, accessed June 14, 2021, https://www.unhcr.org/en-us/figures-at-a-glance.html.

8. "Climate Change and Disaster Displacement," UNHCR, accessed June 14, 2021, https://www.unhcr.org/en-us/climate-change-and-disasters.html.
9. Allison O'Connor, Jeanne Batalova, and Jessica Bolter, "Central American Immigrants in the United States," Migration Policy Institute, August 15, 2019, https://www.migrationpolicy.org/article/central-american-immigrants-united-states.
10. "Syria Refugee Crisis Explained," UNHCR, updated February 5, 2021, https://www.unrefugees.org/news/syria-refugee-crisis-explained/#Where%20are%20Syrians%20fleeing%20to?.
11. Joshua Partlow and David Agren, "Mexico's Presidential Front-Runner, AMLO, Doesn't Want to Escalate the Drug War," *Washington Post*, June 30, 2018, https://www.washingtonpost.com/world/the_americas/mexicos-presidential-front-runner-amlo-doesnt-want-to-escalate-the-drug-war/2018/06/29/f3081f12-7320-11e8-bda1-18e53a448a14_story.html.

CHAPTER 21

A MUSLIM PERSPECTIVE ON REFUGEES

HAMZA YUSUF

Every religion has a cosmology, a foundational narrative of how we arrived here. In all three Abrahamic faiths, Islam being the most recent, we find the arrival of human life on Earth itself to be a refugee narrative.

Adam and Eve were expelled from a garden and given refuge for a time here on our planet: "We said, 'O Adam, live with your wife in Paradise and eat freely from it anywhere you may wish. Yet do not approach this tree lest you become transgressors.' But Satan caused them both to slip from their state and forced them out. We said, 'Go down from here enemies one to the other, and on Earth you shall find refuge and livelihood for a while.'"[1] Thus, the human story begins with banishment and a flight from one place to another: our first parents had refugee status.

Stories of fleeing and migrating as strangers to strange lands—and finding refuge—recur among the shared Abrahamic prophets: the prophet Abraham fled from Nimrod the king to the land of Canaan; the prophet Moses and the children of Israel were enslaved in Egypt and fled from the Pharoah and his forces to Sinai; the Virgin Mary fled with Joseph from Jerusalem to Egypt to protect her child from King Herod; the prophet Muhammad's earliest community fled the persecution of the Meccans and migrated to Ethiopia, and the Prophet himself was a refugee, fleeing Mecca for Medina to avoid assassination and the religious persecution of his people.

The Qur'an reminds us, "There is no refuge from God except to God."[2] The divinely orchestrated conditions of this earthly abode, it seems, often require that the outward physical journey to seek safe haven simultaneously accompanies an inward spiritual journey to seek refuge in God. The prophet Muhammad was once asked, "Who is the true refugee to God?" He replied, "The one who opposes his self and flees from sin." The twelfth-century theologian Abu Hamid al-Ghazali said, "All of humanity are wayfarers moving inexorably toward their Lord."

Islam teaches that all wayfarers deserve food and safety. The Qur'an states, "Let them worship the Lord of this House who has satiated their hunger and freed them from fear."[3] According to exegetes, these two conditions—freedom from hunger and freedom from fear—must be fulfilled as prerequisites for worship; interestingly, in Abraham Maslow's famous "hierarchy of needs" pyramid, physiological needs and safety lie at the base, as they remain our most basic needs. The prophet Muhammad is reported to have said, "The child of Adam has no entitled rights beyond these: shelter from the elements, clothes to cover his nakedness, and grain and water to nourish him." According to Islamic law, these are God-given human rights. Hence, those of us who have must give to those who have not. This obligation manifests clearly in the injunction to pay zakat, a poor tax of 2.5 percent of one's yearly standing wealth, in addition to a percentage of animal stock, and a 5 to 10 percent grain tax, depending on irrigation techniques.

The etymology of the word "refugee" includes the Latin *fugere*, which means "to flee." As for those who flee persecution, whether religious or political, another God-given right applies to them, known as *ḥaqq al-īwa*, "to seek and receive refuge." Service to those in need qualifies believers: "They care for those who have taken refuge with them and have no desire in their hearts for what has been given them, preferring them to themselves, even if it means hardship for them; and those who are preserved from their own avarice are the ones who succeed."[4]

Believers must tend to the needs of those burdened with hardships, no matter who they are or what land or nation they have fled, because the Earth and everything therein belongs to God, and God's servants have the right to travel the Earth seeking provision: "It is God who made the Earth accessible to you, so travel its roads, and eat of what God has provided, but know to God you will return."[5] The fourteenth-century poet Hafez famously asked, "How would you act if you realized that all who inhabit the Earth are God's guests; how would you treat them then?" The first prophetic tradition taught to students of Hadith is "Those who show mercy will be shown mercy by the Merciful Himself; have mercy on those in the Earth, and the One in Heaven will have mercy on you." The scholars agree that "on those in the Earth" in this foundational tradition covers all peoples, regardless of color, creed, or country.

Like its sister religions, Islam has a checkered past but also many glorious examples of the ideals of the faith that call us to serve the Creator by serving

His creation. In 1492, the Ottoman sultan Bayezid II welcomed more than 150,000 Jews fleeing from Spanish persecution to Turkey, granting them citizenship and then building beautiful synagogues—many of which stand to this day—for the newly arrived refugees. In the 1840s, during the Irish potato famine, the Ottoman sultan Abdul-Majid sent not only money but also ships with grain to provide relief for the needy. In 1860, when local Druze attacked the Christian quarter in Damascus, Emir Abdelkader of Algeria saved more than four thousand Christians, including his enemy, the French consul and his staff, by giving them refuge in his compound and defending them with his Algerian troops. During the Nazi occupation of France, Si Kaddour Benghabrit, a Moroccan imam, risked his life by hiding hundreds of Jews at the Paris mosque and saved many others by issuing certificates that allowed them to hide their Jewish identity and claim they were Algerian Arabs instead.

Refugees have always been a part of life on Earth, and tending to their needs has always been a part of the Islamic tradition. It is that noble tradition of service to others that is urgently needed today, as we are overwhelmed with a refugee crisis that the world has not seen since World War II. Our response tests our mettle and reflects our national character, and it will shape our own future for better or for worse. From a metaphysical perspective, this profoundly precarious situation awaits our response. Our Prophet taught us, "There is no leader who closes the door on someone in need or one suffering in poverty except that God closes the gates of the Heavens during his time of need." We shall reap what we sow, and now is the time for sowing seeds of solace. In due time, God willing, we might reap the rewards of the righteous.

Notes

1. Qur'an 2:35–36.
2. Qur'an 9:118.
3. Qur'an 106:3–4.
4. Qur'an 59:9.
5. Qur'an 67:15.

CHAPTER 22

MIGRATION AND REFUGEES

A Christian Perspective

DANIEL G. GROODY

In the wake of the Arab Spring in 2013, a large group of migrants and refugees hopped onto a boat on the North African coast and launched into the open sea.[1] Like others who had gone before them, many were fleeing poverty, political instability, and numerous human rights violations. They left their homeland in the hopes of finding shelter and protection on European shorelines, but in the course of their journey, their vessel capsized in the middle of the Mediterranean, and most of them drowned. Eight survived the shipwreck by clinging desperately to the fishing nets of a nearby Tunisian-flagged boat. Seeing the boat in the distance, the refugees pleaded desperately to be saved. When the fishermen saw them holding on to their lines, however, they ignored their cries, severed their nets, and left them to die in the middle of the ocean.[2]

One of those who heard about this story was the newly elected Pope Francis. The plight of these refugees moved him so deeply that, within a week's time, he chose to make his first pastoral visit outside of the Vatican to the small and isolated Italian island of Lampedusa. Located in the middle of the waters between Africa and Europe, with less than eight square miles in area, it is little more than a rock in the middle of the ocean. As more and more people have perished in these waters, Lampedusa has come to symbolize the global refugee crisis and all those who are forcibly uprooted and cast into the sea of a merciless world.

When the pope arrived on the island, he celebrated mass in the open air, near the harbor, next to a "boat graveyard," where the remains of migrant ships pile up. From the wooden remnants of these vessels, a local carpenter crafted the altar from a migrant boat's hull, the lectern from ships' rudders, and the chalice from the driftwood of capsized refugee boats. As they were used during the liturgy, the pope made an urgent plea both for those perishing at sea and for those "drowning" in a "globalized indifference" to the last and the least among us.

In light of this event, this chapter explores the integral connection between the global migrant and refugee crisis and Christian ethics in action. First, I will offer some context on the global refugee crisis. Second, I will examine, in the words of Francis, the growing "globalization of indifference" to those on the move. Third, I will speak about promoting a "globalization of solidarity" that leads to strengthening the bonds of the human community. I will conclude by looking at the theme of reconciliation. My thesis is that, in a world that increasingly ignores and discards refugees as "no bodies," the Church's mission is not only to help each refugee discover their dignity as "some-body" but also to reveal that they are in fact connected to "every-body." The Eucharist highlights our essential interconnectedness and calls us to move from "otherness" to "oneness" and from alienation to communion.

The Global Refugee Crisis

Although migration has been happening since the dawn of human history, never has the world seen the scope and scale of human movement as in our own times. At the time of writing, there were 68.5 million forcibly displaced people worldwide.[3] Forty million are internally displaced, 25.4 million are refugees, ten million are stateless people, and 3.1 million are asylum seekers. Forty-four thousand a day are forced to flee their homes because of conflict and persecution, and nearly one person in the world is displaced every two seconds.

Included within these numbers is a growing list of thousands of migrants and refugees who die each year in the mountains, in the deserts, and in open seas.[4] It is difficult to assess just how many perish in the course of their journeys, but since 2014, the International Organization for Migration has recorded more than fifteen thousand migrant deaths in the Central Mediterranean alone, although the actual number is most certainly higher.[5] In the environs of Lampedusa, one in thirty-six migrants attempting to cross the Central Mediterranean route perished in 2017.[6]

By the time this book is printed, however, these figures will already be obsolete, and, if trends continue, the numbers most certainly will be higher. This means that the issue of migration is not going away anytime soon. We are living in the "age of migration"[7]: It has not only shaped our past and present but will also profoundly affect and influence our future. Because how we respond to it will determining who we become—individually and collectively—the issue of ethics in action is absolutely essential.

The Globalization of Indifference

What challenges do migrants and refugees pose to the human conscience? How can the religions of the world make a life-giving response to those whose lives are so threatened? And in what ways can Catholic Christian faith inspire action that promotes a more just and humane society? These questions, no doubt, were at the heart of Pope Francis's message and his motivation to go to Lampedusa.

When he received word about the refugees who had died at sea, he said the message reached him "like a painful thorn in my heart."[8] In response, he wanted to make "a gesture of closeness" and to challenge the conscience of the world, "lest this tragedy be repeated." He wanted to remember those the world has ignored and forgotten, those who are poor and marginalized, those whose deaths are often not even acknowledged or recognized in the public eye.

While to the world they are often regarded as "no-bodies," Francis chose to highlight their plight in order to bring out the ways they are "some-body" to God. The heart of Catholic social teaching remind us that their worth and merit rest not on their economic assets, their social standing, or their political status. By virtue of their creation by a loving God, they have an inherit value and dignity rooted in the fact that they are created in God's image and likeness. Since all share in that dignity by virtue of creation, we all share inseparable bonds that connect us to each other as a human family.

The current disorders of the world, however, are inseparable from the disorders within the human heart, which strain and can even sever these human bonds. The fishermen who ignored the pleas of migrants and cut their fishing nets make visible the invisible rupturing of the bonds of human connection. The sundering of these divine-human bonds is what the Church calls sin. While we are born into this condition by view of original sin, the Church also names as true our personal sins through our actions or inactions, which are further compounded by the social sins structured into the systems of our current social, political, and economic order. The fruit of such sin is injustice and indifference. At Lampedusa, Francis wanted the world to see that such indifference to the last and the least not only deprives those refugees desperately crying out for help but also dehumanizes those who ignore their pleas: Indifference keeps us from becoming the human beings God created us to be!

The Globalization of Solidarity

The global refugee crisis presents an opportunity to look not only at the journey of migrants and refugees but also at the larger horizon of our own human journey. Ethics in action asks not only what is happening in the world and what are we to do about it but also who are we in this world before God and what is my responsibility to my neighbor? While the economic, political, and social forces that influence our current world order are reshaping our interactions like never before, underneath the sea of changes are human constants that are as old as the oceans. In the depths of the human heart are the anthropological and spiritual truths of who we are before God and who we are called to be in relationship to one another. These spaces confront us with undeconstructable questions: Who am I? Who is God? Who am I to my brother and sister?

While the forces of globalization are dividing us in many ways, John Paul II has spoken about the need to foster a "globalization of solidarity."[9] This is another way of saying that the first and primary task of life concerns our relationships. The four primary relationships that shape our lives are those with ourselves, creation, other creatures, and, ultimately, the Creator. The just ordering of these relationships leads to peace and communion. The fragmentation of these relationships leads to disorder, injustice, and alienation.

The ethical challenge posed by the global refugee crisis is not only to help those being dehumanized by the current crisis but also to determine what kind of human beings we become as a result of our response to the crisis. Indifference to the poor not only deprives the poor but also dehumanizes the rich because it alienates a person from God, from other people, and even from oneself. Even if we have arrived on different ships, as Francis put it, "We are all in the same boat and headed to the same port!"[10]

Conclusion: The Eucharist and Our Call to Communion

The reason Jesus Christ came was to reconcile the world to God. This means that the Christian mission is about the ordering of our relationships in such a way that they lead toward union with God and communion with one another. The Eucharist, as the source and summit of the Church's activity, both celebrates and promises that work of reconciliation. The global migrant and refugee crisis presents enormous challenges to the human community, and the pope's visit to Lampedusa highlight many of its important ethical dimensions and the call to committed action.

The narratives and symbols richly embedded in religious traditions like Catholicism have much to say about the challenges of the present moment, such as the migrant and refugee crisis. At their core, spiritual truths give rise to ethical challenges that call us to become more human together. They offer a dual role of denouncing the dehumanizing operative political and social narratives of migration and announcing the reign of God and the call to reshape our lives in life-giving ways that are constant with that Kingdom. The Liturgy of the Eucharist reveals an alternative narrative, one that reminds us of the centrality of relationships, our solidarity with the "no-bodies" of the world, and our call to communion with God and one another.

This chapter has sought to unpack the theological significance of Pope Francis's visit to Lampedusa and its connection to the global migrant and refugee crisis. It has also provided an opportunity to explore how Christianity can encourage us to examine in more depth who "they" are, who God is, and who we are in our common journey together. Putting ethics in action is the work of reconciliation, which means, as Pope Francis summarized it, "welcoming, protecting, promoting and integrating migrants and refugees."[11]

Notes

1. This chapter is drawn in part from a previously published article. See Daniel G. Groody, "Cup of Suffering, Chalice of Salvation: Refugees, Lampedusa, and the Eucharist," *Theological Studies* 78, no. 4 (December 2017): 960–87.
2. Barbie Latza Nadeau, "Pope Prays for Lost Refugees on Visit to Mediterranean Island," *CNN*, July 8, 2013, http://www.cnn.com/2013/07/08/world/europe/pope-lampedusa-refugees/index.html; Martin Barillas, "Pope Francis Condemns 'Indifference' to Illegal Migrants Plight," *Spero News*, July 8, 2013, http://www.speroforum.com/a/ ANXKLSYPJA54/74160-Pope-Francis-condemns-indifference-to-illegal-migrants-plight#.WF0qY7G-J04.
3. For updated statistics, see the UNHCR website: https://www.unhcr.org/figures-at-a-glance.html.
4. The International Organization for Migration estimates more than sixty thousand documented migrant deaths globally since 1996, although the actual number is most certainly higher. See International Organization for Migration, *Fatal Journeys: Tracking Lives Lost During Migration* (Geneva: International Organization for Migration, 2014), 11. See also International Organization for Migration Global Migration Data Analysis Centre, *The Central Mediterranean Route: Migrant Fatalities January 2014–July 2017* (Berlin: International Organization for Migration, 2017), https://gmdac.iom.int/sites/gmdac/files/C%20Med%20fatalities%20briefing%20July%202017.pdf. For the latest global statistics on migrant deaths, see the website of the Missing Migrants Project: http://missingmigrants.iom.int/en/latest-global-figures.
5. For current statistics, see the website of the Missing Migrants Project: http://missingmigrants.iom.int. See also International Organization for Migration Global Migration Data Analysis Centre, *The Central Mediterranean Route*.
6. See International Organization for Migration Global Migration Data Analysis Centre, *The Central Mediterranean Route*.
7. Stephen Castles and Mark J. Miller, *The Age of Migration: International Population Movements in the Modern World*, 5th ed. (New York: Guilford, 2013).
8. Francis, "Visit to Lampedusa: Homily of Holy Father Francis," Vatican website, July 8, 2013, http://w2.vatican.va/content/francesco/en/homilies/2013/documents/papa-francesco_20130708_omelia-lampedusa.html.
9. John Paul II, *Ecclesia in America* (1999), no. 55, https://www.vatican.va/content/john-paul-ii/en/apost_exhortations/documents/hf_jp-ii_exh_22011999_ecclesia-in-america.html. Here he states, "The Church in America is called not only to promote

greater integration between nations, thus helping to create an authentic globalized culture of solidarity, but also to cooperate with every legitimate means in reducing the negative effects of globalization, such as the domination of the powerful over the weak, especially in the economic sphere, and the loss of the values of local cultures in favor of a misconstrued homogenization."

10. John Paul II, *Evangelii gaudium* (Vatican City: Vatican, 2013), no. 99, http://w2.vatican.va/content/dam/francesco/pdf/apost_exhortations/documents/papa-francesco_esortazione-ap_20131124_evangelii-gaudium_en.pdf.
11. Francis, "Message of His Holiness Pope Francis for the 104th World Day of Migrants and Refugees 2018," Vatican website, January 14, 2018, https://w2.vatican.va/content/francesco/en/messages/migration/documents/papa-francesco_20170815_world-migrants-day-2018.html.

PART VI

AN ETHICAL CONSENSUS ON SUSTAINABLE DEVELOPMENT

Businesses as Agents of Sustainable Development

ETHICS IN ACTION

Businesses as Agents of Sustainable Development

The corporate sector has an essential role to play in advancing sustainable development. All religious and ethical traditions represented by Ethics in Action affirm that business is—in the words of Pope Francis—a "noble vocation" with the capacity to improve quality of life in many dimensions. They also affirm the immense achievements of the market economy in innovation, poverty reduction, and increased standards of living.

Today, the urgent challenges of sustainable development make clear that business fulfils its true purpose as a noble vocation when it produces "goods that are truly good and services that truly serve"[1] by orienting its activities and directing its ends toward the common good and acting according to the requisite personal and social principles, values, and virtues. Yet the market system also provides the possibility of pursuing profits through "goods" that are actually "bads" (such as goods that pollute or addict) and "services" that are actually disservices (such as human trafficking and modern slavery).

One of the tenets of a market economy is that market prices guide the efficient allocation of resources to sectors where the social benefits of goods and services, measured by the price that consumers are willing to pay, are equal to social costs, measured by the price that businesses charge for production. Society thereby uses its scarce resources effectively. Yet the tenet utterly fails if market prices fail to reflect their true benefits or costs. This failure occurs in many contexts: monopoly power, financial fraud, environmental pollution, and the use of physical and psychological force, such as in modern forms of slavery.

The market principle also utterly fails when the market excludes the weak and the vulnerable from the benefits of gainful employment, human dignity, and adequate living standards. The market system can leave the poor literally to starve or to die from being unable to afford health care. No naive blindness to market principles can justify a market economy in which the rich luxuriate in great wealth while the poor struggle for their very survival. Yet such is the grim reality of our rich world economy today. Too many business leaders either stand by idly or even contribute directly to the sin of extreme inequality and neglect of the poor and vulnerable.

In short, the business sector is a noble vocation when it operates in a moral framework. When businesses ensure that they are adding true value to society and meeting the needs of the poor, their vocation and business activities raise living standards, help society to meet basic needs, and innovate to enrich our lives. Yet when businesses exploit their market power, degrade the environment, engage in financial fraud, or criminalize their activities, such as through forced labor, human trafficking, tax evasion, and corruption, they degrade the community and the noble vocation of business itself. They do the same when they commit idolatry of the market even as the poor suffer and die within their midst.

In this context, it must be frankly acknowledged that some elements of the institutional framework in which businesses operate today—especially the cynical and sometimes criminal pursuit of profits at all costs, which causes dire harm to others, and relentless lobbying by business for special narrow interests at a cost to society—thwart human dignity and the common good. Pope John Paul II, in the wake of Pope Paul VI in *Populorum progressio*, defined these destructive factors as "structures of sin." These structures are harmful not only because of their distorted purposes and activities but also because of their corroding effect on other social institutions that need to operate within their sphere.

These are not idle warnings. The business sector has not only achieved great heights in reducing poverty in recent times, but it has also plumbed great depths by contributing directly to global warming and other environmental destruction, and to the recent global financial crisis and the epidemic of human trafficking and modern slavery.

A mere one hundred large companies account for around 70 percent of all greenhouse gas emissions since 1988, the fossil-fuel companies chief among them.[2] These corporations not only impose negative externalities on society and know it, but, even worse, some attempt to hide or blur the truth and lobby aggressively to prevent regulations to end their negative externalities. Some multinational companies also have a dismal record of degrading the environment of developing countries in order to extract resources and then fight aggressively against legal actions for restitution—such corporate leaders destroy natural capital and social capital and undermine the well-being of future generations for short-term profits and short-term bonuses. This is organized irresponsibility—in religious language, it is sin.

Similarly, in the case of the 2008 financial crisis, which was in fact a "crisis of ethics," financial decision-makers engaged in rampant fraud, insider trading, and other illegal practices. Ethical lapses also included reckless risk-taking, which was facilitated by weakened financial regulation and supervision. The expectation that those responsible would be bailed out by governments in the event of losses allowed for a privatization of benefits and socialization of losses.

Many multinational companies have exploited globalization and its weak institutional framework to avoid and evade taxes and to demand weak regulatory standards from host governments. Too many degrade human dignity across supply chains by denying workers just wages and by subjecting them to inhumane work conditions—most notoriously and shamefully through new forms of slavery, including forced labor and human trafficking.

All too often, multinational companies are willing partners in widespread corruption—on the other side of every bribe taken is a bribe given. Systemic corruption degrades social capital, undermines the common good, and can lead to civil strife and conflict. And in many countries, corporations lobby extensively to make sure that their own narrow interests take precedence over the common good.

We should not depict business today with an overly broad brush, either positive or negative. We salute the millions of honest businesspeople around the world who manage their businesses according to honest, legal, and moral precepts. We applaud the more than twelve thousand companies that have joined the UN Global Compact, committing businesses to sustainable development and honest business practices. Yet we must also bemoan business-led environmental destruction, financial fraud, market monopolization, human trafficking, modern slavery, tax evasion, and reckless financial speculation, all of which are deeply embedded in today's market economy. Without a moral framework guiding business, the recent gains in productivity and profits will prove fleeting in the face of poverty amid plenty, financial destabilization, social unrest, and pervasive environmental destruction.

In terms of fixing these dysfunctions, a first step is to take aim at the flawed assumption that the sole goal of the business enterprise is to maximize profits and shareholder wealth no matter the costs imposed on others. Profit based on causing harm to others and to nature is utterly unacceptable. Businesspeople who pursue such activities may gain short-term profits but cause long-term disasters for their businesses and society. As Jesus asked, "What good is it to gain the whole world yet forfeit your soul?"

As a matter of priority, Ethics in Action calls for a reorientation of business activity around the common good. This involves a two-fold strategy. The first dimension is the personal spiritual transformation of businesspeople, especially business leaders, predicated on change from within, through love in action. The individual human being, irrespective of institutional setting, must be personally accountable as a moral actor. Senior management has a particular responsibility to define and demonstrate the right values and to set "moral enrichment targets" within the corporation to promote and reward virtuous behavior.

Yet as Pope Francis notes, "self-improvement on the part of individuals will not by itself remedy the extremely complex situation facing our world today" (*Laudato si'*, 219). The pursuit of sustainable development requires a reform of the societal frame, creating greater harmony among business success, environmental sustainability, and social fairness. This effort also requires institutional reforms to dismantle the embedded structures of sin, including the criminalization of business and the relentless pursuit of profits at all costs, including harm to others. Businesses must aim for true social value rather than selfish profits at the expense of others. Governments must regulate against environmental destruction, tax evasion, corruption, and human trafficking.

Business ethics can never be optional, and corporate social responsibility can never be an add-on, mere window dressing, or worse—a cynical exercise in public relations. For such change to take root, it must begin with moral education at different levels—for children, in universities, and among CEOs and managers. A particular challenge lies in changing the curriculum of economics programs and business schools toward a healthier and more realistic vision of human nature, and one that aims at the common good.

Business leaders should take a professional oath, similar to the Hippocratic oath taken by medical practitioners. This must start with "first, do no harm" (*primum non nocere*), but it must also mirror the other elements: "second, act cautiously" (*secundum cavere*), and third, "heal" (*tertium sanare*). Turning to the duties of businesses, Ethics in Action proposes the following "Ten Corporate Commandments."

General Recommendations: The Ten Corporate Commandments

(1) Produce Goods and Services Not Only for Markets but Also for the Common Good

Produce goods and services that facilitate human flourishing and dignity rather than "bads" that cause addictions, diseases, environmental degradation, and other harms. Meet real human needs rather than false needs created by advertising.

(2) Promote Sustainable Development

Be aligned with the SDGs, which are the globally agreed goals to be reached between now and 2030. Invest in sustainable development solutions, which, as Pope Francis notes, may also prove to be highly profitable (*Laudato si'*, 191).

(3) Extend Responsibility and Accountability to All Stakeholders

Adopt a fiduciary duty not only to shareholders but also to stakeholders including workers, suppliers, customers, the environment, and society at large. Do not seek profits from harm to others; do not lobby against the common good; and support countervailing institutions such as unions and civic organizations.

(4) Act in Accordance with the Universal Destination of Goods

Private property rights are not inviolate; they must accord with the common good and dignity for all. Firms should defend human dignity, for example by prioritizing decent wages and work conditions over extra profits, and by employing the marginalized and excluded, including minorities, refugees, and Indigenous communities.

(5) Eliminate All Forms of Modern Slavery

Eliminate all forms of modern slavery, especially forced labor, prostitution, and organ trafficking within the company's and the sector's supply chain. Develop and disseminate innovative models, best practices, and indicators to help end these crimes. Provide jobs and training to former victims of modern slavery proportionate to workforce size.

(6) Ensure Environmental Sustainability

Embed environmental sustainability in core business models. Accept that the main cause of climate change is human activity based on the use of fossil fuels (*Laudato si'*, 23), and pay the full social cost of using shared environmental resources, in line with Pope Francis's teaching that environmental sustainability is a necessary condition for economic activity to be considered ethical (*Laudato si'*, 195). Commit to becoming carbon neutral by using clean energy sources and to eliminating all forms of environmental degradation, including pollution and oil spills.

(7) Link Profit to Social Benefit

Internalize the ideas that profit-making at the expense of others is illegitimate and that legitimate profits must be linked to the provision of a true social benefit. In this context, end negative externalities of the firm, and consider the allocation of a clear part of the company's profits to support the common good, especially the poor.[3]

(8) Commit to Responsible CEO Compensation

Set compensation for CEOs and senior executives based on factors such as avoiding social harms, respecting the law, ending self-seeking lobbying, and contribution by the company to the SDGs. Keep a reasonable balance of the pay of the CEO and the workers in the firm, and publish the income differentials between management and workers.

(9) Do Not Seek or Exploit Monopoly Power

Do not seek to create or exploit monopoly power and monopoly rents, accept and cultivate a healthy balance between corporate competition and collaboration, and support antitrust efforts at the national and international levels to ensure a competitive environment in which market power is limited.

(10) Do Not Bribe, Evade Taxes, or Commit Financial Fraud

Do not avoid or evade taxes, take bribes, offer bribes, or engage in any kind of financial corruption or malfeasance. Support the elimination of tax havens all across the world, and promote greater transparency, stricter reporting standards, and stronger criminal penalties for corporate managers and board members in cases of corruption and criminal fraud.

Ethics in Action Working Group Action Items

Ethics in Action pledges to disseminate these recommendations among its networks and to engage in partnerships with stakeholders to further these goals. In particular, it pledges to work with the European Union in supporting corporate ethics and good governance.

Notes

1. Pontifical Council for Justice and Peace, *Vocation of the Business Leader: A Reflection* (Vatican City: Dicastery for Promoting Integral Human Development, 2018), https://www.humandevelopment.va/en/risorse/archivio/economia-e-finanzia/la-vocazione-del-leader-d-impresa-una-riflessione.html.
2. Paul Griffin, *The Carbon Majors Database: CDP Carbon Majors Report 2017* (London: CDP, 2017), https://b8f65cb373b1b7b15feb-c70d8ead6ced550b4d987d7c03fcdd1d.ssl.cf3.rackcdn.com/cms/reports/documents/000/002/327/original/Carbon-Majors-Report-2017.pdf?1499691240.

3. "Give alms from your possessions. Do not turn your face away from any of the poor, and God's face will not be turned away from you" (Tobit 4:7). For example, the economy of communion asks businesses to divide profits in three ways: reinvesting in the business, giving to those in need, and funding infrastructure to promote a culture of giving and reciprocity. See Luigino Bruni and Tibor Héjj, "The Economy of Communion," in *Handbook of Spirituality and Business*, ed. Luk Bouckaert and Laszlo Zsolnai (London: Palgrave Macmillan, 2011): 378–86.

CHAPTER 23

TOWARD A *LAUDATO SI'* COHERENT CORPORATE RESPONSIBILITY MANAGEMENT

KLAUS M. LEISINGER

Successful entrepreneurial engagement and the resulting business activities in the enterprise and supply chain are the most important drivers of economic development.[1] There are no "good" companies, and there are no "bad" companies—the ethical quality of a corporation's business conduct is as high as its leaders want it to be. If and when business activities are pursued in a way that is compatible with the vision of *Laudato si'*, business will also be an important force for good, as it is both responsible and sustainable: responsible in the sense of *responding* to the questions posed by the environmental and social state of the world, and *sustainable* in the sense of not diminishing the ability of future generations to live their lives with freedom and dignity by making peace, justice, and the preservation of creation integral elements of the new business as usual.

A *Laudato si'* Coherent Corporate Responsibility Architecture

Business leaders can support a *Laudato si'* coherent human development by building up a corresponding corporate responsibility architecture. A first necessary step is for corporate top decision-makers to familiarize themselves with the explicit content and implicit values expressed in the Church's social teaching. A next step is to evaluate ways and means to operationalize such values—and then use them as guideposts for developing the corporate strategy and business model. Value-based management is key to a successful business enterprise competing with integrity. Most of the values woven into the message of the Church's social teaching are also part of management approaches that incorporate enlightened values.

The management guru Peter Drucker many years ago pointed out the central fact that "business enterprises . . . are organs of society. They do not exist for their own sake, but to fulfill a specific social purpose and to satisfy a specific need of a society, a community, or individuals. They are not ends in themselves, but means. The right question to ask in respect to them is . . . What are they supposed to be doing and what are their tasks?"[2] Drucker expected business nearly fifty years ago "to make fulfillment of basic social values, beliefs, and purposes a major objective of their continuing normal activities rather than a social responsibility that restrains or that lies outside of their normal main functions."[3] Today, enlightened business leaders consider it more important to ask *why* and *what for* than the traditional *what* and *how*.[4]

The way to translate values into daily business practices starts with reflection and a commitment to values and continues with an analysis of the appropriateness of existing corporate guidelines and codes of conduct, as well as their amendment in light of the chosen values and the available knowledge. Compliance monitoring and auditing of adherence to this additional internal legislation are to be conducted with the same degree of rigor as would be done with external law.

The new spirit is also represented in corporate target setting and the definition of key performance indicators. The design of performance appraisals and promotion criteria reflect the values defined by top management and are adjusted if and when new relevant knowledge becomes available. Moral enrichment targets and performance appraisals create a level playing field internally: Employees doing the right thing for the right reasons can do so while also pursuing their self-interest. Moral heroism, putting people who prioritize values at a competitive disadvantage, is no longer necessary.

If and when a business strategy compatible with *Laudato si'* becomes effective and efficient, it will still emphasize the conventional responsibilities of good companies—but in a manner compatible with the UN Global Compact:

- *Economic responsibility*, measured in terms of value added in areas such as core competence, securing financial liquidity, cost management, profit margins, market shares, customer satisfaction, and innovation—but not in isolation from other areas of responsibility.
- *Social responsibility*, measured by fair labor norms, fair living wages, basic health care, old-age insurance, employability, and opportunities for life-long learning, as well as by ensuring none of the following: child labor, forced labor, and discrimination (e.g., based on gender, race, or age); all of these factors should also be ensured in the supply chain. Social responsibility also

includes the just distribution of wealth, described by the Pontifical Council for Justice and Peace as follows: "As creators of wealth and prosperity, businesses and their leaders must find ways to make a just distribution of this wealth to employees (following the principle of the right to a just wage), customers (just prices), owners (just returns), suppliers (just prices), and the community (just tax payments)."[5]

- *Environmental responsibility*, measured in terms of such factors as integrated environmental management, minimizing emissions and waste, minimizing the use of nonrenewable resources, encouraging sustainable mobility, and encouraging environmentally friendly innovation; these considerations should also be made in the supply chain.
- *Human rights responsibility*, measured through the implementation of the UN Guiding Principles on Business and Human Rights, which include due diligence.
- *Anticorruption responsibility*, measured by efforts to work against corruption in all its forms, including extortion and bribery.

A sustainable corporate responsibility architecture is not something one can download from a website and transplant into a company. Companies in different sectors have different risk and benefit profiles, and their states of profitability allow for varying degrees of freedom to make constructive investments beyond what is legally demanded. For any responsibility architecture to be effective and efficient, it must be tailor-made and "owned" by all employees at all levels of the corporate hierarchy. Open dialogue with employees and invitations for proposals help to establish each company's unique architecture. Ultimately, to enable a sustainable corporate responsibility architecture, special kinds of leadership personalities are needed.

Enlightened Leadership Personalities, Not Functional Management Experts

Human beings, not legal institutions, make decisions and have ethical considerations. Respective responsibility must be attributed to and demanded from individuals at all levels of the corporate hierarchy, not only top management. Employees at all levels possess professional knowledge, work experience, and social competence, and all have moral talent that can be developed into moral competence. All have something to contribute. Anyone who wants to exert moral influence can do so and is therefore in a position to raise the ethical quality of group decisions.[6]

Many decisions within a corporation having an appropriate responsibility architecture do not require ethical reflection; however, decisions that pose complex dilemmas do. By virtue of their executive power, members of top management have the decisive influence on the value structure and moral culture of the company. It is the nondelegable responsibility of leaders to ensure the following[7]:

- The *corporate purpose and mission* are defined appropriately and comprise more than the economic aspects of the business enterprise. Determining the corporate purpose is the single most important task of leadership: "A justifiable answer to the question of the why, the goal and aim of a task, and also the legitimization of a business model, an entire company and even the structure of a market economy is one of the biggest challenges of leadership. An aspiration to provide leadership without providing a convincing answer to the question of the contribution to a greater whole runs the risk of being implausible and arbitrary."[8]
- *Normatively enriched corporate guidelines and codes of conduct* are developed, taught, and disseminated so that it is clear beyond any doubt what values employees are expected to live and work by, with compliance monitored as if the guidelines and codes of conduct were a legal prescription.
- *Target setting, the content of performance appraisals, and promotion criteria* cover not only financial and technical criteria but also criteria reflecting aspects of a humanistic ethic.
- The *definition of the corporate sphere of influence* within which corporate responsibility rules are valid and compliance monitored comprises the relevant elements of outsourced activities.
- The *dilemmas faced by leaders are articulated*, and the values hierarchy within which dilemmas are resolved is transparent.
- *Values are visibly lived by, setting an example for all employees.* Doing so, leaders act as role models and demonstrate that the values of the corporation are nonnegotiable. A key aspect of leadership positions is to make a targeted influence on others. In the words of Albert Schweitzer, "Example is not the main thing in influencing others. It is the only thing."

Business people competing with integrity do not hide behind insufficient law. Good leaders do not support lobbying for weaker law in areas that affect the common good and human dignity. Leaders managing with integrity base their decisions on human values that have been recognized by all religions and cultures across time.[9]

The Ethics Component of the Responsibility Management Process

The reflection of a given status quo is a complex process and must be properly prepared and executed. Ethics, as defined here, is the evaluation of *actual* human behavior and conduct in light of what they *ought to be* if known universal religious values and principles were followed. Translated into the present context, a corporate responsibility process compatible with *Laudato si'* ought to evaluate corporate conduct in light of the values and principles expressed by *Laudato si'* and other relevant wisdom developed by the Church's social teaching. Within this context, answers to the following questions must be found:

- In light of the normative statements and requests of *Laudato si'* and other documents of the Church's social teaching, what are the main weaknesses and vulnerabilities of the current corporate status quo?
- Which of today's weaknesses and vulnerabilities result from corporations having a societal responsibility different from those of religious communities? Which of these weaknesses and vulnerabilities can be "repaired" or "tamed" by investing additional resources in training, education, integral human development, and technology? Which are not reparable and consequently must be given up?
- How do we define the corporate sphere of influence within which we feel accountable for human rights–related social and environmental standards compatible with the spirit of *Laudato si'*, and where do we draw the limits?
- How do we proceed if the requests of *Laudato si'* and the expectations of civil society conflict with those of the financial community? What exactly does it mean when we say that the economic subsystem has to serve society and not the other way around?

Other questions must also be addressed, such as those included in the Discernment Checklist for the Business Leader, compiled by the Pontifical Council for Justice and Peace.[10]

Enlightened leaders embed their performance values in a net of other values such as social values (e.g., fair labor norms, nondiscrimination, employability), cooperation values (e.g., reliability, flexibility, willingness to collaborate), communication values (e.g., accuracy, materiality, completeness, timeliness), and sustainability values (e.g., integrating the seventeen SDGs and their 169 targets).

Legality is a nonnegotiable must but represents only the ethical minimum. A reliance on law alone triggers legalistic, compliance-based attitudes, and this creates risks: If and when the quality of national law is inadequate, and where applicable law does not (yet) represent state-of-the-art knowledge, an exclusively legal corporate conduct might be illegitimate—if not irresponsible. To live up to the recommendations of *Laudato si'*, business leaders choose *legitimacy* as their base of reference: doing the right thing because it is the right thing to do.

Such requests represent not only the view of the Church's social teaching but also those of established management gurus. According to Peter Drucker, "leadership groups have to take social responsibility, have to think through the values, the beliefs, the commitments of their society, and have to assume leadership responsibility beyond the discharge of the specific and limited mission of their institutions."[11] Business exists for the sake of society; therefore, "free enterprise cannot be justified as being good for business. It can be justified only as being good for society."[12]

For sustainable development to be realized, business must be part of the solution. There are no "good" companies, and there are no "bad" companies—the ethical quality of a corporation's business conduct is as high as its leaders want it to be. Good leaders do not hide behind insufficient law.[13] Corporate leaders with integrity are characterized by values such as wisdom, justice, fortitude, and temperance; leaders who cultivate a reverence for life, mindfulness, and respect are likely to more successfully support sustainable development in the long run.[14]

As the modern business world is characterized by volatility, uncertainty, complexity, and ambiguity (VUCA), corporate leaders' value orientation and moral imagination make a decisive difference. It would, however, be naïve to accept that the moral will of top business leaders alone is sufficient for doing the right thing, as their capacity to act under the structurally different rules of the societies that constitute their markets is to a certain extent limited. This is why it is important to reiterate that *Laudato si'* and the 2030 Agenda are blueprints for comprehensive societal reforms—business must be included but not singled out.

Notes

1. Klaus M. Leisinger is the founder and president of the Foundation Global Values Alliance (https://www.globalvaluesalliance.ch). He is also a professor of sociology at the University of Basel, covering business ethics, corporate responsibility, and sustainable development topics, and serves United Nations institutions as a senior advisor. Between 1983 and 2013, he headed the not-for-profit foundations for development of Ciba and Novartis. This chapter is a shortened version of a longer essay on the subject; see Klaus M. Leisinger, *Corporate Responsibility as if Christian Values Mattered: The Role of Business Leaders in a Global Reform Process in the Spirit of* Laudato si' (Basel: Foundation Global Values Alliance, 2017), http://www.globalewerteallianz.ch/wp-content/uploads/2017-08-01-Klaus-Leisinger-Corporate-Responsability-as-if-Christian-Values-mattered.pdf.
2. Peter F. Drucker, *Management: Tasks, Responsibilities, Practices* (New York: Harper and Row, 1973): 32.
3. Drucker, *Management*, 30.
4. Leipzig Graduate School of Management, *The Leipzig Leadership Model* (Leipzig: HHL, 2016).
5. Pontifical Council for Justice and Peace, *Vocation of the Business Leader: A Reflection* (Vatican City: Pontifical Council for Justice and Peace, 2012), sec. 55, http://www.iustitiaetpax.va/content/dam/giustiziaepace/VBL/Vocation,%202nd%20ed,%20ENG.pdf.

6. B. Z. Posner, "Individuals' Moral Judgment and Its Impact on Group Processes," *International Journal of Management* 3, no. 2 (1986): 5–11.
7. Klaus M. Leisinger, "Business Ethics by Manager Ethics," in *Manifesto for a Global Economic Ethic: Consequences and Challenges for Global Businesses*, ed. Hans Küng, Klaus M. Leisinger, and Josef Wieland (Munich: dtv, 2010).
8. Leipzig Graduate School of Management, *Leipzig Leadership*, 76.
9. See "The Global Ethic Project," Global Ethic Foundation, accessed April 26, 2021, https://www.global-ethic.org/the-global-ethic-project/; and Hans Küng, *Project Weltethos* (Munich: Piper Global Responsibility, 1990), published in English as *Global Responsibility: In Search of a New World Ethic* (New York: Crossroad, 1991).
10. Pontifical Council for Justice and Peace, *Vocation of the Business Leader*.
11. Drucker, *Management*, 18.
12. Drucker, *Management*, 18.
13. See "The Global Ethic Project"; and Küng, *Project Weltethos*.
14. Klaus M. Leisinger, *Die Kuns der verantwortungsvollen Führung* (Bern: Haupt, 2018), published in English as *The Art of Leading* (Minneapolis: CRT, 2020).

CHAPTER 24

SUSTAINABLE INVESTMENT AND ETHICS IN ACTION

KERRY KENNEDY

Sustainable investing is a way to create long-term wealth while supporting the Sustainable Development Goals. It is not an exaggeration to assert that the SDGs will not be achieved unless the entire investing superstructure is modified to become sustainable.

More and more, people are demanding that the investment world better align with their values. One in three millennials seek brands making a positive impact on the world, and 90 percent of consumers would switch to sustainable brands given equal price and quality. This results in four times global growth for brands consumers perceive as sustainable.

Increasingly, investment funds are being directed toward sustainable investments. For example, the Ford Foundation recently announced that it is investing US$1 billion in sustainable enterprises. APG, the largest sovereign wealth fund on Earth, has announced that 100 percent of its assets will be invested in sustainable companies. And Bain recently reported that 10 percent of European investors incorporate environmental, social, and governance (ESG) criteria into their financial analysis, while 60 percent consider ESG factors in helping to manage risk.

Institutional investors are finding that sustainable, or SDG-aligned, investing can increase their risk-adjusted returns. As detailed in the strategy reports of the largest and most successful investment firms, there are many reasons investors are adopting sustainable investing. As Cara Kennedy-Cuomo puts it,

> A mixture of ethical, commercial, strategic, and policy-related factors all motivate investors to tailor their investments towards the SDGs. The SDGs can

help investors align their investments with their stakeholders: a shifting market of beneficiaries and consumers, and internal leadership among fund managers and trustees. Investors are also using the SDGs to align with predicted shifts in government policy, including taxes and regulations. Furthermore, some development finance institutions (DFIs) and national governments are employing blended finance and creating low-cost vehicles for investors to contribute to SDG-aligned development projects, providing a financial incentive for SDG-investing through de-risking investments with a first loss facility.[1]

This brings us to human rights investing. This is important not only because it's the morally right thing to do but also because human rights violations are real perils that impinge on investment risk management.

At Robert F. Kennedy Human Rights, we convene about 150 people twice a year who collectively control $7 trillion in managed investments. The worldwide economy is about $75 trillion, so this is a small group with enormous power to influence all of our lives. They are the chief information officers of sovereign wealth funds, pension systems, university endowments, and large family offices, along with a dozen or more private equity companies. We talk about the most effective tools currently available in the sustainable investment arena, specifically in three tranches:

Supply chain issues

Women and underrepresented minorities as investors, in management and on boards of directors

In the tech industry, how to protect privacy, free expression, and access to information and how to address false and misleading information

Sustainable investing is commonly thought of as investing in a way that takes into consideration the three ESG risk factors. "E" is a company's impact on the environment, "S" is the social impact of the company's policies up and down the supply chain and across all its stakeholders, and "G" is the rigor of its governance mechanisms—whether it has strong systems and is transparent about how those systems operate.

A lot of work has been done on the E and G of sustainable investing. Representing a human rights organization, we are concerned more specifically about the S.

As an example of what we are up against, in 2009, the United States government issued sixty-four safety violations based on the precarious ventilation system at Massey Energy's Upper Big Branch coal mine. A year later, the ventilation system failed, and the mine collapsed. The E (environmental damage) was horrific—the water table was so damaged that it will take generations to recover. The S (social impact), was even more horrific—twenty-nine miners lost their lives. And the G (governance) was causal—the official investigation into the disaster revealed that, to cut costs, Massey had undermined safety procedures at the mine and used its considerable power to corrupt the political system.

The failure of ESG factors in this example led to shareholders enduring more than $10 billion in litigation and impaired the retirement savings of millions of police officers, firefighters, and teachers whose pensions had been invested in Massey.

For investors, the Massey disaster is both a question of social values and investment protocols. Investors who failed to take account of Massey's worker safety practices and transparency issues missed what turned out to be a critical element of risk management when they considered risk-adjusted returns.

The lesson is clear: Investors must look beyond traditional financial algorithms and into the reality on the ground if they are to protect their assets. That is what sustainable investing is all about.

Human rights investing is not a recent concept. In the 1700s, based on the theory of "do no harm," the Quakers divested from companies involved in the slave trade. At the height of the Vietnam War, Nick Ut's photo of a young woman running naked toward the camera, her body seared by napalm, spurred a call for divestment from the chemical manufacturer Dow and other companies profiting from the conflict in Southeast Asia.

"Socially responsible" investing perhaps reached its apotheosis with the anti-apartheid movement in South Africa. That was an extraordinarily important moment, when the people of the world came together and heeded the call for support from Archbishop Desmond Tutu, Nelson Mandela, and other heroes of the anti-apartheid movement. Across the world, students flooded administrative offices, and citizens marched on town squares. They demanded that their governments and their corporations divest holdings that were propping up the White supremacist rulers in Johannesburg. That movement exposed the apartheid regime and firmly established it as a pariah among nations.

The negative flow of investment inspired a business coalition, representing 75 percent of employers in South Africa, to call for an end to apartheid. It hobbled the economy, brought the government to its knees, and helped bring democracy to the country. Moreover, it established socially responsible investing as a legitimate alternative to business as usual.

What happened in South Africa holds lessons for investors everywhere.

Today, sustainability takes center stage. Corporations and investors must ask themselves a fundamental question: What is the impact of environmental degradation, human rights violations, or corruption on investment outcomes for shareholders?

The answer is simple. You might get away with abuse for a few days, or months, or maybe even years, but if you make an investment and fail to consider these issues, you are investing in a ticking time bomb, and, at some point, that bomb is going to explode.

It exploded in Bangladesh in 2013 in a garment factory collapse that resulted in more than 1,300 deaths. Customers around the world were thus forced to take an unvarnished look at where their Mango, Auchan, and Benetton cardigans were made—and the human cost paid by people, often entire families, risking their lives for $40 a month.

It exploded for Nike in 2001, when the company admitted to knowingly and actively employing children as young as ten years old in life-threatening factory conditions in Pakistan and Cambodia.

It exploded in Italy with Parmalat, when fraud and irresponsible corporate behavior led to Europe's largest-ever bankruptcy, and thousands of people lost their jobs or portions of their retirement accounts.

It exploded for gun manufacturers across the United States when the teachers' union in Newtown, Connecticut, learned that its pension fund was invested in the same company that sold the military assault rifles that killed twenty of their six- and seven-year-old students and six of their fellow teachers in the December 2012 mass shooting at Sandy Hook Elementary School. Cerberus, the owner of the company, announced they were selling their holdings in the gun manufacturer weeks later.

It exploded for WorldCom, Enron, and BP. And it exploded for Volkswagen, the world's biggest automaker, when executives admitted to rigging diesel emissions tests in America and Europe. Credit Suisse estimates that the total cost to the company could hit $87 billion.

As each of these disasters shows us, we can no longer ignore the link between human rights and investment practices. This is especially urgent when we consider that, of the hundred largest economies in the world, seventy are no longer countries—they are corporations. As such, corporations are under growing scrutiny. As corporations gain power, they are increasingly being held to the same standards for protecting rights to which governments are currently held.

This shift is a monumental one, and it will change the fundamental structure of our capital markets going forward. As pensioners become savvy about how their hard-earned money is used, they will no longer tolerate investments against their interests. For example, United Auto Workers pensioners demanded an end to its pension fund investing in Kia, a nonunion automobile manufacturer whose cars were in direct competition with those made by the pensioners. As foundations become savvier about their investments, they are beginning to pay less attention to the 5 percent of their wealth committed to programming and more attention to the 95 percent of their wealth that is invested, asking, Do these investments enhance our programming goals, undermine those goals, or are they neutral with respect to those goals? Universities, states, and nations are asking similar questions.

While one-off progress can be made, it has become clear that none of the SDGs will be achieved until all of them are achieved. We will not end violence against women until there is universal education for girls and equal pay for equal work. Equally, so long as women and underrepresented minority-owned firms control less than 2 percent of the $75 trillion investment industry, the SDGs will remain a dream deferred.

We can no longer surrender community excellence and community values to the mere accumulation of material things, as Robert Kennedy said in a speech at the University of Kansas in 1968. The solutions are within our reach. The time to act is now.

Note

1. John Morris, Jamil Wyne, Cara Kennedy-Cuomo, and Michael Cosack. *Institutional Intentionality: Aligning Institutional Investors with the Global SDG Agenda* (Princeton, NJ: 17 Asset Management, April 2019).

CHAPTER 25

THE CASE FOR BUSINESS IN ACHIEVING THE SDGs

JACQUELINE CORBELLI

Over the past 15 years, business has adopted a greater sense of social responsibility, with increasing emphasis on global economic development. This is happening despite the broad range of irresponsible actions that have produced tremendous waste and dangerous imbalances in the world—in terms of human dignity, health, and wealth disparity. Especially in light of the existential threat we face as humans on planet Earth, it has never been a more crucial time for business to be a major source of hope and a key driver and engine of change.

And there are, indeed, important reasons for hope. Consider the notable business contributions made during from 2005 to 2015 under the global umbrella of the United Nations' Millennium Development Goals. The demonstrated commitment during that time frame suggests a potential for business to act as a catalytic and scalable platform for the type of transformative change that is required to face the daunting challenges that weigh increasingly upon us.

By 2015, the MDGs became a galvanizing force for businesses to elevate their social responsibility, thereby setting the stage for the UN's even broader and more comprehensive goals, the Sustainable Development Goals. The SDGs represent an opportunity for businesses to build on the success of the MDGs by applying an integrated framework for implementing industry-wide and cross-industry innovations specifically designed to reach these goals.

The MDGs brought different levels of commitment to address extreme poverty. Global support for these goals stretched across governments, nongovernmental organizations (NGOs), and indeed all stakeholders—spurring a unique energy and motivation for business to act. Supporting the goals created business opportunities for powerful win-wins that not only contributed to a global good but also unlocked new investments for development more broadly. Major companies implemented broad-based initiatives that drove major progress on the MDGs.[1]

This is an important backdrop to the business case for the SDGs. Many of the initiatives implemented in the ten-year MDG time frame stimulated important changes in the investment and operational practices of businesses, which, since 2015, have been refocused on implementing sustainability measures and rebuilding global supply chains. The stage has been set for business to contribute to even greater good.

There is also tangible evidence of the potential for building on these changes. Recent research conducted by the U.S. Coalition on Sustainability sheds an important light on the commitment to the SDGs demonstrated by the largest global corporations. The study, completed in March 2020, assessed the current state of business commitment to sustainability among U.S. Fortune 100 corporations. A core finding is that accomplishing the SDGs will require sweeping and fundamental change in the way businesses operate. Additionally, change of a sufficient magnitude to achieve the goals will require the active participation of the world's most socially responsible business leaders, who can have a pronounced impact on their own *and* affect the scale and speed of necessary cross-industry change.

The results of the study paint a hopeful, if mixed, picture. Among Fortune 100 companies:

- Sixty percent of the largest U.S. businesses support sustainability as a goal.
- Twenty-four percent have sustainability efforts underway that specifically align with the SDGs.
- Forty-four percent of companies pursuing sustainability initiatives are also measuring their progress, and 36 percent have publicly reported on outcomes of these efforts.

In sum, the alignment of companies around sustainability is in full swing, and those efforts represent a meaningful starting point for pursuing focused, measured, scalable outcomes.

Without question, achieving the SDGs will require fundamental change on a global scale. In this vein, businesses have the unique capacity to act on that scale by making comprehensive changes to the design and management of business supply chains. Succeeding will take a special kind of cooperation and collaboration among companies and across industries. This does not require completely uprooting the profit motive, but it does require better aligning it with social purpose.

Looking ahead, we need to build on existing businesses' commitments, integrate and accelerate current efforts, and hold business accountable on the alignment of their goals with the SDGs. This in turn requires a disciplined process of applying principles of transparency in terms of which businesses lead and which lag, why this is the case, and spotting and managing opportunities for making leaps toward deeper and more extensive change. Transparency will give businesses a chance to put their best foot forward. Accountability is key to ensuring that actions taken contribute to the achievement of the SDGs.

There is much room for growth here. The level of commitment and action taken by businesses still misses out on their massive potential in terms of how they can help achieve the SDGs.[2] There is much room for improvement on transforming longer-term commitments into actions. Yet during the time frame of the MDGs, businesses showed their capacity to bend the curve on progress.[3] They can do the same for the SDGs. And they should be supported and facilitated in doing so.

There is, then, great reason to hope that business will persist in its commitment to the social good. As Klaus Leisinger has said, there are virtuous individuals in all organizations, and that bell curve also exists in business. If that virtue can be harnessed, it offers a source of operational leverage that is essential to achieving the SDGs.

Looking ahead, the key to success is a more integral, holistic, and productive dialogue among governments, NGOs, and businesses, combined with a shift in the focus of leadership within the business community in terms of what can be accomplished both individually and collectively. Through targeted actions that include improved cooperation between the public and private sectors, better communication and sharing among corporations within and between industries; improved collaboration among small, medium, and large businesses; and targeted education initiatives, businesses can work hand and hand with other stakeholders to achieve the SDGs. As the public benefits from improved social protections, infrastructure, and environmental conditions, so will businesses benefit down the road.

Indeed, leading companies are in the process of mapping corporate social responsibility, sustainability, and supply chain initiatives to the SDGs. CEOs, investors, and other business leaders are seeking out opportunities to align their activities and targeted outcomes with these goals. Of course, delivering on sustainable development requires more than good intentions, and a significant number of practical realities need to be addressed. For one, businesses must commit to embedding sustainability into the very core of their models and business and decision processes, beyond the executive level. In addition, there is a need for improved business practices, in particular on evolving corporate culture, which itself can take years to accomplish. But more than anything, success will require an unprecedented degree of coordinated and concerted action. Platform solutions such as SustainChain, launched by the U.S. Coalition on Sustainability in 2020, are using the latest in digital technologies, machine learning, and data science to support the private sector with these changes.

To conclude, we learned a valuable lesson with the MDGs: that addressing global challenges cannot and will not come solely from the public sector. There is a specific and fundamental role with immense promise for the private sector—business—that will be paramount to achieving the SDGs on a global scale. At the same time, businesses should not lose out on the opportunity to be a pivotal part of the critical changes coming to our increasingly global world. Next-level positive change and progress are achievable, and business has a crucial role to play.

Notes

1. "Sustainable Sourcing for Suppliers," Unilever, accessed March 14, 2018, https://www.unilever.com/about/suppliers-centre/sustainable-sourcing-suppliers/; "Transforming Healthcare with 5G—Ericsson," Ericsson, accessed March 14, 2018, https://www.ericsson.com/en/cases/2016/5gtuscany/transforming-healthcare-with-5g.
2. United Nations Global Compact, *2017 United Nations Global Compact Progress Report: Business Solutions to Sustainable Development* (New York: United Nations, 2017).
3. Elizabeth Stuart, "How Business Can Help Deliver the Global Goals," Business Commission, April 2016, http://businesscommission.org/our-work/our-work-how-businesses-will-determine-the-success-or-failure-of-the-sustainable-development-goalshow-businesses-will-determine-the-success-or-failure-of-the-sustainable-development-goals.

PART VII

AN ETHICAL CONSENSUS ON SUSTAINABLE DEVELOPMENT

Education

DECLARATION OF THE ETHICS IN ACTION MEETING ON EDUCATION

Education is both a fundamental need and right of all children. Because so many children around the world lack even a basic education, the world community bears an urgent moral responsibility to ensure quality education for all children. We ardently support SDG 4 and underscore its insistence on education *for all*, including the poorest, the marginalized, girls as well as boys, the vulnerable, and those with special needs. More than half a century ago, the Church expressed its deep commitment to education for all in *Gravissimum educationis*.[1] Today, *Laudato si'* devotes its chapter 6 to educating for the covenant between humanity and the environment.

Justice demands that new efforts be preferentially dedicated to those at the bottom of the income distribution so that even the poorest children have the opportunity to reach their full potential through quality education.

We also need a new kind of education to promote integral and sustainable human development. As Pope Francis writes in *Laudato si'*, "There needs to be a distinctive way of looking at things, a way of thinking, policies, an educational programme, a lifestyle and a spirituality which together generate resistance to the assault of the technocratic paradigm."

For the low-income countries, where hundreds of millions of young people currently lack access to education, increased international funding for SDG 4 is a sine qua non. Since Africa lies at the center of the funding challenge, we will do our utmost to support African leaders' plans to establish a special fund for African secondary education to help channel donor financing to African primary and secondary education, with full accountability in the use of these funds.

To achieve quality education for all children, the incremental international funds needed are very modest, roughly US$40 billion per year, or 0.1 percent of the national income of the high-income countries. Yet even such modest flows can do much to ensure a more peaceful, harmonious, and productive world. Such incremental flows are a tiny fraction of the annual incomes of the world's two thousand billionaires and a tiny fraction of the world's annual outlays on armaments.

Investing sufficiently in education may seem demanding, but the costs of ignorance are much higher. Domestically and internationally, peace is endangered by uneducated masses who are not integrated into society. Incremental funding should be raised from governments, philanthropists, religious organizations, and businesses. We support an initiative to mobilize large-scale funding from the world's high-net-worth individuals and corporations, as well as development financing through other means. We will seek support in China and in other newly leading countries in the world today.

In today's world, we need new ways to teach and to train teachers:

- To engage the young brain
- To inspire young students to be interested and open to learning
- To build solidarity, ethical reflection, and moral skills, especially by enabling children to teach children and live in harmony with one another
- To foster global citizenship, cosmopolitan responsibility, and a culture of peace
- To emphasize the goals of critical analysis and empathy to diminish the susceptibility of youth to demagoguery, fake news, and demonization of the other
- To use sport, the arts, music, and culture to inspire the young mind
- To use new technologies for multilingual education, connectivity, global literacy and numeracy, coding, and the new potentialities of instruction enabled via information and communications technology (ICT)
- To allow adults to develop new skills and continually learn

First and foremost, teaching requires on the part of teachers a high level of knowledge so that students, who learn through the process of instruction, may achieve a standard of education that they would not obtain on their own.

Education should develop the ability to observe, to reason, to synthesize, and to develop a sense of justice, respect, tolerance, and compassion for others. We need to teach universal and ecumenical values and virtues as a core purpose of education. These include empathy, fraternity, commitment to justice and human rights, and pursuit of the common good. The challenges of globalization can be met only by a cosmopolitan education, rooted in a world ethos shared between religious and nonreligious people across all cultural divides. It is from this overlapping moral and multireligious consensus, rooted in human dignity and the common good, that the formation of character and the instruction of minds ought to begin. Education must teach respect for all religions, cultures, and traditions.

We need an educational system that promotes a fair, inclusive, and sustainable world—without slavery and exclusion, where we take care of our common home, and where all have access to land, housing, work, education, and the foundations of a dignified life.

We need to teach creativity, critical thinking, independence, and social empowerment as a fundamental preparation for truly open, diverse, and democratic societies.

We need to support all families to nurture and educate their children, from early childhood through adolescence, and to empower parents to understand and pursue healthy and nurturing practices.[2]

Universities have a leading role to play in mobilizing action for the full and timely implementation of the SDGs and the Paris Agreement on climate change—in reorganizing their missions, humanistic values, curricula, and research to support sustainable development and the common good; in supporting teacher training colleges in poorer countries; and in encouraging global and regional academies of science to support such efforts.

Corporations also need to provide training and educational opportunities to foster lifelong learning, to inculcate virtue, and to support the integral development of all workers.

Recognizing that education also takes place outside the classroom, we must consider the impact of politics, large corporations, and media on democracy, which can engage or discourage citizen participation.

Ethics in Action Working Group Action Items

(1) We commit to doing our part—individually, through our respective institutions, and as part of Ethics in Action.

(2) We will explore ways to help the Congregation for Catholic Education and similar educational institutions of other religious faiths open thousands of new schools, especially secondary schools, throughout Africa and other regions in need.

(3) We will support the new initiative Move Humanity (in partnership with the UN Sustainable Development Solutions Network [SDSN]) to reach out to high-net-worth families to help close the education funding gap.

(4) The University of Notre Dame, the SDG Academy, and the Global Master's in Development Practice degree program will join forces to promote and advance a new twenty-first-century sustainable development curriculum for universities around the world, including free online instruction through the SDG Academy. The organizing committee of the World Meeting of Popular

Movements will promote these online courses among the leadership of grassroots organizations.

(5) The University of Pennsylvania, Columbia University, New York University, Yale University, the University of Notre Dame, and other U.S. universities will join forces to advance SDG 4 within the United States and will encourage other university networks around the world to do the same.

(6) Ethics in Action, Religions for Peace, the SDSN, and other interested partners will commit as a group to prepare a new course on the virtues of a sustainable society, engaging the world's leading religious and philosophical traditions. They will seek out best practices of virtue education across the world's traditions. The SDSN and the SDG Academy will support a new global online course on this topic. The Foundation for Amazon Sustainability will test this course in public schools in the Amazon on a pilot basis.

(7) The SDG Center for Africa, on behalf of Ethics in Action, will convey our enthusiastic encouragement to Africa's leaders for a bold partnership of Ethics in Action with the African Union to achieve SDG 4, including through the fund for African secondary education.

(8) SDSN Youth will organize young people around the world through its Global Schools Program to take up the challenge of education for sustainable development at the primary and secondary levels—not only as advocates but also through training, curriculum design, ICT applications, and other skills that young people will need to achieve the SDGs.

Notes

1. "Children and young people must be helped, with the aid of the latest advances in psychology and the arts and science of teaching, to develop harmoniously their physical, moral and intellectual endowments so that they may gradually acquire a mature sense of responsibility in striving endlessly to form their own lives properly and in pursuing true freedom as they surmount the vicissitudes of life with courage and constancy." Vatican II Council, "Decree on Christian Education: *Gravissimum educationis*," Vatican website, October 28, 1965, sec. 1, http://www.vatican.va/archive/hist_councils/ii_vatican_council/documents/vat-ii_decl_19651028_gravissimum-educationis_en.html.
2. As Pope Francis notes, "Today we consider the vocation of families to educate their children, to raise them in the profound human values which are the backbone of a healthy society." Francis, "General Audience, St. Peter's Square," Vatican website, May 20, 2015, http://www.vatican.va/content/francesco/en/audiences/2015/documents/papa-francesco_20150520_udienza-generale.html.

CHAPTER 26

THE CHALLENGE OF EDUCATION

JEFFREY D. SACHS

Quality education for all (SDG 4) should be regarded as the queen of the seventeen Sustainable Development Goals. Without an educated population, there is no realistic prospect of achieving the other SDGs. Education imparts job-market skills that enable individuals to escape from poverty. Increasing the educational attainment of society as a whole is perhaps the most important contributor to national economic development. Education is also a prerequisite for social and political stability. When society is divided between education "haves" and education "have-nots," income inequalities and social disparities are large, and demagogues and populists may successfully appeal to less educated segments of the population. In *The Wealth of Nations* (1776), Adam Smith called for basic education for all for this reason: "The more [the common people] are educated, the less liable they are to the delusions of enthusiasm and superstition, which, among ignorant nations, frequently occasion the most dreadful disorders."[1] Moreover, when societies confront complex sustainable development challenges such as pandemic diseases, climate change, and the loss of biodiversity that require behavioral changes (such as getting vaccinated), a better educated public will respond more rapidly and with less friction. When the public is poorly educated, fake news and propaganda are more readily believed and transmitted.

The targets for SDG 4 are suitably bold. Targets 4.1 and 4.2 call for universal completion of pre-kindergarten through upper-secondary education for all children. Targets 4.3, 4.4, and 4.5 call for universal access to affordable vocational, technical, and tertiary education without gender bias. Target 4.6 calls for ensuring youth and adult literacy and numeracy. Target 4.7 calls for education for all in sustainable development and global citizenship, including education to foster a culture of peace and toleration for cultural diversity.

Even though education is vital for both individual and national success, access to affordable schooling remains one of the greatest problems facing poorer children throughout the world. For hundreds of millions of children today, there is no practical way to access even a rudimentary education, much less higher education or professional training. The COVID-19 pandemic has made the situation far worse, with hundreds of millions of children having experienced school closings for part or all of the 2020 and 2021 school years. More affluent children were often able to continue their education online, whereas students in poorer communities suffered educational setbacks as schools were unable to move classrooms online.

Education and Economic Development

Organisation for Economic Co-operation and Development (OECD) and academic economists have demonstrated a strong relationship between economic performance and scholastic attainment in reading, science, and math at the national level.[2] Economic development is highly correlated with the strength of the knowledge capital of the society, as determined in large part by its investments in education. Countries that score high on standardized international tests, such as the OECD's Programme for International Student Assessment, also show higher economic growth, controlling for other determinants of economic growth.[3] According to the standardized international testing carried out by the OECD, the highest-performing schools are now in East Asia, notably in Singapore, mainland China, Taiwan, Korea, Malaysia, and Hong Kong, and the weakest performing schools are generally in low-income countries.

This linkage is becoming stronger over time, making quality education for all even more crucial. Fifty years ago, there were many decent and remunerative jobs that required just a few years of schooling. During the past fifty years, however, with increasing robotics and automation, smart machines have increasingly taken over the tasks once performed by less-educated workers. Rapid technological change is thereby driving down the wages of less-educated workers or driving them out of the labor force altogether (and perhaps onto social assistance).[4] In the future, decent and well-remunerated work will tend to require at least an upper-secondary education. It has recently been estimated that every additional year of schooling increases an individual's earnings by around 10 percent and that each additional year of schooling increases a country's average annual GDP growth by 0.37 percent.[5]

Calculating the Budget for Education for All

Education for all is of top priority for every nation, yet it is still not being achieved. The main reason is simple: Poor countries cannot afford to provide quality education for all. Here is some very simple arithmetic to explain the point.[6]

The main inputs to schooling are a teacher and a school building with safe water, sanitation, electricity, and connectivity to the internet. Based upon these needs, we can roughly estimate the costs of providing schooling in a low-income setting. First, let us consider the teacher's salary. Suppose we assume that a teacher in a low-income country earns around US$7,500 per year and teaches thirty students; therefore, that teacher's pay is around $250 per student per year. Teacher wages compose roughly half of the overall school costs, which also include the cost of the school building, infrastructure, and management, which would suggest that the total cost of schooling per student is two times $250, or $500 per year.

A cost of $500 per student per year seems eminently affordable, especially since rich countries often spend $10,000 or more per student per year. Yet even $500 per student per year is out of reach for many governments of low-income countries. Suppose that school-aged children make up 40 percent of the population, as is the case in many sub-Saharan African countries with very young populations. In that case, the cost per capita of providing universal education would be 40 percent of $500 per student, or $200 per capita.

Let us now look at the budget. Consider a typical low-income country with a GDP per capita of $1,000 per year and tax revenues of 20 percent of GDP. The government's total tax collections would be a mere $200 per person per year (20 percent of $1,000). Yet $200 per person is the amount needed for the education sector alone. In other words, for a low-income country with $1,000 GDP per capita, 20 percent of GDP collected in taxes, and school-aged children constituting 40 percent of the population, the education budget needed to keep all children in school would roughly equal the total taxes collected by the government! Obviously, governments are not able to devote their entire tax collection to education, as they must also cover other essentials such as health care, infrastructure, environmental management, public administration, social protection, and public security.

The bottom line is that education for all children is too costly for low-income countries unless they receive help from richer countries. If the government of a low-income country could devote one-quarter of its budget (or 5 percent of GDP) to education, then it would be able to provide schooling for roughly one-quarter of its school-aged children. This is what occurs in the poorest countries, which struggle just to keep most children in school through the primary level. These countries suffer a drastic decline in schooling for lower-secondary and upper-secondary education. Only 44 percent of children in sub-Saharan

African developing countries (excluding high-income countries) completed lower-secondary school as of 2018,[7] and only 27 percent completed upper-secondary education. By the time that children reach lower secondary school, governments demand school fees and dues to cover costs. Families must choose which children to send to lower-secondary school and beyond. A family may select only the eldest son for further schooling, causing the other children in that family to end their schooling; this choice means that girls are often forced to leave school.

The Global Education Financing Gap

As of 2019, some 258.4 million school-aged children and adolescents worldwide were out of school.[8] Only when a country reaches a GDP of about $2,500 per capita can it come close to affording a comprehensive education budget. (Note that countries at that higher income level also tend to have somewhat older populations with smaller shares of the population at school age, making the cost of education more affordable for the society.)

Let us now consider the overall financing gap for education for all developing countries combined. For low-income countries, the education financing gap is around 15 percent of GDP. With a total GDP of $531 billion for the low-income countries, the financing gap is around $75 billion per year. If we suppose that middle-income countries need about the same financing to achieve education for all, the financing gap in total for both low- and middle-income countries would be on the order of $150 billion per year. In fact, when the UN Educational, Scientific and Cultural Organization (UNESCO) carefully studied the budgets of developing countries, it came up with an annual financing gap of $148 billion, which is similar to our rough estimate.[9] What counts here is the rough magnitude; when we examine that gap in comparison with the $50 trillion per year GDP of the rich countries, we see that the education financing gap for the low- and middle-income countries is around 0.3 percent of the GDP of the rich countries. That small amount is all that it would take to put all children everywhere in school and achieve SDG 4.

Addressing the Education Financing Gap

It's hard to understand why education is not a higher international financing priority, especially because the financing gap for education is rather small. Nonetheless, education continues to lack funding.

There are some important proposals on the table for how to address the education financing gap. Gordon Brown, the UN special envoy for global education, has called for an International Finance Facility for Education (IFFEd), which would be a loan program to help developing countries obtain low-interest loans to cover at least part of the education gap. Alongside IFFEd, I have recommended a new Global Fund for Education that would be modeled on the successful Global Fund to Fight AIDS, Tuberculosis and Malaria.

We must also mobilize financing from the world's richest individuals, either through wealth taxation or through philanthropy. As of March 2021, *Forbes* magazine calculated that there are 2,755 billionaires worldwide with a combined wealth of $13.1 trillion. A 1 percent wealth tax on these billionaires would essentially close the education financing gap. Alternatively, the world's billionaires could agree to donate 1 percent of their wealth each year.

New information technologies and expanded broadband connectivity will also make possible innovative approaches to improve the quality of education and to expand access to quality education at a lower cost. Teachers and students no longer have to be in the same classroom or even the same country for effective quality schooling. Teachers can create a classroom experience using online conferencing and other digital technologies, as long as all children are given access to the needed digital devices and connectivity. We must mobilize digital technology leaders to assist in the large-scale deployment of online learning where needed in low-income settings.

Let us also remember that some of the challenges to ensuring universal education are social and cultural, not financial. Even in some societies that can afford universal education, girls or children in minority groups are often not in school, or not receiving a quality education, owing to factors including gender, racial, and ethnic discrimination, threats of physical and sexual violence at school, child marriage, and inadequate water and sanitation facilities at schools. There are also huge logistical and political obstacles in ensuring access to education for refugees and for disabled people, with both groups requiring special attention and commitment.

Finally, to secure a quality education for all, we must promote the content of sustainable development in school teaching, as called for by SDG Target 4.7. We should teach a holistic sustainable development curriculum, including the science of sustainable development and the cultural dimensions of tolerance and peace. Of course, this can be done only when all children are in school, in person or online, in safe and healthful conditions. Access to quality schooling for all children remains our most pressing and urgent educational need.

Notes

1. Adam Smith, *The Wealth of Nations*, book V, chapter 1, part III, article III.
2. OECD, "Relationship Between Skills and Economic Growth," in *Universal Basic Skills: What Countries Stand to Gain* (Paris: OECD, 2015).

3. Eric A. Hanushek and Ludger Wössmann, "Education and Economic Growth," in *International Encyclopedia of Education*, vol. 2, ed. Penelope Peterson, Eva Baker, and Barry McGaw, (Amsterdam: Elsevier, 2010), 245–52.
4. OECD, "Relationship Between Skills and Economic Growth."
5. "Education Data Highlights," Global Partnership for Education, accessed June 15, 2021, https://www.globalpartnership.org/data-and-results/education-data.
6. For a far more detailed calculation, see UNICEF, "Chapter 3: Barriers to Education Progress and Learning," in *The Investment Case for Education and Equity* (New York: UNICEF, 2015).
7. "Lower Secondary Completion Rate, Total (% of Relevant Age Group)," *World Bank DataBank*, https://data.worldbank.org/indicator/SE.SEC.CMPT.LO.ZS last modified September 2020.
8. UNESCO, *New Methodology Shows that 258 Million Children, Adolescents, and Youth Are Out of School*, Fact Sheet no. 56 (Paris: UNESCO, September 2019), http://uis.unesco.org/sites/default/files/documents/new-methodology-shows-258-million-children-adolescents-and-youth-are-out-school.pdf.
9. UNESCO, *Act Now: Reduce the Impact of COVID-19 on the Cost of Achieving SDG 4*, Policy Paper 42 (Paris: UNESCO, September 2020), https://en.unesco.org/gem-report/COVIDcostSDG4.

CHAPTER 27

WHAT WILL IT TAKE TO MEET THE SUSTAINABLE DEVELOPMENT GOAL FOR EDUCATION?

RADHIKA IYENGAR

Worldwide, 263 million children are not in school.[1] These children include sixty-one million children of primary school age, sixty million adolescents of lower secondary school age, and 142 million youth of upper secondary school age. In 2015, the Sustainable Development Goals succeeded the Millennium Development Goals, but many millions remain out of school. Education indicators on enrollment and learning are far behind expectations. Globally, progress in education has been pathetically slow. Millions are still deprived of their fundamental human right to be enrolled in an age-appropriate grade and learn at grade level. While countries around the world are already in their fourth year of implementing the SDGs, it is evident that we need to do more.

What is our moral responsibility as international researchers and practitioners toward the field of education? How can we ensure that all children are enrolled in school at the right age and are learning at grade level? It is time we acknowledge that "business as usual" is leaving many millions behind. Education need to be done right, and change has to come now. Let us first examine the current status of education.

Where do things begin to go wrong in education? For many children, very early on. Internationally, access to free early education would ensure that all children are equipped to succeed in primary school and beyond, regardless of background. Evidence from cognitive neuroscience suggests that pre-primary education is key to appropriate brain development, a precondition for acquiring subsequent skills.[2] However, only forty-nine of 206 countries and territories mandate compulsory pre-primary education, and only thirty-six countries mandate free compulsory education. UNESCO's *GEM Report* provides an analysis of early childhood education programs globally, including their policy impact. In 2014, just 44 percent of children were enrolled in pre-primary education worldwide, and just 21.5 percent of children in sub-Saharan Africa.[3] The 2016 report notes that developing countries that invested in early childhood interventions, including Bangladesh and Timor-Leste, reduced child mortality by at least two-thirds in advance of the Millennium Development Goal target date.[4] Pre-primary education is the foundation of all the years of education that will come later. This foundation needs to be rock solid. Three- to five-year-olds need to be paid special attention as their brains are just starting to make the neurological connections necessary to become a fully functional, physically and mentally healthy human being.

In most developing countries, however, the reality is that by the time a child reaches grade 1 of primary schooling, they are developmentally behind in every learning and anthropometric measure. Entering grade 1 should be a stepping stone to success, but many schools are failing to make children ready even for this grade. The learning lag continues to widen throughout primary education. Among children who attend school, 130 million cannot read or do basic math after four years. In recent assessments in Ghana and Malawi, more than 80 percent of students at the end of grade 2 were unable to read a single familiar word such as "the" or "cat."[5]

A slow start in learning impacts all future learning but often goes undetected. And even if it is detected, children often do not get a chance to attend remedial classes to catch up with grade-level competencies, thus keeping them lagging behind. There should be a focus on early literacy and numeracy programs for young learners, especially in low-resourced environments. In many cases, when children arrive in grade 1, they get to hold a book for the first time. Governments and donors must prioritize a scaling up of early literacy and numeracy education through ongoing professional development support for teachers and integrating these topics in pre-service teacher training.

At the secondary level, sixty million children of lower secondary school age and 142 million of upper secondary age are out of school, and 617 million youth worldwide lack basic mathematics and literacy skills.[6] In Western Asia, a disproportionate share of these youth are girls: 20 percent of adolescent girls of lower secondary school age are out of school compared to 13 percent of boys. In sub-Saharan Africa, 36 percent of adolescent girls are out of school compared to 32 percent of boys.[7] These figures are many times higher in conflict-affected countries with high rates of international refugees and displaced peoples. In such places, the youth who are fortunate enough to attend secondary school feel they need additional support. For example, a study conducted in sub-Saharan Africa on the career aspirations of youth suggests that many would like to become doctors, nurses, teachers, and engineers, among other highly skilled careers. However, a majority feel they need additional help from their schools and communities to attain their goals.[8]

Countries that have recently eradicated secondary school fees have seen a steep increase in secondary enrollment,[9] leading to new challenges to maintain quality under strained capacity. As secondary enrollments increase, the problem of a mismatch in skills learned compared to those in demand by rapidly changing economies continues to exist.[10] Some countries are addressing this issue by revising their curricula to be more practice oriented; for example, Rwanda unveiled its new competency-based curriculum unveiled in 2016. Despite such improvements, girls continue to be left behind in secondary education, the worst impact of which is on girls living in rural areas, those who are economically marginalized, and those living in low-income countries.

Social emotional learning (SEL) concepts also need to be added to curricula to attend to the needs of internally displaced peoples and international refugees. Schools in temporary refugee camps are not equipped with the curricula needed to ensure that children's learning needs are met. Given the increasing number of climate refugees and countries experiencing political unrest, integrating SEL concepts into curricula and learning materials is a high priority.

Globally, youth unemployment is three times the adult rate.[11] In 2016, seventy-one million youths around the world were without a job, while another 156 million young workers in developing countries earned less than US$3.10 per day.[12] Globally, around 40 percent of the youth labor force is still unemployed or working yet living in poverty.[13] Some projections estimate that more than two billion jobs will be obsolete by 2030, and half our youth will lack the necessary skills and innovation required for employment.[14] Many graduates, mostly women, slip into the informal sector or become part-time workers. Increased support for creating and implementing educational programs that meet the needs of both youth and the economy is needed. A recent study in Bhopal, India, revealed a mismatch between technical, vocational, and education training (TVET) programs offered by the government and the populations they serve.[15] Previous studies have been highlighting this misalignment for years, resulting in TVET policy documents being revised to address the issue. Unfortunately, however, TVET practitioners and programs have been slow to reflect this change. Focus groups with potential TVET candidates suggest women find it challenging to attend these courses owing to the timing of classes. Moreover, women need programs that incorporate both SEL programming and technical skill enhancement. Finally, more TVET courses are needed that align with the requirements of "green economy" jobs, an area of the economy that is growing quickly.

How can we resolve the education crisis? First, we need citizen-led movements in education to make citizens aware of the learning crises around the world. We need to support governments to ensure that all our children are learning. A powerful force here is the People's Action for Learning (PAL) Network,[16] a multicountry, citizen-led group that develops learning assessments and shares results with their respective communities. Such movements are the seeds of the future of education and can help demystify learning for the masses.

Second, we need to use cognitive science–based approaches, such as Helen Abadzi's one-hundred-day literacy program[17] and the Pratham Teaching at the Right Level program,[18] to ensure 100 percent literacy for all children. While Abadzi's approach focuses on children who struggle with letter recognition,[19]the Pratham approach is geared toward ensuring children progress to their next level of learning. Such literacy approaches are essential to making every child literate.

Third, we need to raise funds to address critical issues in education. Based on UNESCO's costing models, it is estimated that an annual total cost of $340 billion is needed to achieve universal pre-primary, primary, and secondary education in low- and lower-middle-income countries by 2030.[20] The average annual per-student cost of quality primary education in a low-income country is estimated to be $197 in 2030. UNESCO estimates an annual gap of $39 billion between 2015 and 2030. Professor Jeffrey Sachs of Columbia University raised the issue of financing education issue in his keynote speech at the 2019 Comparative and International Education Society conference in San Francisco.[21] Sachs also established a fund for ensuring education in Africa, which will ensure that countries have the required monetary support to ensure that all children receive quality education through the secondary level and will help SDG 4 to be achieved in Africa. However, for the fund to be viable, action will be required from private sector donors, philanthropists, and international financial institutions.

The current education crisis requires our educators to raise their voices. With growing economic disparity, the literacy disparity is growing exponentially.[22] As citizens, we must demand that the basic right of every child to attend school is met. As education experts, we need to facilitate this process and collaborate with national governments. Let us no longer deny any child the right to go to school at the appropriate age and learn what they need to learn.

Notes

The author would like to thank Ms. Haein Shin and Ms. Tara Stafford Ocansey for providing research assistance for this chapter.

1. "263 Million Children and Youth Are Out of School," UNESCO, July 15, 2016, http://uis.unesco.org/en/news/263-million-children-and-youth-are-out-school.
2. Helen Abadzi, "Efficient Learning for the Poor: New Insights into Literacy Acquisition for Children," *International Review of* Education 54, no. 5 (November 2008): 581–604.

3. UNESCO, *Education for People and Planet: Creating Sustainable Futures for All*, Global Education Monitoring Report, 2016 (Paris: UNESCO, 2016), https://unesdoc.unesco.org/ark:/48223/pf0000245752.
4. UNESCO, *Teaching and Learning: Achieving Quality for All*, EFA Global Monitoring Report, 2013–2014 (Paris: UNESCO, 2014), https://unesdoc.unesco.org/ark:/48223/pf0000225660.
5. World Bank, *World Development Report 2018: Learning to Realize Education's Promise* (Washington, DC: World Bank, 2018), http://www.worldbank.org/en/publication/wdr2018.
6. "Sustainable Development Goals: 4. Quality Education," United Nations, https://www.un.org/sustainabledevelopment/education/.
7. "Closing the Gender Gap," UNESCO, March 3, 2017, http://uis.unesco.org/en/news/closing-gender-gap.
8. UNESCO, *Relevance of Post-Basic Education to the Realities of Youth in Sub-Saharan Africa: Exploring the Perspectives of Young People* (Paris: UNESCO, forthcoming).
9. Godwill Arthur-Mensah and Kwamina Tandoh, "FREE SHS So Far Records 83.9% Enrolment," *Modern Ghana*, January 12, 2017, https://www.modernghana.com/news/819996/free-shs-so-far-records-839-enrolment.html.
10. Ruby Sinam Nutor, "IMANI Alert: Youth Employment Prospects and the Educational System of Ghana," IMANI Center for Policy and Education, September 25, 2015, https://imaniafrica.org/2015/09/25/youth-employment-prospects-the-educational-system-of-ghana/.
11. "The Global Youth Unemployment Rate Is Three Times Greater than Between Adults," Youth Employment Decade, January 25, 2018, http://www.youthemploymentdecade.org/en/repor/la-tasa-mundial-desempleo-juvenil-tres-veces-mayor-la-los-adultos/.
12. See International Labour Organization, *Global Employment Trends for Youth 2015: Scaling Up Investments in Decent Jobs for Youth* (Geneva: ILO, 2015), https://www.ilo.org/wcmsp5/groups/public/---dgreports/---dcomm/---publ/documents/publication/wcms_412015.pdf.
13. "Poor Working Conditions Are Main Global Employment Challenge," International Labour Organization, February 13, 2019, https://www.ilo.org/global/about-the-ilo/newsroom/news/WCMS_670171/lang--en/index.htm.
14. "Global Education," Office of Gordon and Sarah Brown, accessed April 26, 2021, https://gordonandsarahbrown.com/global-education/.
15. Radhika Iyengar and Matthew Witenstein, "Amplifying Indian Women's Voices and Experiences to Advance Their Access to Technical and Vocational Education Training," in *Interrogating and Innovating Comparative and International Education Research*, ed. C. Manion, E. Anderson, Supriya Baily, Megan Call-Cummings, Radhika Iyengar, Payal Shah, and Matthew Witenstein (London: Brill Sense, 2019).
16. "PAL Network: People's Action for Learning," PAL Network, accessed April 26, 2021, https://palnetwork.org.
17. Helen Abadzi, *Literacy for All in 100 Days? A Research-Based Strategy for Fast Progress in Low-Income Countries*, GPE Working Paper Series on Learning No. 7 (Washington, DC: Global Partnership for Education, May 30, 2013), http://documents.worldbank.org/curated/en/257651468326683654/pdf/797850WP0ENGLI0B0x0379789B00PUBLIC0.pdf.
18. "Teaching at the Right Level: Strengthening Foundational Skills to Accelerate Learning," TaRL, accessed April 26, 2021, https://www.teachingattherightlevel.org.
19. Helen Abadzi, Radhika Iyengar, Alia Karim, and Florie Chagwira, "Malawi: Why It's Important Children Learn to Read in Their Mother-Tongue," World Education Blog,

October 14, 2014, https://gemreportunesco.wordpress.com/2014/10/14/malawi-why-its-important-children-learn-to-read-in-their-mother-tongue/.

20. UNESCO, *Pricing the Right to Education: The Cost of Reaching New Targets by 2030*, Education for All Global Monitoring Report Policy Paper 18, updated July 2015, https://unesdoc.unesco.org/ark:/48223/pf0000232197.
21. Jeffrey Sachs, "Education for Sustainability" (George F. Kneller keynote lecture, Comparative and International Education Society conference, San Francisco, CA, April 15, 2019), https://cies2019.org/keynote/.
22. Federico Cingano, *Trends in Income Inequality and Its Impact on Economic Growth*, OECD Social, Employment and Migration Working Papers, no. 163 (Paris: OECD Publishing, 2014), http://www.oecd.org/els/soc/trends-in-income-inequality-and-its-impact-on-economic-growth-SEM-WP163.pdf.

CHAPTER 28

"ONLY CONNECT"

Neuroscience, Technology, and Global Literacy

MARYANNE WOLF

Literacy changes the brains of literate individuals, which changes the trajectory of their lives, which changes the society in which they live, which changes the health and economies of our world. Yet, there are sixty-three million children without schools or teachers, and there are another 120 million to 150 million children who attend inadequate schools or whose classrooms have just one teacher for sixty to one hundred children.[1] With this as our world's reality, the Kantian questions are three: What can we know? What should we do? What may we hope? Knowledge, action, and hope are the major principles in our approach to changing literacy in our lifetimes.

What Can We Know?

Let us begin with something most of us know: that without literacy, the "good society" is immeasurably changed. In Aristotle's view of the good society, there are three lives: the life of knowledge and productivity, the life of leisure and entertainment, and the life of contemplation and reflection.[2] Without literacy as the foundation in today's world, none of these lives will be achieved, and our children will never reach their full potential.

We also know where this happens most. So much of the illiteracy and non-literacy in the world is found in sub-Saharan Africa, India, and pockets of our own backyard (the United States).[3]

We also know why this happens. Nelson Mandela said, "Like slavery and apartheid, poverty is not natural. It is man-made and it can be overcome and eradicated by the actions of human beings."[4] So also with literacy and the foundation it gives to the education of every human being. Nonliteracy may begin as an accident of poverty, but it continues under the full glare of all members of our connected world. Leaving millions of our world's people illiterate need not continue. We have the knowledge and ability to change the situation, if we choose to do so.

From my perspective in the cognitive neurosciences, we know a great deal about the reading brain, knowledge that can help to ameliorate the high rates of illiteracy, particularly in remote regions where there are no schools. The implementation of this knowledge, however, is complicated by the fact that the very technologies that will aid our efforts have themselves both promise and threats to literacy's development. The application of this knowledge into practice, therefore, will require from all of us a new kind of informed, ethical imagination. Thus, in this chapter, I will be connecting information not only about the reading brain's development but also about the impact of digital culture on all of us.

To begin: The human brain was never born to read or multiply numbers or program. How then does the human brain learn how to perform any of these relatively "new" cognitive functions? The little discussed capacity within our species' genetic gifts involves design principles that allow humans to make *new circuits* for cultural inventions like literacy and numeracy. It does so by rearranging the connections among its older, more established parts. It is important to underscore that the reading brain does not simply unfold naturally, in the way that oral language does. Rather, the new reading circuit in the brain needs to be built up and elaborated over time through the connection of a group of components that are genetically programmed, including vision, cognition, and language. These new circuits can be very basic or scant and impoverished, or they can be elaborated over time by adding the neuronal networks for some of our most sophisticated processes, including affective processes like empathy and other emotions, our ability to evaluate truth, and our individual ethical and aesthetic values. The circuitry of the brain not only elaborates with our own personal development; indeed, it also reflects our development. In the process of its elaboration, reading acts as both an intellectual discipline for our thoughts and a moral laboratory for human development.[5]

Within this context, there is no one ideal reading brain. There are many reading brains that reflect the writing system used, the individual's educational history, and even the medium. To give a small example of the differences among reading brains, think of the differences among writing systems. Chinese and Japanese reading brains have circuits that place heavier requirements on areas responsible for visual memory.[6] Thus, the visual areas of the right hemisphere of these reading brains are far more activated than for the readers of alphabets.

In other words, just as all learning is local, so too are all reading brains local. They reflect the environment around them, particularly in the early stages of reading development, and not just after a child begins school. Society members need to ensure that among infants and children up to five years of age, the component parts of the reading circuit are developing, beginning with language and cognition. Additionally, we must ensure that we are providing influences that will help to make moral, ethical children.

We who are the teachers help to make these connections happen. But in places where there are no teachers, what can we do? Some of my colleagues and I are attempting to use models of the reading brain as the basis for creating apps on digital platforms to teach children the precursors of reading. These models help us understand how to create or curate apps that support not only the basic decoding parts of the brain but also the expert's "deep reading processes," like inference and critical thinking. When most of us read, few of us are just decoding; rather, we are deploying those deep reading processes that help us to infer, analyze, feel, and discern the truth and value of what we read.

Essential to this process is the cumulative building of background concepts in the young, and indeed in us all. If we don't possess the necessary background knowledge, we can't go on to infer and comprehend. We humans are analogy makers. We make analogies from what we know to what we read. Based on this millisecond-long transmission, we can go on to make inferences. There is much more as well. For example, in the words of the theologian John Dunne, we can "pass over" to the perspective of another person.[7] And we leave our own perspectives in order to try on another's thoughts and feelings. In the process of passing over, we return enriched. Perspective-taking is, therefore, one process of many that allows us to make a critical analysis of what we read. Ultimately these interactive deep reading processes can, upon occasion, lead to generativity, novel thought, and insight, the furthest point of the deep reading brain.

Deep reading has implications far beyond the individual. If readers in a society never learn to develop and use critical analysis and empathy, we will not have the means to sustain a truly democratic society. We will instead have citizens who are susceptible to fake news and the falsely fueled fears and falsely raised hopes of demagoguery. The last thing I ever thought when I began work on neuroscience and the reading brain is that I would be making a statement about why true, deep literacy is essential for a democratic society, such that children, youth, and adults know how to discriminate truth from falsehood and how to understand the perspectives of those who have been deemed "other."

One of the most important but less described aspects of deep reading concerns the development of perspective-taking that leads to understanding and empathy for others. As discussed earlier, Dunne used the term "passing over" to describe what is an essential gift of reading. Passing over is what makes us able to enter the lives, historical epochs, and understandings of others. Whether from a psychological or theological perspective, reading provides us with a moral laboratory for our thoughts. As Dunne wrote, "Here is the Golden Key. It is the capacity to pass over to others and come back to ourselves. We all have the capacity, but we do not all discover it, come to use it, learn to pass over."[8]

This essential aspect of deep reading was captured in a conversation between the Pulitzer Prize–winning novelist Marilynne Robinson and the former U.S. president Barack Obama. They spoke about the novel's role in developing empathy, to which Robinson remarked that seeing others as "the sinister other" was the "greatest threat to our democracy."[9] Along similar lines, the novelist Jane Smiley remarked that she did not believe that the novel would die but that it could be sidelined, that we could have leaders who do not read, who do not understand the viewpoints of others, which would "coarsen and brutalize our society."[10]

What Should We Do?

Within this broader context, what should we do with this information about reading development and the larger problem of global nonliteracy? I want to suggest that we can use information from the reading brain's development, from our instruction to children, and from technology to help form and teach the circuit over time so that children can become expert readers, capable of the most sophisticated thought. The brain can make various types of circuits, which is both heartening when it elaborates over time and worrisome when it remains primitively basic. The onus upon researchers and educators (whether in the classroom or teaching remotely via digital technology) is to teach the many components of the circuit, using reverse engineering of our models of the reading brain, so we don't stop at just decoding and decoding precursors in early childhood development but provide the foundation for developing deep reading processes.

And now we come to the great conundrum of current research on reading development in the twenty-first century. Although most people around the world have begun to use digital technology to great advantage, there are impediments in using this technology that were never suspected. The unintended collateral damage of digital culture represents an unexpected part of my recent work. I have become both an advocate of digital devices in providing literacy precursors to nonliterate children with no schools or with inappropriate educational environments and a reluctant but fierce critic of the dominant use of digital screens for all reading. It is imperative to understand that the reading brain reflects not only the writing system but also the medium. We are only now beginning to understand that digital reading, with its emphases on fast processing and multitasking, can disrupt the allocation of time to deep reading processes. New research indicates that changes in attention and memory form when the reader expects a constant, immediate stream of information from external platforms of knowledge. A recent European meta-analysis of more than 171,000 subjects found that the ability of young readers to understand the same material was superior when read in print versus on a digital screen.[11]

We're all involved in what I describe as the digital chain hypothesis.[12] Within this chain, how much we read (how much we're inundated with daily) affects how we read. Eye-movement research indicates how much we've already changed. We're skimming, we're browsing, we're word spotting. We're using an "F" or "Z" pattern by and large, and then we triage. The effects on the beginning of the circuit are inevitable. Attention is split among competing stimuli, leading to what is sometimes called "continuous partial attention" in young people. This should come as no surprise, since by reflex, human beings have an ancient novelty bias that causes us to look at whatever is new to any of our senses. This reflex kept our ancestors alive, but now it is posing a problem to children. They literally can't look away when bombarded by multiple competitors for their attention. By most measures, all people's attention spans, not just children's, are about half of what they were ten to fifteen years ago.[13] A reduced attention span affects memory, which changes the consolidation of what we read into background knowledge, which affects our future ability to make analogies and inferences.

Most importantly, this chain of changes affects the allocation of time to critical analysis and empathy, which influences what we read, what is published, and why we read at all. With more to read and less time to invest in sophisticated analyses, many people resort to the comfort of familiar silos of knowledge that challenge little, confirm what we thought before, and do not give us the perspective of others who think differently. The unsettling paradox of our time is that the more information we have, the more likely many of us are to turn to unchallenging, undemanding, less dense, less syntactically complex information. This is the stuff of threat to any democracy. Complexity is reduced, beauty vanishes, and citizens and leaders become more and more susceptible to simplistic reasoning and false information.

Conclusion

How can this cumulative knowledge—about the plasticity of the reading brain and about the advantages and disadvantages of technology—inform approaches to achieving global literacy? By looking at the reading brain and by studying the impact of different mediums, we have the opportunity to make better use of "digital wisdom" when we make tablets for nonliterate children in remote areas of our globe and for literate children in our backyards. The ongoing work of my colleagues at Curious Learning, Global Learning XPRIZE, MIT, Georgia State University, the University of California San Francisco, and UCLA concerns both of these goals. My own recent work involves a developmental proposal to build a biliterate brain that will know how to read deeply regardless of medium. Within this proposal is the hope that every new reader will be taught to decode fluently and learn to read with empathy, critical reasoning, and novel insights all their own.

Ultimately, my colleagues and I hope that global literacy initiatives around the world will join efforts so that we can reach one hundred million children and reduce world poverty by 12 percent.[14] I believe, like John Dunne, that an ethical imagination requires the knowledge that each of us has in each of our disciplines. It is when we "only connect" our knowledge and our ethical concerns *in action* that we can achieve the world Pope Francis describes: "Children are a sign. They are a sign of hope, a sign of life, but also a 'diagnostic' sign, a marker indicating the health of families, society, and the entire world. Wherever children are accepted, loved, cared for and protected, the family is healthy, society is more healthy, and the world is more human."[15]

Notes

1. UNESCO, *One in Five Children, Adolescents and Youth Is Out of School*, Fact Sheet No. 48 (Paris: UNESCO, February 2018), http://uis.unesco.org/sites/default/files/documents/fs48-one-five-children-adolescents-youth-out-school-2018-en.pdf.
2. Aristotle, *Aristotle's Politics: A Treatise on Government* (London: G. Routledge, 1895).
3. UNESCO, *Mapping the Global Literacy Challenge*, Education for All Global Monitoring Report, 2006 (Paris: UNESCO, 2006), http://www.unesco.org/education/GMR2006/full/chapt7_eng.pdf.
4. Nelson Mandela, "In Full: Mandela's Poverty Speech," *BBC News*, February 3, 2005, http://news.bbc.co.uk/2/hi/uk_news/politics/4232603.stm.
5. Maryanne Wolf, *Reader, Come Home: The Reading Brain in a Digital World* (New York: HarperCollins, 2018).
6. H. S. Huang and J. Richard Hanley, "Phonological Awareness and Visual Skills in Learning to Read Chinese and English," *Cognition* 54, no. 1 (January 1, 1995): 73–98, https://doi.org/10.1016/0010-0277(94)00641-W.
7. John S. Dunne, *Reading the Gospel* (Notre Dame, IN: University of Notre Dame Press, 2000).
8. John S. Dunne, *The House of Wisdom: A Pilgrimage of the Heart* (Yorktown Heights, NY: Meyer-Stone, 1988).
9. Barack Obama and Marilynne Robinson, "President Obama and Marilynne Robinson: A Conversation in Iowa," *New York Review of Books*, November 5, 2015, https://www.nybooks.com/articles/2015/11/05/president-obama-marilynne-robinson-conversation/.
10. Jane Smiley, *Thirteen Ways of Looking at the Novel* (New York: Knopf, 2005).
11. Pablo Delgado, Cristina Vargas, Rakefet Ackerman, and Ladislao Salmerón, "Don't Throw Away Your Printed Books: A Meta-analysis on the Effects of Reading Media on

Reading Comprehension," *Educational Research Review* 25 (November 1, 2018): 23–38, https://doi.org/10.1016/j.edurev.2018.09.003.

12. Wolf, *Reader, Come Home.*
13. Wolf, *Reader, Come Home.*
14. UNESCO, "One in Five Children, Adolescents and Youth Is Out of School."
15. Francis, *The Blessing of Family: Inspiring Words from Pope Francis* (Cincinnati, OH: Franciscan Media, 2014).

PART VIII

AN ETHICAL CONSENSUS ON SUSTAINABLE DEVELOPMENT

Climate Justice

ETHICS IN ACTION

Climate Justice

Laudato si' calls on us to redress a world beset by structures of injustice. We can point to five main challenges of injustice that threaten human dignity, undermine the common good, subvert democracy, and endanger our very survival and health owing to environmental degradation:

- Global promises by governments that are not kept
- The persistence of unequal distributions of income and wealth
- Unjust redistributions from poor to rich
- Bullying associated with corporate lobbying
- States that do not accept migrants while adding to the distress causing the migrations

In addition, for the first time in history, we are faced with the grave threat of human-induced climate change, the moral dimensions of which are a core feature of *Laudato si'* and the Sustainable Development Goals.

We have examined the challenge of climate justice in light of these structures of injustice. We acknowledge that scientific evidence can *attribute* the harms of environmental degradation to identifiable agents, such as the major producers of fossil fuels and governments that fail to regulate economic activities for the public good. Such "attribution" is in the form of a probability or likelihood that human actions have caused a specific loss or damage. For example, scientists have shown that recent heat waves, extreme rainfalls, and floods have been made more likely to occur (and therefore to occur more frequently) because of human-caused global warming.

Global warming is the result of the dangerous and unconstrained use of fossil fuels. As Pope Francis points out in *Laudato si'* in discussing climate as a common good, "A number of scientific studies indicate that most global warming in recent decades is due to the great concentration of greenhouse gases (carbon dioxide, methane, nitrogen oxides and others) released mainly as a result of human activity."[1] Moreover, thanks to recent investigations, it is possible to identify the companies that have contributed most to the historic rise in carbon dioxide levels in the atmosphere. For example, according to one study, ExxonMobil, Chevron, Saudi Aramco, and Gazprom (the two largest investor-owned and two largest state-owned producers, respectively) together produced emissions constituting nearly 10 percent of the rise in global mean surface temperatures between 1880 to 2010.[2]

Ethics in Action identified several ways to address climate justice. These include public actions to regulate carbon dioxide emissions under the Paris Agreement on climate change; behavioral and attitudinal changes by the lead managers of major companies; responsible investing by universities, foundations, insurance companies, pension funds, and others (sometimes called environmental, social, and governance investing); divestments from the fossil-fuel industry; consumer boycotts of companies that befoul the environment with impunity; legal challenges to the behavior of companies and governments, including compensation paid to those suffering "losses and damages" from climate change and other environmental degradation (such as air and water pollution); public awareness campaigns by religious leaders, climate scientists, energy engineers, public health officials, and others; and other means.

Ethics in Action members agreed that there is no single "magic bullet" to overcome climate injustice. Many tools should be deployed, and those tools should be tailored to the specific conditions and needs of different communities and parts of the world. A comprehensive and effective global effort should be top-down as well as bottom-up; through moral suasion as well as public policy; and through cooperation as well as litigation. All pathways will be needed, and urgently, given the grave and potentially irreversible consequences of unchecked global warming.

In this spirit, Ethics in Action identified eleven action items that it would carry out during 2018 and beyond in order to contribute to global solutions. In this work, Ethics in Action will seek partnerships and participation in global forums and processes at the United Nations and UN agencies, as well as other venues. Ethics in Action will report on its activities and progress in this regard at the end of 2018, with recommendations for further steps in the following years.

Ethics in Action Working Group Action Items

(1) Ethics in Action outreach to the top ten oil and gas companies: Gazprom, Rosneft, ExxonMobil, Petro China, BP, Royal Dutch Shell, Chevron, Petrobras, Lukoil, Total, Statoil, ENI
(2) Legal analysis for strategies on losses and damages, public trust, and government accountability (Cesar Rodriguez, Lisa Sachs, Dan Galpern, Yann Aguila)
(3) Meeting of universities and investment funds on ethical investing in, and divestment from, the oil, gas, and coal sectors (Erin Lothes, Lisa Sachs, University of Notre Dame)
(4) Support for the Niger Delta bishops and local Niger Delta communities to meet with Royal Dutch Shell under the auspices of Ethics in Action (Jeffrey Sachs, Cardinal Onaiyekan, Bill Vendley)
(5) At the United Nations headquarters in New York, convene climate-impacted small-island states (under the auspices of the UN Sustainable Development Solutions Network) to discuss strategies for climate justice, including public health (Jeffrey Sachs, Dan Galpern)
(6) Develop massive open online courses and organize workshops on *Laudato si'* and the ethics of climate change (all participants of Ethics in Action, SDG Academy, Fetzer Institute)
(7) Introduce the Global Pact for the Environment project to the G20 process (Yann Aguila, Jeffrey Sachs)
(8) Convene the Interfaith Rainforest Initiative and rainforest country representatives at the UN and in regional meetings (Amazon, Congo Basin, Southeast Asia) to discuss the adoption of zero-deforestation goals
(9) Convene leaders of faith communities to develop strategies for inspiring, educating, and mobilizing congregations and faith communities at the local level in support of comprehensive action to address climate change (Fetzer Institute)
(10) Support World Health Organization (WHO) initiatives, including the First Global Conference on Air Pollution and Human Health, October 30 to November 1, 2018, and communicate Ethics in Action support to the WHO director-general
(11) Support efforts to price carbon emissions, especially through carbon taxes, to make sure polluters are held financially accountable for the costs they impose on society

Ethics in Action also identified the following "ten commandments" to mark the path toward climate change mitigation.

General Recommendations: The Ten Climate Change Commandments

(1) No new coal plants
(2) No new oil and gas
(3) No new fracking
(4) No new pipelines
(5) No new deforestation
(6) Shift to battery electric vehicles by 2030
(7) Reduce beef eating
(8) Divest from greenhouse gas companies
(9) Sue oil producers
(10) Connect renewables

Notes

1. Francis, *Laudato si'* (Vatican City: Libreria Editrice Vaticana, 2015), sec. 25, http://www.vatican.va/content/francesco/en/encyclicals/documents/papa-francesco_20150524_enciclica-laudato-si.html.
2. B. Ekwurzel, J. Boneham, M. W. Dalton, R. Heede, R. J. Mera, M. R. Allen, and P. C. Frumhoff, "The Rise in Global Atmospheric CO2, Surface Temperature, and Sea Level from Emissions Traced to Major Carbon Producers," *Climate Change* 144 (2017): 579–90, https://doi.org/10.1007/s10584-017-1978-0.

CHAPTER 29

CLIMATE DISRUPTION

A Personal Journey Into the Ethical and Moral Issues

VEERABHADRAN RAMANATHAN

"Oh mother Earth, mountain breasted, ocean girdled,
forgive me for trampling on you"

—Three-thousand-year-old Sanskrit prayer on sustainability

From Climate Change to Climate Disruption

The planet has warmed by one degree Celsius, ushering in a change in climate not experienced in the last fifty thousand years or more. In its 2017 annual bulletin, the American Meteorological Society linked the new weather extremes experienced during the last two decades to this one-degree warming. Based on sophisticated statistical analyses, they declared, "We are experiencing new weather extremes because we have created a new climate. These extremes have claimed the lives of 605,000 people over a twenty-year period. It is now scientifically defensible to refer to climate change as climate disruption."

In another ten years, there is a probability of more than 50 percent of this warming exceeding 1.5 degrees Celsius. If that happens, the planet will be warmer than at any time during the last 130,000 years. All of the climate change–related extremes we are witnessing such as fires, hurricanes, and floods could intensify by another 50 percent. If we ignore the problem, the warming could exceed two degrees Celsius in thirty years. If that happens, 1.7 billion people will be subject to deadly heat; several hundred million will be subject to vector-borne diseases like Lyme disease, the Zika virus, and dengue; and tens of millions will be forced to emigrate. If we continue to ignore the problem, the gaseous greenhouse blanket will become so thick that the planetary warming by the year 2100 is likely to be in the range of three degrees Celsius (60 percent probability) to six degrees Celsius (5 percent probability). A six-degree warming would lead to existential threats to *Homo sapiens*, rich and poor. The effects of warming on glaciers, the sea level, and other aspects of the environment may last for thousands of years or more.

The Ethical and Moral Issues

For the world's poorest three billion people, who live in rural areas, an increase in temperature of two degrees Celsius would be enough to pose existential threats raising an unprecedented intragenerational equity issue. For the middle three billion, living in urban areas, a warming of three to four degrees Celsius would be sufficient to pose systemic risks to their social systems. With the possibility of a warming of six degrees Celsius, a world population of ten billion, and a mass extinction of species, we should ask whether civilization as we know it can survive beyond this century. This is an unfathomable question of intergenerational equity.

Let me give you a graphic metaphor of the unprecedented ethical issues posed by such catastrophic climate disruptions. If there is a 5 percent probability that the plane we are about to board will crash, none of us would board the plane. Yet we are putting our children and grandchildren on such a plane—a plane called planet Earth. I deal with this moral and ethical issue each time I visit my grandchildren.

Climate disruption poses yet another vexing ethical and moral issue. In May 2014, I had the singular opportunity to bring such issues to the attention of Pope Francis in a parking lot just outside his residence at the Vatican. I said the following to Pope Francis: "We are all concerned about climate change. About 50 percent to 60 percent of the climate pollution is due to the wealthiest one billion on the planet; the poorest three billion, who have a negligible impact on the climate pollution, will suffer the worst consequences of climate change caused by the wealthiest."

In my view, Pope Francis's encyclical published in 2015, *Laudato si'*, and Rachel Carson's *Silent Spring* are the two most influential documents to guide humanity toward a sustainable pathway, a pathway that elevates human dignity and natural wealth at least to the same heights as material wealth. In addition, *Laudato si'* is unequalled in its appeal to the ethical and moral issues of climate change. In one famous passage in the encyclical, Pope Francis pleads for us to "hear both the cry of the earth and the cry of the poor."

In my opinion, the most influential scientific and social thought in the encyclical is the call for an "integral ecology" approach, according to which the impacts of environmental degradation on natural systems should be studied on equal footing with the impacts on social systems.

With the help of the Pontifical Academy of Sciences, Pope Francis followed the encyclical by forming a global alliance of scientists, policy makers, and faith leaders. Members of the Pontifical Academy of Sciences include eighty of the world's leading experts in natural sciences, twenty-one of whom are Nobel laureates in biology, physics, or chemistry. This academy reports directly to the pope and briefs him on all matters related to nature, ecology, and human affairs.

In 2015, the Pontifical Academy of Sciences organized a series of meetings with the United Nations, including its secretary-general, and with mayors from around the world to explore the links among climate change, social exclusion, slavery, and poverty, among other social issues. At this summit, the pope came over to me and hugged me—my first papal hug since 2004, when I had been elected to the academy by Pope Saint John Paul II. My first thought after the hug was that the pope was signaling to me that he had done his part and it was now for the rest of us to do ours, at the grassroots level.

Forming a New Alliance

How urgent is the situation?

Take a child who is five years old today. When that child reaches eighteen years of age, the planet will have crossed the threshold of dangerous climate change (a warming of 1.5 degrees Celsius). By the time that child is forty years old, catastrophic climate disruptions are likely to be the norm, given the likelihood of a warming of two degrees Celsius by that time.

Is there still time to avoid such catastrophes? The answer is yes. I led a study with fifty researchers from across all ten campuses of the University of California system. The study, "Bending the Curve: Ten Scalable Solutions for Carbon Neutrality and Climate Stability," was commissioned in early 2014 by the president of the university at that time, Janet Napolitano. Our task was to propose ten solutions to the climate change problem. One of our top solutions was for society to go beyond the technological solutions that are currently at the forefront of mitigation discussions and bring societal transformation to the fore.

We concluded that education, from kindergarten to adulthood, must form one of the pillars of societal transformation. Toward this objective, at the University of California we have established the "Bending the Curve: Climate Solutions" protocol, which outlines scalable climate solutions across different fields.[1] In addition, through a partnership between the University of California and the

California State University system, we are developing recommendations to build climate literacy among all elementary and high school students in California.

The developments at the Vatican summarized earlier convinced me that scientists and policy makers must form an alliance with religion. Further, this alliance must include health care providers since one of the biggest impacts of climate disruption is on human health. Such an alliance has the transformative potential to garner the sort of public support we need for drastic actions such as redirecting the massive fossil fuel subsidy of trillions of dollars and decoupling the economy from fossil fuel consumption by a total switch to solar and wind energy. An alliance among science, policy, and religion can have a significant impact on slowing down climate change. My reasons for believing in the positive impact of such an alliance are as follows:

- The alliance would provide a forum in which climate change could be discussed in an apolitical manner. The apolitical nature of the forum would address the politicization of climate change science in the United States and likely in other countries.
- Through this alliance, we could provide unbiased information about climate change and its solutions in every church, every synagogue, every mosque, every temple, and every other place of worship in the world. Doing so would be one of the most expeditious way to bring climate awareness to the world and would have a good chance of garnering massive public support for solving the problem.
- Religion and science both desire the protection of creation, or nature. This synergy in the ultimate goal would prevent the usual conflicts between science and religion.
- Because of the delay in taking meaningful actions to slow climate change, it has become a major ethical and equity issue. Places of worship offer the most natural setting to discuss and remediate such ethical and moral problems.

In seeking an alliance with religion, we have to be aware that many scientists have rejected religion as faith based, not fact based. We often forget that religions have given us tool kits for survival through the ages, as well as numerous cultures and forums to contemplate our humanity and what it is to be human. In this moment of existential threat, we need to put aside our differences and work together for the common good of protecting future generations.

However, in forming such an alliance, we must address the following questions:

- Can science and religion find a common language?
- Can science and religion work together to contemplate the sustainability of nature and humanity?
- How do we deal with the pace of technological change?

- How do we change our attitude toward nature and toward one other?
- What do we owe our children, grandchildren, and generations yet unborn?
- How do we improve people's trust in science?
- How do we educate scientists on the value of religion in protecting nature and humanity?

An alliance between science and religion will enable us to address the fundamental ethical and moral questions related to climate disruption.

Note

1. "Bending the Curve: 10 Scalable Climate Solutions," University of California San Diego, https://btc10solutions.ucsd.edu.

CHAPTER 30

THE RELIGIOUS CASE FOR ENVIRONMENTAL CARE

EMMANUEL ADAMAKIS

The encyclical of the Holy and Great Council of the Orthodox Church, convened in Crete in June 2016, states the following:

> The roots of the ecological crisis are spiritual and ethical, inhering within the heart of each man. This crisis has become more acute in recent centuries on account of the various divisions provoked by human passions—such as greed, avarice, egotism and the insatiable desire for more—and by their consequences for the planet, as with climate change, which now threatens to a large extent the natural environment, our common "home." The rupture in the relationship between man and creation is a perversion of the authentic use of God's creation. The approach to the ecological problem on the basis of the principles of the Christian tradition demands not only repentance for the sin of the exploitation of the natural resources of the planet, namely, a radical change in mentality and behavior, but also asceticism as an antidote to consumerism, the deification of needs and the acquisitive attitude. It also presupposes our greatest responsibility to hand down a viable natural environment to future generations and to use it according to divine will and blessing. In the sacraments of the Church, creation is affirmed and man is encouraged to act as a steward, protector and "priest" of creation, offering it by way of doxology to the Creator—"Your own of your own we offer to You in all and for all"—and cultivating a Eucharistic relationship with creation. This Orthodox, Gospel and Patristic approach also turns our attention to the social dimensions and the tragic consequences of the destruction of the natural environment.[1]

Environmental care and sustainable development constitute challenges that pertain to justice and responsibility in a world in crisis. The word "crisis" in Greek means "judgment." These issues point to the consequences of our economic systems and challenge our lifestyles. Speaking of a challenge means to be invited to think further and to act! It means to recognize the problem by seeking a solution. A challenge calls out for a solution, stimulates, puts things in motion. How, then, can we think of environmental care as a challenge pertaining to justice and responsibility, and a challenge that is connected to religion?

A Challenge for Our Confession of God, of the World, and of Humanity

Today's ecological challenge is not only related to globalization; it is also (geo-) political, economic, and philosophical. What about the spiritual challenge? Are we not misled by seeing ourselves as masters and possessors of nature, a nature the sole purpose of which is to serve us? According to the Scriptures, we confess the world as God's creation, in which life thrives and one can sense the divine.[2] We therefore believe that we are an integral part of this good creation and recognize that the destinies of nature and humanity are intimately interrelated. In this regard, the biblical texts teach us that God has given us the "faithful and prudent" stewardship of creation.[3] To protect humanity is to protect creation, and to protect creation is to protect humanity.

No Environmental Care Without Conversion

As the Ecumenical Patriarchate never ceases to repeat with other Christians, and especially with His Holiness, Pope Francis, along with other religious leaders, protecting the environment must be a common goal. While some political leaders of the world have rejected the 2015 Paris Agreement on climate change, faith-based institutions have taken on the crucial task of raising awareness of the dangers related to the destruction of the natural environment. An ecological spirituality should be a spirituality of conversion. Conversion requires the transformation of the inner self as the starting point of an external change. Scientists emphasize tirelessly the need for a radical change in our lifestyles in order to reduce the polluting activities that contribute to climate change. This is a reality that Christianity calls *metanoia*, a reversal of the whole being. In the patristic tradition of the Desert Fathers—those seekers of the spiritual forged through centuries of ascetic experience—this *metanoia* encourages a clear-sighted way of regarding humanity. It is precisely this vision that was imagined by Saint Isaac the Syrian, a mystic of the seventh century who considered the goal of the spiritual life to be the acquisition "of a merciful heart that burns with love for all creation . . . for all of God's creatures."

To meet the challenge of environmental care, we must also take on a spiritual challenge: to change our lifestyles. The spirit of conversion in Christian spirituality calls for an in-depth mutation, a conversion of the being that simultaneously touches on and exceeds environmental issues. Love for one's neighbor—present and future—will override selfishness. The collective action of believers will apply pressure on world leaders and decision-makers. Sobriety will respond to the appetites of over-consumerism. Sharing will reduce inequality. Finally, charity will encompass the political and social spheres. Through prayer and commitment, we can be led to a new life, to the possibility of a sustainable, just, and peaceful society. As Ecumenical Patriarch Bartholomew has said, "We cannot separate our concern for human dignity, human rights or social justice from concern for ecological protection, preservation and sustainability. These concerns are forged together, comprising an intertwining spiral that can either descend or ascend."[4]

Our Lifestyles: From Conversion to Transfiguration

A structural transformation of our lifestyles and consumption patterns cannot be achieved unless all stakeholders commit themselves to this purpose, which includes citizens, nongovernmental organizations, social movements, businesses, and communities, as well as church and spiritual movements. This transformation involves identifying which levels of decision-making and organization, from the local to the regional, from the national to the international, are most appropriate for engaging and address particular environmental and social challenges. The evidence of the climate emergency is forcing us to develop a new imagination, new alternatives, and new forms of commitment. We must question policy-makers, get involved in local projects, call for the accountability of political and economic actors, and develop local and communitarian solutions. Politics must be put back in the service of community, and citizens must be put at the center of politics and decision-making mechanisms, with the common good as the ultimate goal. Faced with the global crisis of climate change, religions must renew the word they preach. They must remind the world of fraternity as the essential basis of unconditional inclusivity, of justice and solidarity for the most vulnerable, of peace for all as a condition for good living, of simplicity and gratitude as seeds of a renewed hope.

Climate Change: A Challenge that Will Open Our Eyes

Climate change is a sign of spiritual blindness. But the scales may fall from our eyes, as we see ourselves as part of the creation and acknowledge God in

His beauty. In our societies, which have become increasingly urbanized and oppressed, we are invited to take time out, to go out to meet creation and to attend to the Scriptures—and why not combine the two by reading our Bible in nature! There, through contemplation and meditation, with the help of the Holy Spirit, we can be led to a spiritual rebirth. Feeling gratitude for what has been given to us will lead us to the realization of our responsibility to protect creation and to share its gifts. It will also make us more humble, for our understanding of creation is not exhausted by scientific knowledge, nor can it be mastered by technology. Its forces and its mystery go well beyond us.

It is therefore our moral obligation to engage actively in favor of environmental care as the manifestation of a universal ethos that is non-negotiable with special interests. It is not too late to act, but we cannot allow ourselves to put off until tomorrow what we can do today. We are all in agreement about affirming the need to protect the natural resources of our planet, which are not infinite and should not be considered commodities to be traded. The unity of humanity concerning the protection of creation, which we are calling for, at the same time obliges us to place our acts in accordance with our words. I therefore conclude with the words of His All-Holiness Ecumenical Patriarch Bartholomew in his Message for World Oceans Day on June 8, 2016:

> If we have created the dire conditions that we now face, we are equally accountable for and capable of remedying the health of our environment. Each of us must learn to appreciate the way in which our individual and collective lifestyles—our choices and priorities—impact the environment.[5]

Notes

1. "Encyclical of the Holy and Great Council of the Orthodox Church," Holy and Great Council, June 2016, sec. 14, https://www.holycouncil.org/-/encyclical-holy-council.
2. See Genesis 1:1 to 2:25.
3. Luke 12:42.
4. Bartholomew, "Reflections by Ecumenical Patriarch Bartholomew," Orthodox Fellowship of the Transfiguration, November 4, 2015, https://www.orth-transfiguration.org/creation-care-and-ecological-justice/.
5. Bartholomew, "Message by His All-Holiness Ecumenical Patriarch Bartholomew for World Oceans Day," Orthodox Fellowship of the Transfiguration, June 8, 2016, http://www.orth-transfiguration.org/wp-content/uploads/2016/11/Message-By-His-All-Holiness-Ecumenical-Patriarch-Bartholomew-June-8-2016.pdf.

CHAPTER 31

HEALTH JUSTICE IS CLIMATE JUSTICE

TED SMITH AND CHRISTINA LEE BROWN

Times of crisis demand that we be as clear as possible: Our global community is fundamentally unhealthy. We have let other values, such as profit through cheap labor or unpriced natural resources, take the front seat. This has given rise to an obsession with economic power and limitless growth. We must set a new course immediately.

Health, as originally conceived by the World Health Organization in its founding charter, is "a state of complete physical, mental and social well-being and not merely the absence of disease or infirmity."[1] This definition can be further elaborated to include considerations of our physical health, psychological health, intellectual health, environmental health, financial health, nutritional health, cultural health, and spiritual health. Dating back to Aristotle, it has been considered that the good life of each person and the good society depend on this kind of multidimensional and interdependent understanding of health and well-being. Achieving a new understanding of health as our highest value will guide us most directly to the many important changes that must be made.

Our Biosphere: Air, Water, and Soil

First-time travelers to space generally report that the thin film of soil, water, and gas that supports all life appears fragile and miraculous. What do astronauts see that strip miners, fracking technicians, and rainforest loggers fail to see? This biosphere supports all life on Earth. A healthy environment, by definition, is one that supports life. Yet we have been extracting and dumping, all over the planet, under the false hope that we will never have to experience consequences. For many, the irreversible perils of climate change are in fact just that—instead of just pockets of human suffering making occasional headlines—such as those caused by the Great Smog of London, the Love Canal disaster, the Deepwater Horizon oil spill, and mountaintop-removal coal mining—we have ruined it all.

Clean Air Is Much More Than Carbon

The World Health Organization has identified air pollution as the number-one health issue of our time, with nine out of ten people in the world breathing unhealthy air each day. To the astronaut floating above the earth, it is plain to see that your air is my air and that in a closed system in which air is recirculated, air pollution in one place will find its way around the planet or be exchanged as water or soil pollution. Unlike the naive industrialists of the last century, we know that pollution does not go away. Air pollution has been associated with a long list of expensive and disabling conditions, from heart disease to depression to infant mortality. More broadly, the environmental injustice of air pollution compresses the financial health of the people living in close proximity to sources of pollution. In the United States, it is most often people of color who live near coal-burning power plants and other sites of dirty industry. In the developing world, too, wealth buys healthier surroundings. It is no surprise, then, that those living under a smokestack cannot accumulate wealth from the land they live on, as its value has been destroyed by pollution. Climate change for the people in these places is not a problem for the future; it has been, and is, a reality now with awful and immediate consequences. The injustice is all around us, and, for perhaps the first time ever, the fates of the rich will join those of the poor. If we continue to avoid making decisions through the lens of health, it is not likely to be the moral imperative to right a wrong but the terror of having nowhere to hide that will motivate action. If we commit ourselves to considering the health injustices that our public, private, collective, and personal values are creating, we will achieve climate justice through decisions to rely on renewable energy, less-extractive industries, and circular economies.

Toward a Culture of Health

With a foundation of healthy air, water, and soil, we can re-form our human society. To be just, we must understand how other aspects of health can be out of balance in inequitable circumstances. The earlier example of financial health inequities that come from living alongside pollution is just the start. At the core of the term "sustainability" is the notion that the outputs of one activity or process become the inputs for another. It is in this way that oxygen naturally enters the atmosphere and cycles through water, animals, and plants to ultimately be liberated again as oxygen, cycling over and over, maintaining life.

Many of the forms of health that we celebrate in human society are in deficit when we don't respect each life around us as necessary and of value. The financial health of a community must be aligned with the financial health of each individual within the community. There can be no sustainable prosperity in a system that creates profound inequity. As income gaps widen in the Western world, there is a visible strain as other forms of health collapse under the pressure. For example, for the last three years in the United States, the average human life span has begun to decrease owing to "diseases of despair," which include suicide, drug overdose, and alcoholism. Despair preys on those who have given up hope of having a good life, withing for freedom from unending suffering and burden.

Similarly, a failure to value nutritional health has given rise to a "food science" industry in which we no longer teach nutrition or how to cultivate our food. It is "agri*culture*," not "agri-*industry*," that made sure food ended up as compost that fertilized the soil to grow the next crop. Treating food cultivation like a factory process has given us an obesity epidemic, exposed workers to toxins, imperiled much of the biodiversity of the planet, and, again, endangered our own resiliency.

Another form of health we know should not be out of balance is our intellectual health, the value we place on learning about ourselves and our surroundings. Again, a singular lesson of climate change has been the very high and broad levels of illiteracy about the realities of how we are living, as well as a disregard for science. When we don't respect the capacity that we have as individuals and as communities for lifelong learning, the resulting ignorance leads to an unlivable climate, conflict, and the logic of moral hazard.

Ever since humans began painting on cave walls, there have been music and dancing and so many other truly human creative acts that can be described as examples of our cultural health. Here, too, injustice takes many forms, with a departure from creative culture as a shared social value being the beginning of unsustainability. The most resilient cultures have been those that work together, play together, and pray together. As segregation enters these domains, cultural health inequities erode our capacity to work together to tackle existential threats such as plagues, dictators, and natural disasters. A consumption culture is not a shared culture. It depends on winners and losers. Healthy cultures convey ancient wisdom and shared values. We would all benefit from placing the right value on cultural health to have a people to save.

Spiritual Health

More than a quarter-century ago, a diverse group of faith leaders gathered to celebrate the things that all faith traditions share in common. Annually, Louisville's Festival of Faiths continues to celebrate these different faith traditions for the universal wisdom they share. The Ethics in Action initiative has created something similar as it calls on faith leaders from all traditions to bring their faith values to the many injustices of our time. Much like the plight of intellectual health and the associated "war on science," spiritual health has had an analog in a celebration of secular ethics without faith. Here, too, lies the erosion of our ability to ask, Why am I here? What is my purpose? What will happen when I die? These are not questions for science. They are questions for spirituality. When the faithful are put at a great distance from the faithless, there can be tension and unhealthy dialogue. Here the injustice is the failure to see the common heart within us all. Our religious leaders know that as the least of us suffer, we all suffer. Our diseases of despair have been treated for centuries by many different faith systems. The act of valuing every human life happens reliably when we value spiritual health.

As the World Health Organization suggested long ago, let's value health as much more than the absence of disease. Let's try to create a world where each one of us can flourish.

Note

1. "Constitution," World Health Organization, accessed August 20, 2021, https://www.who.int/about/governance/constitution.

CHAPTER 32

ENVIRONMENTAL INJUSTICE

How Treaties Undermine the Right to a Healthy Environment

LISA SACHS, ELLA MERRILL, AND LISE JOHNSON

Our planet faces unprecedented threats, including irreversible global warming, loss in biodiversity, and water pollution and scarcity. The impacts of these environmental crises also threaten human rights and exacerbate inequality within and between countries and populations.[1] Slowing or halting these worsening environmental trends—and addressing the impacts of environmental change on populations—will require cumulative policy responses at the national and international levels. Fortunately, efforts to align globally around these challenges are underway: The 2030 Agenda emphasizes and elaborates on the environmental crises of and goals for the coming decade, including specific targets for governments, private-sector entities, and other actors to work toward. The proposed Global Pact for the Environment aims to codify a right to a healthy environment and to create a framework for the international protection of that right.[2]

In the race to take action to protect the environment and ensure humanity's sustainable use of the planet's resources, many policy-makers and advocates have overlooked how provisions buried in thousands of bilateral and multilateral investment agreements have stymied such critical policies and threaten to do so at an increasing scale. Specifically, the investor-state dispute settlement (ISDS) mechanism, found in most international investment agreements, including both bilateral investment treaties (BITs) and the investment chapters of free trade agreements (FTAs), allows foreign investors to sue host governments for government measures that undermine their profitability, no matter the objective of the challenged measure.

This private dispute mechanism limits governments' ability to adopt and enforce policies meant to protect citizens and the environment from potentially harmful corporate activities and thus has important implications for the future of environmental justice. The impacts of these treaties and of ISDS have only recently become clear; most of the more than 3,300 treaties have been signed within the past forty years or so.[3] Disputes brought under the ISDS mechanism of these treaties are even more recent; as of the end of 2018, 944 known disputes had been filed, with more disputes brought confidentially[4] and others threatened and then settled, formally (23 percent of all cases brought between 1987 and 2018)[5] or informally, before an award was issued.[6] Half of all known disputes were brought between 2012 and 2018,[7] and, in 2018, 70 percent of the fifty substantive decisions by tribunals were made in favor of the investors.[8]

Investor-state dispute settlement allows investors to enforce vaguely worded treaty protections against discrimination and unfair treatment, bringing the disputes outside of domestic administrative and judicial processes to be decided by an ad hoc, party-appointed, party-paid tribunal, unconstrained by domestic rules and procedures and often without consideration or deference to other obligations of the host government or the investor under domestic or international law.[9]

Investors have increasingly—and overwhelmingly—used the ISDS mechanism to challenge government measures of general applicability, including those taken in the public interest to protect the life, health, and security of people or the planet, and have been awarded substantial sums by the tribunals, out of public coffers, to compensate investors for economic impacts on their investments, including, in many cases, future lost profits from an impacted investment. Alarmingly, alongside national and international efforts to protect the environment from mounting risks—many of which are caused by the very investors that these treaties protect—these awards are quietly undermining environmental conservation, governance, rights, and justice. This chapter highlights several areas of environmental justice that have been impacted by ISDS cases, including in particular measures related to climate action,[10] protection of water resources,[11] antipollution regulations,[12] and efforts to protect communities' rights to representation and access to justice.[13]

Implications for Climate Action

The international scientific community's assessment that the world needs to strand 80 percent of proven fossil fuel reserves and transition to zero-carbon energy systems[14] in order to avoid the most disastrous consequences of global warming has enormous implications for global investments. While trillions of dollars of new investments will be required to meet growing demands for clean energy, significant existing investments in fossil fuel extraction, transmission, and processing will have to be urgently phased out. In line with goals and commitments agreed to by the international community, individual countries are increasingly adopting a range of policy tools to shift energy generation and transmission, including phasing out coal-fired power plants, adopting mechanisms such as carbon-pricing schemes, or employing other regulatory tools, such as fuel efficiency standards, methane leakage, and emissions controls. Additional measures will be necessary, including the revocation or modification of permits for exploration for, extraction of, and power generation from fossil fuel resources.

Each of these measures will impact the profitability (and in some cases, the viability) of investments related to carbon-intensive energy, the types of economic impacts that have triggered ISDS claims alleging unfair treatment, regulatory expropriation, de facto discrimination, and other breaches of treaty protections.

Indeed, we've already started seeing the first cases. In 2015, following years of delay, the United States government rejected TransCanada's construction permit to build the Keystone XL Pipeline, stating that the pipeline and associated activities "would not serve the national interests of the United States." President Obama announced that the pipeline would undercut efforts on the part of the government to make the United States a leader in climate action.[15] In its US$15 billion North American Free Trade Agreement (NAFTA) suit against the United States, the company alleged it had borne substantial expenses over the course of the delay and sought compensation for future lost profits it allegedly expected to earn from the development and operation of new fossil fuel infrastructure.[16] Ultimately, the case was withdrawn by TransCanada after President Trump was elected and approved the resubmitted permit application.[17]

Also in 2015, the new provincial government of Alberta, Canada, announced that it would phase out its coal-fired power plants by 2030 under its Climate Leadership Plan.[18] The plan included compensation of $1.4 billion to three companies owning coal-fired plants for the early closure of their plants. In response, Westmoreland, whose mine-mouth operations supplied coal to the majority of phased-out operations, alleged that Alberta's compensation of the coal-fired power plants but not of the coal mines themselves, which by design were able to serve only the associated plants, breached NAFTA treaty obligations to provide fair and equitable treatment. Westmoreland claimed $357 million in damages.[19]

Another investor, Lama Energy Group, also sued Canada, under the Canada-Czech Republic BIT, alleging that the Government of Alberta was unduly delaying approvals for their oil sands project, in light of the government's environmental concerns.[20] The policies that triggered these actions, however, were subsequently reversed; after the election of conservative Premier Jason Kenney in June 2019, Lama received regulatory approval for its project despite pushback from First Nations groups who claim the activities will put sacred lands and drinking water at risk.[21]

In all three cases, changes to climate-related policy, made in the hands of less environmentally minded governments, led to the withdrawal of arbitration by the investor, indicating a discouraging trend in global policy-making.

As cities, states, and countries continue to adopt policies in line with their international commitments and in response to domestic pressures and climate-related incidents to address the mounting climate crisis, the threat of additional ISDS cases mounts. Most recently, in response to the Dutch government's announcement in 2018 that it would shut down all coal-fired power plants by 2030, Uniper, the owner and operator of one of the country's largest power plants, threatened arbitration under the Energy Charter Treaty if the legislation were passed into law.[22] The risks of ISDS are both that governments will be deterred from taking such critical measures (or will reverse measures in response to a filed suit) and that governments will provide free risk insurance to investors for fossil fuel–related investments, distorting investment decisions and incentives and increasing the public costs of climate action.

Implications for Water Access

Treaty protections, as they have been interpreted by investment tribunals, have also allowed investors to challenge governments' ability to protect water resources from contamination and ensure that water is accessible and affordable to all. In Romania, Gabriel Resources was developing the Rosia Montana gold and silver mine when conflicts arose surrounding the potential threats the mine posed to nearby water sources. Pressured in part by environmental groups, Romania did not issue necessary environmental permits for the mine for fear of cyanide pollution, particularly given recent memory of the cyanide spill and subsequent environmental disaster in 2000, in which Romania was found by the European Court of Human Rights to be liable by failing to conduct an adequate environmental impact assessment (EIA).[23] Gabriel Resources filed a request for arbitration in 2015, contending that the rejected EIA breached the fair and equitable treatment standard, among other treaty provisions.[24] As of 2017, they had invested $700 million to finance the project.[25] The company is seeking $4.4 billion in damages from Romania, roughly 2 percent of Romania's GDP.[26]

Echoing the Gabriel Resources case but on the other side of the Atlantic Ocean, a company, Lone Pine, which holds permits for petroleum and natural gas exploration in the Utica basin in Canada, claimed $100 million from Canada after a Quebec moratorium on fracking below the St. Lawrence River, in response to concerns about the impacts of fracking and other development activities on the water source.[27]

In Colombia, a series of cases were brought against the country after the Constitutional Court of Colombia ordered the prohibition of mining activities in *páramos* (wetland) regions in 2016 in order to preserve an important source of the country's water supply. As a consequence of this decision, the National Mining Agency reduced the size of several mining companies' concession areas in order to exclude areas located within conservation zones. Eco Oro Minerals, Red Eagle Exploration, and Galway Gold all brought cases against Colombia in response. Noting that none of the three cases has been settled, combined, their claims amount to almost $1 billion under the free trade agreement between Colombia and Canada.[28]

Implications for Environmental Protection

Environmental impact assessments are a widely accepted and relied upon feature of licensing processes to inform governments and other stakeholders of anticipated environmental impacts from projects seeking approval; the assessed impacts are intended to guide decision-making, including the gateway decision of whether to approve a proposed project and then whether and how to influence its design. While domestic institutions have pathways for challenging any determinations made on the basis of an EIA, investors have used ISDS to bypass those domestic processes, including domestic rules and procedures, alleging that EIA processes and outcomes have breached more favorable treaty protections.

Bilcon, an American mining company, sought to develop a mining and marine terminal project in Canada and was required to obtain various approvals from provincial and federal authorities. As part of the EIA, an expert panel was assembled with the mandate to provide a nonbinding opinion on whether the project should proceed in light of its potential social and environmental impacts. The panel proposed that the project be rejected in light of its anticipated impacts, including that the project was inconsistent with "core community values." Nova Scotia and the federal government then rejected the project, based on the panel's recommendation. The Government of Canada stated that the mining project was likely to cause negative environmental impacts that were protected against under the Canadian Environmental Assessment Act.[29]

The tribunal ruled in Bilcon's favor, finding that the government's decision to deny the project violated NAFTA's investment chapter. The tribunal ordered Canada to pay $7 million (though the investor had sought hundreds of millions in damages),[30] stating that the "advisory panel's consideration of 'core community values' went beyond the panel's duty to consider impacts on the 'human environment,'" in violation of NAFTA.[31]

Canada unsuccessfully challenged enforcement of the award in federal court; the presiding judge acknowledged that the decision raised "significant policy concerns," including "its effects on the ability of NAFTA Parties to regulate environmental matters within their jurisdiction, the ability of NAFTA tribunals to properly assess whether foreign investors have been treated fairly under domestic environmental assessment process, and the potential 'chill' in the environmental assessment process that could result from the majority's decision," but that the federal court had a very limited scope to review the tribunal's determination.[32] Indeed, the risk of regulatory chill, that is, that the threat of an ISDS dispute will dissuade governments from adopting certain policies, has been increasingly recognized, even though it is difficult to document.[33]

Implications for Representation

The extraordinary rights that ISDS confers on investors comes, in fact, at an even greater cost to the rights of other stakeholders, including domestic citizens who are impacted by the investments. Copper Mesa Mining Corporation's exploration concession in the Junin region of Ecuador, for example, faced great opposition from the community, partly owing to the environmental concerns surrounding potential mining activities. Copper Mesa's response is widely recognized to have violated basic standards of decent corporate conduct, including, among other things, using threats and intimidation. The company hired a private security force, which fired pepper spray and live rounds of ammunition into crowds of protestors. In response to the escalating conflict at the mine site, Ecuador eventually revoked Copper Mesa's concession, citing Copper Mesa's failure to consult the community.

In light of Copper Mesa's egregiously irresponsible and aggressive response to the community, a group of citizens sued Copper Mesa and the Toronto Stock Exchange in Canadian courts, but the case was dismissed for lack of jurisdiction.[34] Copper Mesa, however, successfully sued Ecuador before an arbitration tribunal, which found in Copper Mesa's favor. The tribunal indicated that Ecuador's failure had been in siding with its own citizens and responding to the escalating domestic crisis; the tribunal wrote, "It's of course difficult to say now what [Ecuador] should have done to resolve all the claimants' difficulties and still more so whether anything it could have done would have changed the claimants' position for the better. Plainly the Government of Quito could hardly have declared war on its own people. Yet in the Tribunal's view, it could do nothing."[35] Ecuador was ordered to pay $19 million in damages to the company.[36]

The pattern of these outcomes is, unfortunately, not unique. More recently, in Armenia, local communities protested the Amulsar gold project, owned by Lydian International, because of concerns over the mine's environmental impacts on nearby lakes, mineral springs, and agricultural land, particularly in light of its location in an area with significant seismic activity.[37] Lydian's subsidiaries in Canada and the United Kingdom filed arbitration requests against Armenia after the project was temporarily shut down because of protests, though these have not yet been brought to litigation; Lydian claims already to have invested $500 million in the project.[38] As in so many cases, the threat of arbitration seemed sufficient to change a government's mind; in August 2019, Prime Minister Nikol Pashinyan announced that mining could proceed,[39] saying that the project posed no environmental threat. At the same time, Lydian announced that a domestic court confirmed that criminal proceedings could continue against some of the community protestors.[40] Two weeks after Pashinyan made his announcement, on a call hosted by the Armenian government to discuss a private consulting group's audit of the project and relevant EIAs, the consulting group refused to conclude that the project was ecologically safe, stating that another EIA could be necessary.[41] In response, Lydian stated in a press release that it was "deeply disappointed" in the results of this discussion, complaining of unfair treatment by the Government of Armenia.[42] It is once again unclear whether the Armenian government will allow the project to proceed.

Conclusion and Recommendations

Notably, many treaties, especially free trade agreements, recognize the importance of bilateral or multilateral cooperation on environmental protection and preservation. Treaties include nonderogation provisions, committing treaty partners not to lower environmental standards or enforcement in order to attract or retain investment. NAFTA created the Commission for Environmental Cooperation[43]; the United States–Peru Free Trade Agreement includes an annex on forest governance[44] that initially earned the support of environmental groups (though that support has waned in implementation)[45]; and the Japan-Mexico Economic Partnership Agreement, "recognizing the need for environmental preservation and improvement to promote sound and sustainable development," commits parties to a number of cooperative activities.[46] However, these provisions to promote cooperative environmental protection and governance have had disappointing results[47] for a variety of reasons, including the lack of enforcement mechanisms and the will to enforce them.

By contrast, the dispute-settlement provisions in the investment chapter, which can be enforced directly by investors without the involvement of their home state, have much sharper teeth, and investors and their counsel have demonstrated their eagerness to bite. Vaguely worded or circuitous treaty exceptions for measures taken to protect the environment or other public interests rarely limit the liability of countries; tribunals have found that such measures were applied discriminatorily or arbitrarily or that they may be allowable but still require compensation for an affected investor.

In large part because of the growing number of cases that have successfully challenged public interest measures and awarded substantial damages to investors, governments and especially their citizens are starting to question the legitimacy of ISDS and its suitability for twenty-first-century governance challenges. Research also suggests that in addition to undermining or discouraging important regulatory measures, the procedural and substantive aspects of ISDS can also exacerbate inequality[48] and undermine the rule of law.[49]

As the number of ISDS cases increases year on year, the scope of challenged measures widens, and as the costs of both the litigation and the resulting awards continue to increase, states should strongly consider a moratorium on disputes, if not the outright termination of treaties, while discussions on more suitable, SDG-aligned, and rule-of-law-respecting governance mechanisms are explored and negotiated.[50] Investment has a critical role to play in achieving the Sustainable Development Goals, including those on environmental protection and restoration and access to justice; accordingly, investment governance—including through international commitments and cooperation—has an equally critical role in shaping those investment flows, and their contributions to and impacts on sustainable development. Enough ISDS cases have illustrated the tremendous risks of putting enforceable investor protections at the heart of investment governance. Global investment governance needs to be redesigned for the twenty-first century, with people and the planet at the core.[51]

Notes

1. Greta Reeh, "Human Rights and the Environment: The UN Human Rights Committee Affirms the Duty to Protect," *Blog of the European Journal of International Law*, September 9, 2019, https://www.ejiltalk.org/human-rights-and-the-environment-the-un-human-rights-committee-affirms-the-duty-to-protect/.
2. "Objectives," Global Pact for the Environment, accessed September 6, 2019, https://globalpactenvironment.org/en/.
3. "Primer: International Investment Treaties and Investor-State Dispute Settlement," Columbia Center on Sustainable Investment, updated May 31, 2019, http://ccsi.columbia.edu/2019/06/03/primer-international-investment-treaties-and-investor-state-dispute-settlement/.
4. Not all arbitration rules require disclosure of the legal challenge or of the related case materials. "Primer: International Investment Treaties and Investor-State Dispute Settlement."
5. "Review of ISDS Decisions in 2018: Selected IIA Reform Issues," *UNCTAD IIA Issues Note*, Issue 4 (July 2019).
6. For example, a report listing possible reasons for China's previously low profile in ISDS cases suggests this may be because of the country's preference for settling such disputes informally through diplomatic discussion. Dilini Pathirana, "A Look Into China's Slowly Increasing Appearance in ISDS Cases," *Investment Treaty News*, September 26, 2017.

7. "Primer: International Investment Treaties and Investor-State Dispute Settlement."
8. Either on jurisdictional grounds or on merits. "Review of ISDS Decisions in 2018."
9. "Primer: International Investment Treaties and Investor-State Dispute Settlement."
10. E.g., *TransCanada v. US*, Case No. ARB/16/21 (ICSID. 2016).
11. E.g., *Vattenfall I v. Germany (I)*, Case No. ARB/09/6 (ICSID. 2009); *Lone Pine v. Canada*, Case No. UNCT/15/2 (ICSID. 2013).
12. E.g., *Vattenfall I v. Germany (I)*, Case No. ARB/09/6 (ICSID. 2009); *Renco v. Peru*, Case No. UNCT/13/1 (ICSID. 2011).
13. E.g., *Copper Mesa v. Ecuador*, Case No. 2012–2 (PCA. 2011); *Bear Creek Mining v. Peru*, Case No. ARB/14/21 (ICSID. 2014).
14. James Leaton, Nicola Ranger, Bob Ward, Luke Sussams, and Meg Brown, *Unburnable Carbon 2013: Wasted Capital and Stranded Assets* (London: Carbon Tracker Initiative and Grantham Institute at the London School of Economics and Political Science, 2013).
15. Bill Chappell, "President Obama Rejects Keystone XL Pipeline Plan," *NPR*, November 6, 2015.
16. Ethan Lou, "TransCanada's $15 billion U.S. Keystone XL NAFTA Suit Suspended," *Reuters*, February 28, 2017.
17. Lou, "TransCanada's $15 billion U.S. Keystone XL NAFTA Suit Suspended."
18. Alberta Climate Change Office, *Climate Leadership Plan: Implementation Plan 2018–19* (Edmonton: Government of Alberta, June 2018).
19. *Westmoreland v. Canada* (UNCITRAL, 2018).
20. Damien Charlotin, "Frustrated by Delays in Licensing Process Czech Oilsands Investor Puts Canada on Notice of a Claim Under Bilateral Investment Treaty," *IA Reporter*, April 8, 2019.
21. Canadian Press, "Alberta Oil Sands Project Wins Regulator Approval Despite Indigenous Objections," *Globe and Mail*, June 13, 2018.
22. Uniper, *Annual Report 2017: Financial Results* (Düsseldorf: Uniper, 2017), https://www.uniper.energy/sites/default/files/2018-03/2018-03-08_fy2017_uniper_annual_report_en.pdf, 23; Bart Meijer, "Netherlands to Ban Coal-Fired Power Plants in Blow to RWE," *Reuters*, May 18, 2018, https://www.reuters.com/article/us-netherlands-energy-coal/netherlands-to-ban-coal-fired-power-plants-in-blow-to-rwe-idUSKCN1IJ1PI; Edwin Van Der Schoot, "Claim for Coal Prohibition for State," *De Telegraf*, September 5, 2019, https://www.telegraaf.nl/financieel/1134267479/claim-om-kolenverbod-voor-staat; Damien Charlotin, "Netherlands Poised to Face Its First Investment Treaty Claim, Over Closure of Coal Plants," *IA Reporter*, September 7, 2019, https://www.iareporter.com/articles/netherlands-poised-to-face-its-first-investment-treaty-claim-over-closure-of-coal-plants/.
23. Clovis Trevino, "As Romania Grapples with Mining Regulation in Aftermath of 2000 Environmental Catastrophe, a Foreign Investor Loses Patience with Delays," *IA Reporter*, January 21, 2019; *Tătar v. Romania*, Case No. 67021/01 (ECHR. 2009).
24. *Gabriel Resources v. Romania*, Case No. RB/15/31 (ICSID. 2015).
25. Gabriel Resources Ltd., *Annual Information Form of Gabriel Resources Ltd. for the Year Ended December 31, 2017* (London: Gabriel Resources Ltd., April 30, 2018), http://www.gabrielresources.com/site/documents/AIF_2018_Master_Filing_300418.pdf.
26. "Romania," World Bank Group, accessed August 7, 2019.
27. *Lone Pine v. Canada*, Case No. UNCT/15/2. (ICSID. 2013).
28. *Eco Oro v. Colombia*, Case No. ARB/16/41. (ICSID. 2016); *Red Eagle v. Colombia*, Case No. ARB/18/12. (ICSID. 2018); *Galway Gold v. Colombia*, Case No. ARB/18/13. (ICSID. 2018).

29. *Clayton/Bilcon v. Canada*, Case No. 2009–04 (PCA. 2008).
30. In addition to pre-award interest for the previous twelve years. Luke Peterson, "Canada Ordered to Pay $7 Million for Botched Environmental Review, but NAFTA Arbitrators Reject I.S. Investors' Bid for $400+ Million in Lost Profits," *IA Reporter*, February 25, 2019.
31. *Clayton/Bilcon v. Canada*, Case No. 2009–04 (PCA. 2008).
32. "Judgement of the Federal Court of Canada," *2018 FC 436 (Conclusion)* (2018).
33. UNCITRAL, *Report of Working Group III (Investor-State Dispute Settlement Reform) on the Work of Its Thirty-Seventh Session (New York, 1–5 April 2019)* (Vienna: UNCITRAL, 2019); Lise Johnson and Jesse Coleman, "International Investment Law and the Extractive Industries Sector," *Columbia Center on Sustainable Investment Briefing Note*, table 2 (January 2016); Kyla Tienhaara, "Regulatory Chill in a Warming World: The Threat to Climate Policy Posed by Investor-State Dispute Settlement," *Transnational Environmental Law* 7, no. 2 (July 2018): 229–50; Julia Brown, "International Investment Agreements: Regulatory Chill in the Face of Litigious Heat?" *Western Journal of Legal Studies* 3, no. 1 (2013); Gus Van Harten and Dayna Nadine Scott, "Investment Treaties and Internal Vetting of Regulatory Proposals: A Case Study from Canada," *Journal of International Dispute Settlement* 7, no.1 (2016): 92–116.
34. "Copper Mesa Mining Lawsuit (re Ecuador)," Business & Human Rights Resource Centre, March 3, 2009, https://www.business-humanrights.org/en/latest-news/copper-mesa-mining-lawsuit-re-ecuador/.
35. *Copper Mesa v. Ecuador*, Case No. 2012–2 (PCA. 2011).
36. *Copper Mesa v. Ecuador*, Case No. 2012–2 (PCA. 2011).
37. Raffi Elliot, "PM Pashinyan Approves Amulsar Mining," *Armenian Weekly*, August 19, 2019, https://armenianweekly.com/2019/08/19/pm-pashinyan-approves-amulsar-mining/.
38. Lisa Bohmer, "Armenia Allows Lydian's Gold Mining Project to Proceed After Company's Threat of Investment Treaty Arbitration," *IA Reporter*, August 22, 2019, https://www.iareporter.com/articles/armenia-allows-lydians-gold-mining-project-to-proceed-after-companys-threat-of-investment-treaty-arbitration/; "Lydian Announces Submission of Notices to Government of Armenia Under Bilateral Investment Protection Treaties," Lydian International, March 11, 2019, https://www.lydianinternational.co.uk/news/2019-news/452-.
39. Bohmer, "Armenia Allows Lydian's Gold Mining Project to Proceed."
40. Bohmer, "Armenia Allows Lydian's Gold Mining Project to Proceed."
41. Raffi Elliot, "Armenian Government Still Undecided on Amulsar Issue," *Armenian Weekly*, September 4, 2019, https://armenianweekly.com/2019/09/04/armenian-government-still-undecided-on-amulsar-issue/; "Lydian Is Deeply Disappointed with ELARD's Misleading Comments on Skype Call with the Government of Armenia," Lydian Armenia, accessed September 6, 2019, https://www.lydianarmenia.am/index.php?m=newsOne&lang=eng&nid=241.
42. "Lydian Is Deeply Disappointed."
43. A commission established by Canada, Mexico, and the United States under the North American Agreement for Environmental Cooperation (NAAEC) to support efforts to address environmental concerns, especially those posed by free trade among the three countries. "About the CEC," Commission for Environmental Cooperation, accessed September 4, 2019, http://www.cec.org/about-us/about-cec.
44. "United States–Peru Trade Promotion Agreement," entered into force February 1, 2009, annex 18.3.4, https://ustr.gov/trade-agreements/free-trade-agreements/peru-tpa/final-text.
45. "Implementation and Enforcement Failures in the US-Peru Free Trade Agreement (FTA) Allows Illegal Logging Crisis to Continue," Environmental Investigation Agency,

June 2015, https://www.illegal-logging.info/sites/files/chlogging/Implementation_and_Enforcement_Failures_in_the_US-Peru_Free_Trade_Agreement_(FTA)_Allows_Illegal_Logging_Crisis_to_Continue.pdf; "Peruvian Timber Exporter Excluded from Selling in the United States for Three Years," Center for International Environmental Law, October 19, 2019, https://www.ciel.org/news/peruvian-timber-exporter-excluded-selling-united-states-three-years/.

46. "Agreement Between Japan and the United Mexican States for the Strengthening of the Economic Partnership," entered into force April 1, 2005, art. 147(1), https://www.mofa.go.jp/region/latin/mexico/agreement/agreement.pdf.
47. United States Government Accountability Office, *Four Free Trade Agreements GAO Reviewed Have Resulted in Commercial Benefits, But Challenges on Labor and Environment Remain*, Report to the Chairman, Committee on Finance, U.S. Senate (Washington, DC: United States Government Accountability Office, July 2009), http://www.gao.gov/new.items/d09439.pdf.
48. Lisa Sachs and Lise Johnson, "Investment Treaties, Investor-State Dispute Settlement and Inequality: How International Rules and Institutions Can Exacerbate Domestic Disparities," *Initiative for Policy Dialogue Working Paper* no. 306 (2018), http://ccsi.columbia.edu/files/2017/11/ISDS-and-Intra-national-inequality.pdf.
49. Mavluda Sattorova, *The Impact of Investment Treaty Law on Host States: Enabling Good Governance?* (London: Hart, 2018); Tom Ginsburg, "International Substitutes for Domestic Institutions: Bilateral Investment Treaties and Governance," *International Review of Law and Economics* 25, no. 1 (2005): 107–23.
50. Lise Johnson, Lisa Sachs, Brooke Güven, and Jesse Coleman, "Clearing the Path: Withdrawal of Consent and Termination as Next Steps for Reforming International Investment Law," *Columbia Center on Sustainable Investment Policy Paper* (2018).
51. Lise Johnson, Lisa Sachs, and Nathan Lobel, "Aligning International Investment Agreements with the Sustainable Development Goals," *Columbia Journal of Transnational Law*, forthcoming 2019.

CHAPTER 33

ETHICS IN ACTION AND DIVESTMENT

ERIN LOTHES

Though divestment involves untangling complex investment structures and provokes intense emotional reactions, on one level it is quite simple. If we as a global society wish to have clean, renewable energy accessible to all, we must stop purchasing and investing in fossil fuel energy. Those who want apples should stop planting oranges. In this short chapter I hope to express another dimension to divestment, which is its religious, even sacramental, meaning.

First I indicate some often-heard concerns, offer ethical responses to them, and propose a spiritual interpretation of divestment as a sacramental action. I then reference various resources, as well as some barriers ahead, particularly for the United States Catholic community, which is my own faith community.

Three concerns are often heard: that divestment is risky, impractical, and symbolic. These are legitimate concerns that deserve attention alongside ethical assessments.

Risks. Indeed there are risks to any investment strategy, and all portfolios are subject to losses. Certainly fossil fuels have been profitable. Yet as carbon becomes priced or regulated, or increasingly politically or socially unpopular, stranded assets are becoming a real risk on the horizon. Thus, new definitions of fiduciary responsibility are emerging that may define investments in fossil fuels as impractical and risky themselves. Furthermore, Catholic social teaching excludes seeking maximum returns if doing so means compromising moral standards. In terms of risks and returns, faith-based socially responsible investment guidelines do not ask about the lost profitability from companies barred by other values screens. These products and services are excluded on a moral basis.

Practicality. Divestment is also dismissed as impractical: the act of selling one's shares is seen as hypocritical and futile because others will purchase the shares, and the fossil fuel infrastructure will thus remain unchanged. On the contrary, divesting and reinvesting are eminently practical because reinvested funds can be directed to impact investing, sustainable investments, and the essential scaling up of currently available renewable infrastructure.

Society must invest in renewable energy systems on a massive scale. A March 2017 report from the Columbia University Center on Global Energy Policy points out that wind and solar would benefit from institutional investors (through pensions, endowments, insurance, sovereign wealth funds, and foundations), as some of the capital, tax breaks, and accounting mechanisms available to fossil fuel corporations are unavailable to these renewable sources of energy.[1] This capital is *not* needed to fund fossil fuel systems, which benefit from billions of dollars in subsidies, as clearly documented by the International Monetary Fund and an October 2017 report from Oil Change International, showing $20 billion of U.S. federal and state subsidies to the fossil fuel industry through support for exploration, deductions for drilling costs, reduced royalties for mining on public lands, and deductions for oil spill penalties, among other mechanisms.

These subsidies, along with the fossil fuel industry's well-documented tactics of disinformation, amid active efforts to *expand* the extraction of fossil fuels, represent entrenched opposition to accelerating the renewables revolution.

These are signs that divestment is needed. The interfaith environmental coalition GreenFaith articulates precisely this opposition as one of the criteria employed by many faith communities when discerning a decision to divest. The criteria include large-scale, systematic harm; intractable opposition to change; and the need for religious groups to redefine society's moral code, reject complacency, and spur an appropriate disgust for corporate-sponsored ecocide.[2]

Such intractable opposition must be countered by a strong moral witness that calls people to recognize the seriousness of our problem and the need to interrupt business as usual and to courageously take the opportunity to impact society's moral code.

Symbolism. Divestment is indeed symbolic. In fact, to draw upon the Jesuit theologian Karl Rahner's articulation of a Catholic theology of symbol, divestment is a "real symbol," the very reality of the thing presented within the symbolic action or image: "A symbol is not something separate from the symbolized object. On the contrary, the symbol is the reality, which reveals and proclaims the thing symbolized . . . being its concrete form of existence."[3] As such, divestment is a visible sign of invisible realities, the visible reality of conferring funds to build a healthy and sustainable economy.

Conversely, business-as-usual investments manifest the invisible reality of the power relationships that pervade energy and inequality, power inequities that are plainly visible in energy poverty, disproportionate climate impacts, and ecological debt.

Thus, the symbol of divestment is in no way an empty performance but a practical contribution to building a renewable energy economy. The sacramental realization of love of neighbor enacted by redirecting funds to clean investments, for the sake of a stable climate and fruitful earth, is no less symbolic than providing funds to buy bread for the hungry. It is the actual reality of love of neighbor, future generations, and all living families of the earth.

Love of neighbor requires a preferential option for the poor. In our current era of climate change, enacting that option requires reversing environmental injustice. That theme has been addressed by magisterial social teaching since at least 1972, as emphasized by Pope Saint John Paul II in his 1990 World Day of Peace message, analyzed by Pope Benedict XVI in *Caritas in veritate*, and comprehensively established by Pope Francis in *Laudato si'*, with its integrated critique of culture, economics, technology, and politics. All texts present a deeply compassionate and spiritual call to ecological conversion.

Regarding energy, *Laudato si'* states the following:

> We know that technology based on the use of highly polluting fossil fuels—especially coal, but also oil and, to a lesser degree, gas—needs to be progressively replaced without delay . . . the international community has still not reached adequate agreements about the responsibility for paying the costs of this energy transition . . . Politics and business have been slow to react in a way commensurate with the urgency of the challenges facing our world.[4]

Therefore, Francis explains, "There is an urgent need to develop policies so that, in the next few years, the emission of carbon dioxide and other highly polluting gases can be drastically reduced, for example, substituting for fossil fuels and developing sources of renewable energy."[5]

The excellent text *Energy, Justice, and Peace: A Reflection on Energy in the Current Context of Development and Environmental Protection*[6] anticipates the teaching of *Laudato si'*, applying the preferential option for the poor to energy decisions. The text clarifies that energy should "primarily solve the shortages of the poorest populations"; thus, advanced countries have "the moral duty of developing the use of the most complex and capital-intensive energy technologies."[7]

This development takes place only through research and investment that scale up the individual, institutional, and societal deployment of currently available technologies. And funding the development and deployment of renewable energy requires divestment that reverses the ongoing prospecting, extraction, and expanded production of fossil fuels.

The essay "Catholic Moral Traditions and Energy Ethics for the Twenty-First Century" further articulates criteria from Catholic social teaching, identifying seven principles of energy ethics that empower conversations about *energy decisions as ethical decisions* in a U.S. Catholic context.[8]

Indeed, such conversations are not only an ethical imperative but a spiritual opportunity. As the Protestant theologian Langdon Gilkey wrote, the strength of a sacramental religious vision is the "divinely granted capacity to allow finite and relative instruments to be media of the divine, and endow all of secular and ordinary life with the possibility and the sanctity of divine creativity."[9]

Finite and relative *financial* instruments play an essential role in secular life and cannot be left out of the creativity needed to direct funds appropriately in order to establish a healthy, sustainable, and just economy. Thus, a sacramental view of divestment places finance within a moral vision, seeing the pragmatic exchange of investments within the covenantal economy of God's providence, neighborly justice, and creaturely praise and gratitude.

This sacramental interpretation of the meaning of divestment builds upon the message of *Laudato Si'* and other episcopal and papal teachings, which call us to see the complex social and environmental crisis as a religious summons to ecological conversion.

Resources

What are the resources for further ethical discussion? The interfaith community has given strong witness to the religious meaning of divestment. At COP22, the 2016 UN Climate Change Conference in Marrakech, an interfaith statement on climate change was signed by hundreds of faith leaders worldwide. This statement, a call to accelerate fossil fuel divestment and clean energy investment, was delivered to the office of the UN secretary-general. A collection of essays by religious leaders from the world's major faith traditions on energy ethics, which I had the privilege to edit with GreenFaith, was also released at that time.[10] GreenFaith's Divest and Invest Now movement continues to support faith communities from all traditions.[11]

The Vatican's Dicastery for Promoting Integral Human Development and Catholic Relief Services have hosted three impact investing conferences to share best practices related to investing for the common good and mitigating climate change. Also working in this area is the highly professional Catholic Impact Investing Collaborative, a network of wealth management professionals and financial officers. They are committed to values investing, have links with the Interfaith Center on Corporate Responsibility, and have significant experience with divestment.[12]

Trócaire, the official overseas development agency of the Catholic Church in Ireland, has issued a helpful resource, "Ethical Investments in an Era of Climate Change," which is available on the websites of the Catholic Impact Investing

Collaborative and the Global Catholic Climate Movement.[13] The Global Catholic Climate Movement has supported the divestment of more than one hundred Catholic groups worldwide.

Barriers

The traditional Catholic support of shareholder engagement is an important effort to bring a moral voice to corporate boardrooms. It is effective in many campaigns for social justice. However, shareholder advocacy has had only limited success in obtaining change from fossil fuel companies. The pace of change so far achieved through shareholder engagement does not match the actual need for a faster shift to renewable energy as defined by science. In part, this is because of the *distinction between advocating for change in process and advocating for change in product.*

Changes to a business model that improve the safety, inclusion, health, transparency, wages, well-being, and equitable treatment of workers are *process* changes that a corporation can accept while continuing to make and sell its product. However, it is inherently challenging for a company to comply with *product* changes, that is, requests to put themselves out of business.

Yet many Christian groups will wish to continue their witness through shareholder engagement, and if they are to join the divestment movement, they must have the option to choose portfolios divested from fossil fuels while maintaining relationships with faith-based shareholder engagement. At present, these options are quite limited, which represents a significant barrier to some Catholic divestment efforts. Fossil-free investment vehicles must be developed, and investment advisors must be educated about them to accelerate the divestment movement.

It is the moral challenge of this generation to reverse the fossil fuel economy and fund a clean, healthy, sustainable, inclusive, renewable economy as an imperative of intergenerational justice. Divestment is both an ethically and spiritually meaningful part of the solidarity that works for justice and a stable, sustainable Earth community. In Pope Benedict's words, "If you want to cultivate peace, protect creation."[14]

Notes

1. Travis Bradford, Peter Davidson, Lawrence Rodman, and David Sandalow, "Financing Solar and Wind Power: Insights from Oil and Gas," March 2017, Columbia Center on Global Energy Policy, https://www.energypolicy.columbia.edu/sites/default/files/Financing%20Solar%20and%20Wind%20Power.pdf

2. Fletcher Harper, *Divest and Reinvest. Now. The Religious Imperative for Fossil Fuel Divestment and Reinvestment in a Clean Energy Future* (New York: GreenFaith, 2013).
3. Karl Rahner, *Theological Investigations*, Vol. IV, trans. Kevin Smythe (Baltimore, MD: Helicon, 1966), 224.
4. Francis, *Laudato si'* (2015), sec. 165.
5. Francis, *Laudato si'* (2015), sec. 26.
6. Pontifical Council for Justice and Peace, *Energy, Justice, and Peace: A Reflection on Energy in the Current Context of Developmental and Environmental Protection* (New York: Paulist, 2014).
7. Pontifical Council for Justice and Peace, *Energy, Justice, and Peace*, 88, 111.
8. Erin Lothes Biviano, David Cloutier, Elaine Padilla, Christiana Z. Peppard, and James Schaefer, "Catholic Moral Traditions and Energy Ethics for the Twenty-First Century," *Journal of Moral Theology* 5, no. 1 (2016): 1–36.
9. Langdon Gilkey, *Catholicism Confronts Modernity: A Protestant View* (New York: Seabury, 1975), 196–97.
10. Erin Lothes Biviano, ed., *Light for a New Day: Interfaith Essays on Energy Ethics* (New York: GreenFaith, November 2016).
11. See "Divest and Invest Now," GreenFaith, https://greenfaith.org/take-action/divest-and-invest-now/.
12. See "Catholic Impact Investing Collaborative," CIIC, https://www.catholicimpact.org.
13. Trócaire, *Ethical Investments in an Era of Climate Change* (Maynooth, Co. Kildare, Ireland: Trócaire, 2017), https://catholicclimatemovement.global/wp-content/uploads/2017/11/GCCM_Trócaire-Catholic-Toolkit.pdf.
14. Benedict XVI, "Message of His Holiness Pope Benedict XVI for the Celebration of the World Day of Peace," Vatican website, January 1, 2010, http://www.vatican.va/content/benedict-xvi/en/messages/peace/documents/hf_ben-xvi_mes_20091208_xliii-world-day-peace.html.

PART IX

AN ETHICAL CONSENSUS ON SUSTAINABLE DEVELOPMENT

Modern Slavery, Human Trafficking, and Access to Justice for the Poor and Vulnerable

ETHICS IN ACTION ON MODERN SLAVERY, HUMAN TRAFFICKING, AND ACCESS TO JUSTICE FOR THE POOR AND VULNERABLE

The Sustainable Development Goals call for equal access to justice for all (SDG Target 16.3), and the end of crimes against humanity such as forced labor, human and organ trafficking, child labor, and modern slavery (SDG Target 8.7). For the poor, these fundamental human rights are still not realized. This partly reflects a flawed libertarian mentality that assumes formal consent is all that matters, ignoring issues of power, coercion, and the degradation of human dignity. Unchecked climate change can also contribute substantively to these "crimes against humanity"—especially in light of the displacement of peoples caused by climate change (twenty-four million in 2016).

Ethics in Action met at Casina Pio IV on March 12 and 13, 2018, to promote new ways to put SDGs 5, 8, and 16 into practice around the world, especially for the world's poorest and most vulnerable people.

The discussion pointed to the need for a framework of action that engages several types of actions and interventions simultaneously:

- Strengthened legal frameworks and law enforcement by government
- Prevention of crime through criminal deterrence of predators
- Strengthened administrative and regulatory frameworks
- Application of the UN Guiding Principles on Business and Human Rights
- Real-time measurement and monitoring of abuses and public reporting

- Empowerment of the poor and vulnerable (through access to protection and justice, legal defense, trade unions, and other organized efforts)
- Increased social service provision
- Mobilizing partners of goodwill
- Use of the SDGs and *Laudato si'* as rallying points
- Training of trainers in and public awareness of legal empowerment
- Engagement of the private sector
- Training of corporate actors with vulnerable supply chains

General Recommendations: Achieving SDG Targets 8.7 and 16.2

Ethics in Action provided the following recommendations for achieving SDG Targets 8.7 and 16.2:

- Better supply chain management
- Combatting modern slavery, prostitution, sex trafficking, and gender-based violence
- Systematic management of organ transplantation to prevent organ trafficking and transplant tourism
- Criminal justice for the poor, including the professionalization of local police, prosecutors, and courts and an end to impunity for perpetrators of sexual violence, human trafficking, and slavery

For each of these areas, Ethics in Action heard testimony regarding key initiatives. Ethics in Action will establish a legal access working group to pursue detailed recommendations on best practices that will report back to Ethics in Action at the June 2018 meeting.

In considering the policy approaches, Ethics in Action examined six initiatives:

- Ending human rights abuses in the former Katanga cobalt supply chain
- The Nordic model to combat prostitution and sex trafficking
- The Mexican model of attention to and reintegration of victims of human trafficking
- A new model to regulate organ donation and transplantation
- The International Justice Mission model of legal protection and justice for the poor
- The Move Humanity campaign to help fund equal access to justice for the poor

Here is a summary of the key findings.

Ending Human Rights Abuses in the Former Katanga Cobalt Supply Chain

The Democratic Republic of the Congo holds half the world's cobalt reserves, and demand for the main mineral component of lithium-ion batteries is set to surge as electric cars proliferate. In 2016, the Congo mined 54 percent of the 123,000 tons of cobalt produced worldwide. Yet cobalt mining has led to a social disaster, indeed a vivid and startling case of the "resource curse," characterized by child and forced labor, massive pollution, and extreme poverty in the region. While there is not enough evidence on the exact number of children engaged in the worst forms of child labor in the region, UNICEF has estimated that forty thousand children were working in the mines in 2015.

One highly successful community-based framework in Lualaba Province (in former Katanga) is the Bon Pasteur model, implemented by the Good Shepherd sisters in Kolwezi. The model has rigorously demonstrated the feasibility of a low-cost, community-based development strategy to combat child and forced labor in the cobalt region. In the past five years, the Bon Pasteur model has succeeded in the following:

- Reducing by 91 percent child labor in a cobalt mining community where now 1,674 children are in school and have become advocates for children's rights
- Raising income, food security, and self-confidence for three hundred women and girls through education and alternative livelihoods
- Creating community-based safe spaces for five thousand people to report and prevent human rights violations and mobilize victims to advocate for the change of unjust laws and systems

The Bon Pasteur model's unique approach is based on the following:

- Radical inclusivity: putting the poorest first and designing actions around their basic needs
- Integrating human rights and development: providing education and food security and promoting community-based livelihoods in farming
- Persevering in building long-term human relationships, uplifting the spiritual value of each human being in an extremely materialistic environment
- Focusing on both process and outcomes, resisting the adoption of "prepackaged" donor-driven models of intervention
- Adopting a strategic approach to engaging the powerful and leveraging the moral and ethical credibility of religious women to resist corruption and invest in long-term local capacity development

The Nordic Model to Combat Prostitution and Sex Trafficking

Prostitution reflects a fundamental disrespect of human dignity and is incompatible with a humane society. The Nordic model, first implemented in Sweden in 1999, is based on the principle that no human being should ever be for sale. In this legal approach, prostitution is understood as an institution imbued with harm for the person who is bought as a commodity. Following Sweden, several countries have passed legislation that recognizes prostitution as sexual exploitation: South Korea (2004), Iceland, (2008), Norway (2009), Canada (2014), Northern Ireland (2015), France (2016), and the Republic of Ireland (2017). The abolitionist approach to prostitution means that sex buyers are penalized (as are pimps and traffickers) while the individuals used in prostitution are decriminalized and provided with exit services and job training. Once prostitution is understood as a form of violence against human beings, this legal approach makes sense.

There are five major pillars of the Nordic model: (1) making the buying of sex a criminal offense, since demand is understood to be a fundamental cause of prostitution; (2) the full decriminalization of those who are used in prostitution; (3) high-quality social services for the victims; (4) strengthening laws against procuring, pimping, and sex trafficking; and (5) addressing all factors that drive poor and vulnerable human beings into prostitution. The Nordic model calls for a fairer and more equal society, eliminating the pay gap between women and men, better resources and support for parents and children, and tackling all other factors that trap people in poverty. Achieving these goals requires a holistic approach, including public information campaigns, education programs in schools, and training for police and other civil servants. It also calls for the law to be prioritized and coordinated nationally.

According to an evaluation of the Swedish legislation after ten years, street prostitution in Sweden had been halved—and while there had been an increase in prostitution in neighboring Nordic countries, this was not the case in Sweden. The ban on the purchase of sex had also undermined organized crime, and surveys showed that the ban had had a deterrent effect on prospective buyers. There was no indication that the risk of physical abuse had increased for people used in prostitution—on the contrary, while nearly seventy women in prostitution have been murdered in Germany since prostitution was legalized in 2002, there has not been a single murder of a woman used in prostitution in Sweden since the Nordic model was implemented.

The Mexican Model of Attention to and Reintegration of Victims of Human Trafficking

Efforts are also needed to ensure the reintegration of victims of human trafficking into society. This requires a long-term strategy that includes rescue, shelter, education, legal support, and family and social integration.

Mexican programs are currently implementing such strategies. Owing to the trauma experienced by human trafficking victims, the victims require greater protection and attention than is currently typically given. Short-term restoration is not enough. Accordingly, the nongovernmental organization Comisión Unidos vs Trata has advocated for long-term care, accompaniment, and social restoration, tailored to the individual needs of the victim.

This model puts human dignity at the heart of restoration, as this is exactly what has been taken away from victims, many of whom experience great anger and psychological problems. Every person involved in the victim's care must be trained in restoring dignity from the first moment—in the way she is spoken to, in assuring her that she will be taken seriously, in offering patience while she decides to share her story, in ensuring that she will be fed, clothed, and offered a dignified place of rest and recuperation. Every step of the way, she must feel that she is being cared for and not exploited.

The model, known as the virtuous circle, is predicated on the willingness to accompany the victim by offering her legal, psychological, physical, health, educational, and cultural support for as long as necessary, while providing her with the necessary tools to reach her life's goals. This dignity-centered approach aims at allowing the survivor to obtain a professional or vocational career, a decent job, good health, emotional stability, and dignified housing. It allows her to be proud of what she has accomplished and to become financially and emotionally independent.

It is also vitally important that every aspect of the criminal justice system be trauma informed and victim friendly. Police, prosecutors, judges, courtroom personnel, and social service officials should be trained and supported in such approaches.

A New Model to Regulate Organ Donation and Transplantation

SDG Target 8.7 offers an additional tool to combat organ trafficking. It supports existing global protocols, including the Declaration of Istanbul (2008); the WHO Guiding Principles on Cell, Tissue and Organ Transplantation (2010); and the Council of Europe Convention Against Trafficking in Human

Organs (2015). Most recently, on September 8, 2017, the UN General Assembly adopted Resolution 71/322 to direct countries to establish best practices, including regulations that organ transplants are performed only in the following ways:

- In authorized centers
- With proper regulatory oversight
- With specific procedures for authorizing every organ removal and transplant procedure
- With the use of national registries to ensure the transparency of practices, traceability, and the quality and safety of human organs

Ethics in Action endorses the development of a WHO task force to work with governments in implementing these principles, as requested by member states at the 2017 World Health Assembly. Ethics in Action also recommends that the work of such a task force be addressed at the 2018 World Health Assembly, because the World Health Organization has estimated that 10 percent of the total number of transplants performed annually involve an organ sale. Six WHO regions are geographically designated worldwide, and each region has member states where media reports and professional communications have raised concerns of organ trafficking. These media reports indicate thousands of individuals have recently sold organs in India and Egypt.

Ethics in Action acknowledges the reform that has been accomplished in China prohibiting the use of organs from executed prisoners and the prohibition of foreign individuals undergoing organ transplantation in China (so-called transplant tourists). This reform is illustrative of the WHO Guiding Principles of equity, transparency, and fairness.

The China model has the following features:

- The State Council promulgated its Regulations on Human Organ Transplantation in 2007, and the National People's Congress promulgated the Eighth Amendment of Criminal Law in 2011, which serve as the legal foundation for criminalizing all organ-related crime in China.
- China established a single mandatory national organ allocation computer system (the China Organ Transplant Response System), which is interconnected with four transplant registries (for liver, kidneys, heart, and lungs) to ensure the traceability and fairness of organ distribution.
- The Government of China now authorizes transplant hospitals by licensing policy, which is enforced with unannounced field audits conducted by national authorities.
- China is now implementing a national anti–organ trafficking surveillance system through the joint efforts of health and legal authorities. This data

system may serve as an example of an operational mechanism to combat organ trafficking for the rest of the world, empowered by information integration between health and legal authorities.
- An essential feature of the China model is the resolve of the Government of China to sustain reform, effectively driven by the cooperation of professionals and exemplified by the leadership of Professor Huang Jiefu.

These developments in China represent a new era of organ donation and transplantation that complies with SDG Target 8.7 and emphasizes the need for strong government support to accomplish such SDGs. More than nine hundred thousand individuals in China will require kidney replacement treatment either by transplantation or dialysis in the near future; thus, expanding the source of organs from deceased individuals or through ethical live donation (that does not involve an organ sale) will require government monitoring of transplant practices.

The International Justice Mission Model of Legal Protection and Justice for the Poor

The International Justice Mission (IJM) empowers poor and vulnerable individuals through interventions to ensure that local law enforcement systems protect communities and hold perpetrators to account. IJM notes that the public justice system is the sole service provider of criminal accountability and physical restraint of the aggressor. IJM interventions are based on the following principles:

- A national and global commitment to empowering and equipping local public justice professionals
- Mentorship and accompaniment of local criminal justice officials to bring relief to individual victims, diagnose weaknesses in the existing system, and build the will and capacity to address these weaknesses
- An evidence-based and case-based diagnosis of public justice system gaps
- Local and national implementation of capacity-strengthening mechanisms
- Measurable, case-based improvements to public justice system functioning
- The measurement of impacts in the prevalence of the crime
- Community-based and survivor-based advocacy
- The availability of community justice workers to assist victims and hold local authorities accountable
- Information and communications technologies to assist the identification and rescue of victims and the collection and analysis of crime data for policy response

The IJM model of strengthening justice systems has contributed to the protection of the poor, decreased the prevalence of violent crime, improved system performance, and provided victim restoration in a number of contexts:

- Since 1997, IJM has collaborated with local authorities to bring relief to approximately 14,200 victims of bonded labor, 3,200 victims of commercial sexual exploitation, and 700 victims of child sexual assault.
- Since 1997, IJM has collaborated with local authorities on more than 1,200 cases in which suspects were charged through local criminal justice systems.
- A 2010 study documented a 79 percent reduction in the availability of minors being sold for sex in Cebu, Philippines, since IJM began working with local authorities in 2006.[1]
- In 2016, studies in Manila and Pampanga, Philippines, documented a 75 percent to 86 percent reduction in the availability of minors being sold for sex since IJM began work in those locations in 2009 and 2012, respectively.[2]
- A study found a 73 percent reduction in the availability of minors being sold for sex from 2012 to 2015 in target areas in Phnom Penh, Cambodia, where IJM worked.[3]
- In 2017, 92 percent of IJM clients who had been freed from bonded labor achieved full restoration.

The Move Humanity Campaign to Help Fund Equal Access to Justice for the Poor

Starting in 2018 and continuing through 2030, the Move Humanity Campaign will call upon all billionaires (those with more than US$1 billion in net worth) to direct at least 1 percent of their net worth each year toward the SDGs. Move Humanity will appeal for voluntary giving but will also call on all UN member states to introduce a 1 percent SDG levy no later than 2023 on billionaires who do not give to the SDGs voluntarily.

The Move Humanity Campaign will be organized around 12 guiding principles. These principles will establish the priorities of the SDGs, including the end of poverty (SDG 1), universal health coverage (SDG 3), universal basic education (SDG 4), gender equality (SDG 5), universal access to renewable energy (SDG 7), the end of modern slavery (SDG 8), biodiversity conservation (SDGs 14 and 15), and access to justice for all (SDG 16). Philanthropic funds will allow for the SDG financing gap of the low-income and lower-middle-income countries to be closed.

One target for new philanthropic funding will be access to justice for the poor. One possibility is a new global fund to allocate financial resources to end human trafficking, forced labor, organ trafficking, prostitution, child labor, and modern slavery.

Move Humanity's Guiding Principles

- The seventeen SDGs are the world's global development priorities, constituting the globally agreed framework for the years 2015 to 2030.
- Development assistance should be complementary with domestic financing and conditional on strong national financing efforts.
- Development assistance should be directed to low-income and lower-middle-income countries in order to close the SDG financing gap.
- Official development assistance from each donor country should reach the long-standing target of at least 0.7 percent of gross national income.
- Private development assistance should reach at least 0.3 percent of the donor country's GDP, with assistance from ultra-high-net-worth individuals constituting the largest portion of the assistance.
- Private development assistance from ultra-high-net-worth individuals should equal at least 1 per cent of their net worth per year, with extra giving in one year being allowed to carry over for giving in later years. These individuals should demonstrate SDG leadership by publicly committing to this goal, consistent with other commitments such as the Giving Pledge.
- High-net-worth development assistance should be monitored and reported annually.
- High-net-worth funding should be directed largely toward pooled SDG funds that support national SDG strategies and ensure rigorous monitoring and evaluation of all funding.
- Ultra-high-net-worth giving should be based on voluntary giving supplemented by national SDG levies on high-net-worth individuals who do not contribute voluntarily.

Ethics in Action Working Group Action Items

(1) Ethics in Action will establish a legal access working group to pursue detailed recommendations on best practices, and the working group will report back to Ethics in Action at the June 2018 meeting.
(2) Ethics in Action, supported by the UN Sustainable Development Solutions Network, will establish a working group on modern slavery, human trafficking, and access to justice for the poor and vulnerable.
(3) The working group on modern slavery, human trafficking, and access to justice for the poor and vulnerable will report back to Ethics in Action with recommendations at the October 2018 meeting.
(4) The working group on modern slavery, human trafficking, and access to justice for the poor and vulnerable will seek out additional partners including the

Responsible Cobalt Initiative, the Organisation for Economic Co-operation and Development, the Columbia Center on Sustainable Investment, and the Amazonas Sustainability Foundation, among others.

Notes

1. Andrew Jones, Rhonda Schlangen, and Rhodora Bucoy, *An Evaluation of the International Justice Mission's "Project Lantern": Assessment of Five-Year Impact and Change in the Public Justice System* (Washington, DC: IJM, October 21, 2010), https://ijmstoragelive.blob.core.windows.net/ijmna/documents/studies/Cebu-Project-Lantern-Impact-Assessment_2021-02-05-071021.pdf?mtime=20210204231021&focal=none.
2. Dave Shaw and Travis Frugé, *Child Sex Trafficking in Metro Manila: Using Time-Space Sampling to Measure Prevalence of Child Sex Trafficking in Metro Manila, the Philippines* (Washington, DC: IJM, 2016), https://ijmstoragelive.blob.core.windows.net/ijmna/documents/studies/ijm-manila-final-web-v2_2021-02-05-062616.pdf?mtime=20210204222616&focal=none. Dave Shaw and Travis Frugé, Child Sex Trafficking in Angeles City: Using Time-Space Sampling to Measure Prevalence of Child Sex Trafficking in Angeles City and Mabalacat in the Philippines (Washington, DC: IJM, 2016), https://ijmstoragelive.blob.core.windows.net/ijmna/documents/studies/ijm-pampanga-final-web-pdf-v2_2021-02-05-064344.pdf?mtime=20210204224344&focal=none.
3. Robin N. Haar, *External Evaluation of International Justice Mission's Program to Combat Sex Trafficking of Children in Cambodia, 2004–2014* (Washington, DC: IJM, December 2015), https://ijmstoragelive.blob.core.windows.net/ijmna/documents/studies/2015-Evaluation-of-IJM-CSEC-Program-in-Cambodia-Final-Report_2021-02-05-064542.pdf?mtime=20210204224542&focal=none.

CHAPTER 34

ACTUALIZING JUSTICE FOR THE POOR

MARCELO SÁNCHEZ SORONDO

Our contemporary world is beset by many injustices: broken promises on behalf of governments; unequal distribution of income and wealth; unjust redistributions from poor to rich; bullying of states and individuals on behalf of multinational corporations solely concerned with profit; states that do not welcome migrants while contributing to the causes of mass migration; and global warming caused by human activity, mostly owing to fossil fuel consumption. Concerned with this "globalization of indifference," which tends to exclude whole masses of people from society, Pope Francis, soon after his election, asked the Pontifical Academy of Social Studies to study human trafficking and modern slavery in the forms of forced labor, prostitution, and organ trafficking. On several occasions, he has defined these practices as "crimes against humanity." In order to describe this kind of exclusion, contempt, and, ultimately, nonparticipation in what makes us human, it is necessary to understand the double nature of the violence of these crimes. It is an attack on corporal integrity and a combination of all forms of abuse—torture, repetitive rape, organ harvesting, and forced labor, including that of children—that destroy a person's primary trust in him- or herself and in others. But it is also an attack inflicted on the survivor's soul, which creates wounds deeper and more complex than those already caused by the physical violence.

What kind of psychological and moral violence are we talking about? Friends, or individuals who love each other, approve each other's existence. Trafficking victims often lack true friends, and so, too, they lack the affirmation that makes friendship the "unique good," as defined by Simone Weil, inspired by Aristotle.[1] The humiliation of the victim, perceived as the withdrawal or rejection of that affirmation to exist, harms, first and foremost, at a pre-juridical level. The humiliated person feels looked down on or, worse, completely unvalued. Deprived of this essential approval, the person almost becomes nonexistent. Therefore, the humiliation caused by forced labor, prostitution, involuntary organ harvesting, and rape makes the victim feel they do not exist as an end in itself but merely as an individual's property, or as a means for the benefit of others.[2] This destroys the very core of the trust that one can put in a person. Somehow, the betrayed victim feels worse than someone whose existence has not been recognized. It is for this reason that the protocol of victim rehabilitation is often based on the reconstruction of self-trust, and of trust in one's fellow human beings, as well as on a public admission of the iniquity of having been betrayed and sold. Betrayal by the person in whom we have put all our trust is worse than death itself.

The Struggle for Participation from a Judicial Standpoint

Importantly, as a judicial institution, slavery was banned thanks to the progressive penetration in history of Christ's message of brotherhood and the antislavery struggles that began toward the end of the eighteenth century, which ended with the abolition of this scourge—despite considerable reluctance, as was the case in the American Civil War—in most countries around the world. Modern international agreements (e.g., the 1926 Slavery Convention) reassert the prohibition of slavery, which is considered a crime against humanity. However, slavery is still culturally entrenched in some countries (e.g., India, Sudan, and Mauritania) and has reappeared in new forms, such as forced labor, prostitution, organ trafficking, and child slavery. Naturally, the victims of these new forms of slavery are deprived of both negative and positive civil rights. They are the targets of the most ruthless discrimination.

The struggle for political rights took place in the most developed countries of the world during the nineteenth century and continued in the twentieth century in the context of debates on the representational nature of democratic regimes, once the sovereignty of citizens and their right to express themselves through elections finally began to be recognized.

The biggest concern today is the exclusion and marginalization of the majority with regard to egalitarian participation in the distribution of goods on national and planetary scales. This is true of both market goods and nonmarket goods such as dignity, freedom, knowledge, integration, and peace. The biggest cause of human suffering, and, ultimately, of rebellion, is the alarming and unfair contrast between the theoretical attribution of equal rights to all and the unequal and unfair distribution of fundamental goods for most human beings. Despite living in a world of abundant wealth—a world where economic activity has exceeded US$120 trillion a year—countless people continue to live with poverty and social exclusion, two scourges that facilitate the expansion of the new forms of slavery. This alarming inequality—together with dominance wars and climate change—is the cause of the largest forced migration in human history, which is now affecting as many as sixty-five million people. We should not forget either the growing number of individuals—estimated at fifty million—who have been ravaged by the new forms of slavery and human trafficking, such as forced labor, prostitution, and organ trafficking. These are all veritable crimes against humanity that must be recognized and denounced as such. The fact that the human body should be bought and sold as if it were just another commodity on the market is appalling, and it is a symptom of a profound moral and social decay. Almost one hundred years ago, Pope Pius XI foresaw the entrenchment of inequality and injustice as a consequence of global economic dictatorship, which he called the "internationalism of finance or international imperialism."[3] For his part, Pope Paul VI denounced, almost fifty years later, the "new and abusive form of economic domination on the social, cultural and even political level."[4]

Talking about social participation, I cannot ignore the invitation that we can find in a statement that is ancient and new at the same time and loaded with immense theoretical and practical meaning. Today, every Christian, including us scholars, and every human being destined to the love of Christ, must internalize the words of Saint Leo the Great, who reminds us of the letter of Saint Peter:

> *Agnosce, o christiane, dignitatem tuam*, acknowledge, O Christian, the dignity that is yours! Being made a "participant in the divine nature" (θείας κοινωνοὶ φύσεως), do not by an unworthy manner of living fall back into your former abjectness of life. Be mindful of Whose Head, and of Whose Body, you are a member. Remember, that wrested from the powers of darkness, thou art now translated into the Light and the Kingdom of God.[5]

Insofar as the Lord will reign in us and among us, we will be able to participate in divine life, and we shall be for each other "instruments of grace, so as to pour forth God's charity and to weave networks of charity."[6]

Saint Paul, in correspondence with what we have said about the participation of grace in the Kingdom of Christ, says, "With freedom did Christ set us free: stand fast therefore, and be not entangled again in a yoke of bondage" (Τῇ ἐλευθερίᾳ ἡμᾶς Χριστὸς ἠλευθέρωσεν· στήκετε οὖν καὶ μὴ πάλιν ζυγῷ δουλείας ἐνέχεσθε).[7] So Christ gives grace and freedom to all human beings. In ancient times, many knew that one person could be free, such as a tyrant or the chief of a tribe, or that many were free, such as the citizens by birth and philosophers of Ancient Greece or Ancient Rome. But the idea that all people were free by their essence comes from the grace and message of Christ. All human beings are destined to the utmost grace and freedom, and the Holy Trinity lives inside each human being through the grace of Christ and the collaboration of everyone.

From the theological and empirical point of view, according to Saint Paul, the achievement of freedom and the subsequent abolition of ancient slavery in the course of history depended on and the new forms of slavery today depend on the opposition of sin and grace. This opposition includes and fosters the other antagonisms that fight tenaciously within the human being—truth versus error, good versus evil, virtue versus vice, wholeness versus corruption, and so on—in the impenetrable puzzle of the enigma that is the human heart. It is a transcendental phenomenology of the conflict of the two laws revealed by Saint Paul—good and evil, virtue and concupiscence—that agitate our bodies and obscure our minds, debilitating our will. "I see and approve of the better, but I follow the worse" (*Video meliora proboque, deteriora sequor*), states Ovidius. Saint Paul responds by saying, "For I know that nothing good dwells within me, that is, in my flesh. I can will what is right, but I cannot do it. For I do not do the good I want, but the evil I do not want is what I do."[8] This is the deepest invitation and the most profound revelation that can free us from the social situations of the new forms of slavery, just as in the past it freed us from the juridical institution of slavery.

From Indignation to Dignity, Freedom, and Peace

Since Pope Francis's request to the Academy to deal with modern slavery and human trafficking, we have tried first to establish the facts and then to find models and best practices to restore dignity, freedom, peace, and happiness to the victims.

Establishing the facts before proceeding with the work is important because it describes the extent of the problem. If you cannot count it, you cannot fight it. According to the most serious estimates, there are about fifty million victims of modern slavery and ten million victims of organ trafficking. Many of these individuals are among the sixty-five million refugees currently in the world. Our research has identified models and good practices to combat modern slavery as created by states and individuals. Forced labor is, in a way, easier to fight. We must trust that, once awareness of the severity of the issue has spread, public opinion, states, and multinationals will work to address it, for example by ensuring supply chains are free of forced labor and using a packaging mark that indicates that products have been made without forced labor.

In view of eradicating the new slavery and ensuring the social participation, dignity, freedom, and happiness of each person, we need to work together and across boundaries to create "waves" that will affect society as a whole, from top down and vice versa, moving from the periphery to the center and back again, from leaders to communities, and from small towns and public opinion to the most influential segments of society.

Today we can celebrate a new synergy between the spirit of the United Nations and other international organizations and the spirit of religions. This is testified by the requests we get from the UN to hold meetings together. Although individuals of different religions do not pray at the same altar, they can and should act together to promote human dignity and defend the freedom of each person and promote good relations with the earth, therefore promoting sustainable development.

Along these lines, the Pontifical Academy of Sciences organized its first important meeting to eradicate human trafficking and modern slavery in 2014[9] with Pope Francis and the leaders of many faiths, who all agreed to define "modern slavery, in terms of forced labour, prostitution, and organ trafficking," as crimes against humanity. We have held several meetings since to discuss and combat trafficking in all its forms and continue to reach out to different sectors of society to raise awareness about modern slavery and human trafficking. At the Academy, we recognize that it is not only a case of stating that modern slavery is a crime against humanity but that practical steps must be taken, for example to prosecute traffickers and pimps, as well as the customers who create a market for sexual exploitation, destroying the lives of the victims and their families. In particular, the Academy has identified models to eradicate these crimes and thus recommends following the so-called Nordic model, which was adopted by France last year. For the first time in history, the true cause of the problem is being criminalized—that is, not the victims (the women in prostitution) but the customers.

These new moral imperatives are enshrined in the Sustainable Development Goals promulgated unanimously by the United Nations after Pope Francis's address at the UN General Assembly. SDG Target 8.7 affirms that we must "take immediate and effective measures to eradicate forced labour, end modern slavery and human trafficking and secure the prohibition and elimination of the worst forms of child labour, including recruitment and use of child soldiers, and by 2025 end child labour in all its forms." In addition, SDG Target 5.2 states the need "to eliminate all forms of violence against all women and girls in the public and private spheres, including trafficking and sexual and other types of exploitation," and SDG Target 16.2 states the need "to end abuse, exploitation, trafficking and all forms of violence against and torture of children."

The beatitudes are valid for everyone in every culture and religion. If we follow them closely, we will heal the wounds of humanity, which are also the wounds of Christ in the contemporary world. It is with these intentions that Ethics in Action brought us together. We hope to offer a decisive contribution to a world where dignity, justice, freedom, and peace are a reality for all.

Notes

1. Simone Weil, "Amitié," in *Oeuvres* (Quarto) (Paris: Gallimard, 1999), 755.
2. "So act that you use humanity, in your own person as well as in the person of any other, always at the same time as an end, never merely as a means." Immanuel Kant, *Groundwork of the Metaphysics of Morals*, trans. and ed. Mary Gregor and Jens Timmermann (Cambridge: Cambridge University Press, 2012), 41.
3. Pius XI, *Quadragesimo anno* (May 15, 1931), §109.
4. Paul VI, *Octogesima adveniens* (May 14, 1971), §44.
5. Leo the Great, *Serm. I de Nat.*, P. G. 54, 192.
6. Benedict XVI, *Caritas in veritate*, §5.
7. Galatians 5:1.
8. Romans 7:18 f.; cf., Ovidius, *Metamorfosis*, lib. 7, vv. 20.
9. "Joint Declaration of Religious Leaders Against Modern Slavery," Pontifical Academy of Sciences, December 2, 2014, http://www.pas.va/content/accademia/en/events/2014/jointdeclaration.html.

CHAPTER 35

MULTIRELIGIOUS ACTION AGAINST MODERN SLAVERY AND TRAFFICKING

WILLIAM F. VENDLEY

The Scandal of Modern Slavery and Trafficking

If forced labor, child labor, human and organ trafficking, and modern slavery are out of view too often for too many, the scandal of these atrocities surrounds us, including being woven into the productive webs that provide us "goods and services." These crimes against humanity are an unspeakable burden and humiliation for the estimated 40.3 million people in modern slavery as of 2016, including 24.9 million victims of forced labor.[1] These figures tell us that there are 5.4 victims of modern slavery for every one thousand people. Of these, one in four are children. Women and girls are disproportionately affected by forced labor, accounting for 99 percent of victims in the commercial sex industry and 58 percent in all sectors.[2] These grave crimes are explicitly proscribed in the Sustainable Development Goals (SDG Target 8.7).

The Basic Consensus of World Religions

What then is the position of the world religions on the modern versions of these crimes? Given their far-flung and widely dispersed communities, can we discern a basic consensus of moral concern among the world's religions?

Historical Admission of Complicity

Before attempting to point to the emergence of a moral consensus among the world's religions on modern slavery and trafficking, it is essential to note the relatively recent positive developments within religious communities on questions of slavery. Prior to the historic beginnings of the abolitionist movements toward the end of the eighteenth and continuing through the nineteenth and twentieth centuries, many religions "accepted" slavery as a cultural given. And even in the twentieth century, we can find examples of religious communities contorting their sacred texts in justification of slavery. The potential reasons for the relative slowness of the widespread rejection of slavery can be duly debated. However, with the advent of religious positions that explicitly reject slavery, a dialectic of core religious understandings of the dignity of people and its contradiction in slavery can easily be discerned, leaving many modern-day religious believers incredulous that their religious forebears could have tolerated the scandal of slavery for so long.

A Basic Multireligious Consensus on the Dignity of People as a Basis for Working to Abolish Modern Slavery and Trafficking

Today, there is a basic multireligious consensus on the dignity of people. The term "person" in our use denotes an embodied mystery, expressed differently across religions. Various expressions for "person" labor to make clear the radical mysteriousness of the reality. Buddhist terms like "no-self," "selfless self," and "formless self" are examples, while the many theistic expressions of "kenotic" or "self-emptying" selves are others. Notably, despite these differing notions, people across religious traditions have no difficulty recognizing and acknowledging all human beings as such.

Furthermore, the person is typically understood by today's religious communities as the subject of the rights and implied responsibilities set forth in such seminal and widely accepted documents as the Universal Declaration of Human Rights.[3] Notably, in this regard, diverse religious communities are widely collaborating on the defense of these fundamental rights. Moreover, as part of their respective commitments, each religious community is also engaged in the creative task of rooting these fundamental rights and their respective notions of the person within the soil of their respective religious traditions' experience of sacred mystery.

This multireligious consensus on universal human dignity, grounded in each community's experience and understanding of sacred mystery, is the shared basis for the active rejection of modern slavery and trafficking. Nevertheless, this consensus urgently needs to be applied in concrete actions to free, heal, and restore the people ensnared in the scandal of modern slavery and trafficking.

Method and Mechanisms for Engaging Multireligious Collaboration to Help Resolve the Scandal of Modern Slavery and Trafficking

A Multireligious Method

The method developed by Religions for Peace for multireligious action is practical, powerful, and open to continuous creativity. It marshals, adapts, and engages the remarkable panoply of "assets"[4] that many religious communities have built over millennia.[5]

At its simplest, the method involves assisting religious communities to correlate, or work out a connection, between their capacities for action and the drivers of a given problem or challenge. When applied to the analysis of modern slavery, it can disclose large, underused capacities for action that lie within reach of religious communities and their multireligious cooperative mechanisms.

Vitally, the method also clarifies what kinds of capacity building are needed to turn potential assets into actual assets that can equip religious communities and their multireligious associations for more effective engagement in eliminating modern slavery and trafficking.

Religious and Multireligious Mechanisms

Many, although not all, religious communities exist simultaneously on local, national, regional, and global levels and have linkages across these four levels. Ideally, lower levels are included in successive higher levels so that all within a community are served by its global level with related global action, while local, national, and regional action likewise take place in accord with each community's application of a principle of subsidiarity.

The interreligious mechanisms fostered by Religions for Peace strive to pace the ways religious communities organize themselves from local to global levels. Thus, Religions for Peace has local, national, and transnational regional units, as well as a world council elected at World Assemblies. The multireligious mechanisms on each level are led by representatives of participating religious communities from the given level of organization, with strong participation of religious women's and youth groups on all levels. Each level of multireligious organization—from local to global—relates by a principle of subsidiarity. In this way, Religions for Peace strives to "mirror" the presence of specific religious communities on their respective levels of organization within complementary multireligious mechanisms across the four levels co-led by their own representatives.

More than ninety national multireligious bodies, as well as six continental regional bodies and one elected world council, are solely dedicated to multireligious collaboration for peace understood holistically. These multireligious mechanisms engage the earlier-discussed method in their activities.

Applying the Method and Engaging the Multireligious Mechanisms to Help Eliminate Modern Slavery and Trafficking: Core Steps

Three core steps are essential to applying the method and engaging the multireligious mechanisms to help eliminate modern slavery and trafficking. The first is a detailed analysis of the central drivers of the pathology of modern slavery and trafficking. The second is to identify the potential assets of collaborating religious communities with the capacity to impact selected drivers.[6] The third is to find practical ways and means to help mobilize, equip, and deploy the religious assets.[7]

Drivers of the Problem

Modern slavery and trafficking have many drivers, including overarching forces such as the dislocation, displacement, and poverty of large numbers of people (estimated to be twenty-four million) owing to climate change. Additional factors include widespread massive ignorance with low public awareness of the scale and depth of modern slavery and trafficking, ignorance regarding the intersecting levels of these crimes ranging from local to global, the stigma experienced by victims (especially victims of sexual crimes), weak and under-enforced legal frameworks and criminal deterrence, inadequate forms of monitoring, the inadequate mobilization and training of relevant partners of goodwill, and the lack of critically needed services essential for restoring the humanity of victims.

Each of these complex drivers can become a kind of lens useful for the establishing the inventory of potential assets for multireligious cooperation.

Multireligious Assets Linked to the Dimensions of the Problem

Religious Community Assets

Religious communities have at least three broad classes of religious assets relevant to combatting modern slavery and trafficking: social, moral, and spiritual.

- **Social:** The social assets of religious communities include the vast panoply of religious infrastructures: local churches, mosques, or temples; the women's and youth associations associated with them; their linkages from district to national levels; the national denominational organization; councils of churches, mosques, and temples; related presses; and a large number of religiously affiliated nongovernmental organizations, schools, and hospitals. The scale of religious infrastructure varies from country to country, but in most developing countries, it is the most extensive, interconnected, and locally led social infrastructure, reaching from the smallest village to the capital. Moreover, this national social infrastructure can also be connected to regional and global mechanisms, each with their respective sets of assets. These social assets, which span levels from local to global, are particularly valuable when the drivers of modern slavery and trafficking also span these levels.
- **Moral:** Religious communities have moral assets that build upon and unfold the great strengths of their spiritualities and relate directly to modern slavery and trafficking. These assets include much more than the consensus on human dignity and codes of ethics that relate directly to proscribing slavery and trafficking. They also include mechanisms of inculcating moral visions and sensibility by means of nurturing an intimate grammar of religious identity, with the advantage that new challenges—such as modern slavery and trafficking—can be examined in relation to communal moral memory as it is preserved in a variety of ways, thereby further fostering and strengthening the communal consensus on the need to address modern slavery and trafficking.
- **Spiritual:** The spiritual assets of religious communities are, in the eyes of their communities, their greatest assets. Spiritual assets defy easy description but typically point to what is most interior and experientially grounding regarding the meaning of human life before sacred mystery. Spiritualities can be spoken of as virtues or habitual orientations to the sacred dimension of personal and social experience and are intimately involved in actualizing human potential for flourishing.[8]

Importantly with regard to modern slavery and trafficking, each religion's spirituality can also focus expressly on spiritually motivated ways to confront, transform, and heal this scandal. These ways include a commitment to repair injustice based on unflinching honesty, repentance, restitution, reconciliation, and the restoration of dignity; calls for the transformation of social structures that hurt us or fail to protect us into ones that nourish and protect us; sober calls for self-sacrifice for the sake of others and the common good; calls for the voluntary bearing of innocent suffering; calls for returning good for evil; and calls for forgiveness and unrestricted compassion and love.

Multireligious Assets

Building on the primary assets of the religious communities, the multireligious approach harnesses the complementarity of diverse communities' distinctive assets, provides efficiencies in mobilizing and equipping, and facilitates partnerships with public institutions keen on supporting the positive power of multireligious cooperation while retaining principled neutrality regarding explicit religious claims.[9]

Mobilizing, Equipping, and Deploying Religious Assets for Action: General Considerations to Be Applied Across the Global Religions for Peace Network

Modern slavery and trafficking span the world on local, national, transnational regional, and global levels. A number of site-specific multireligious plans thus need to be developed and loosely coordinated. The following general reflection on method and multireligious mechanisms focuses on four key functions to be adapted and loosely coordinated across the global Religions for Peace network. Actualizing these key functions is to be done by harnessing the relevant religious and multireligious assets mentioned earlier and deploying them broadly into the religious communities accessible through Religions for Peace mechanisms. The core key functions are education, advocacy, special services, and solidarity and partnership.

Education

The half-hidden scandal of modern slavery and trafficking must be brought into the light. Religions for Peace—ideally in partnership with qualified experts and groups, including the Pontifical Academy of Social Sciences and the UN Sustainable Development Solutions Network—needs to prepare accurate, pedagogically appropriate, multireligious educational materials that can be translated into numerous languages, easily downloaded, and used across all levels (local to global) of its multireligious coalition. In addition to providing accurate information, these educational materials need to clarify the moral and spiritual imperative for multireligious action and include materials relevant to local religious activities.

Advocacy

Advocacy for the victims and for the transformation of the drivers of modern slavery and trafficking is essential. Typically, Religions for Peace advances global advocacy campaigns led by its World Council and adapts them for participation across all levels of the Religions for Peace movement. These include highly targeted forms of advocacy aimed at actors and institutions implicated in the scandal of modern slavery and trafficking, as well as to sympathetic groups and people. Importantly, these campaigns are designed to facilitate each regional and national multireligious structure to participate on its own level. Advocacy kits are developed to be adaptable at all levels and thus to reach to the relevant public.

Special Services

In modern slavery and trafficking, there is a need for specialized organizations competent in protecting potential victims and healing and restoring the dignity of victims. At the same time, religious congregations in some parts of the world can also provide frontline services. Working with experts, Religions for Peace can develop training kits and trainer programs appropriate for congregations in the areas of highest need.[10] In addition to providing needed services, these programs have the added advantage of engaging local communities of faith.

Solidarity and Partnership

Solidarity with the victims of modern slavery and trafficking needs to be cultivated across all sectors of society, especially religious communities with members of all sectors. Specially targeted activities to build solidarity need to be advanced. These include efforts to intentionally foster solidarity in all educational, advocacy, and special services initiatives.

In addition, wide-ranging partnerships need to be initiated. While this chapter has noted the potential work of Religions for Peace, special mention must be made of His Holiness Pope Francis and the Vatican agencies he directs, including notably the Pontifical Academy of Social Sciences. Religions for

Peace deeply respects the leadership of Pope Francis and the related work of the Pontifical Academy of Social Sciences and the UN Sustainable Development Solutions Network and commits to a principled partnership with them.

Turning the Wheel

This brief reflection has described a method and set of multireligious mechanisms to address the scandal of modern slavery and trafficking. Putting this method into action across the Religions for Peace mechanisms, ideally in partnership with the Pontifical Academy of Social Sciences, the UN Sustainable Development Solutions Network, and others, will require significant leadership in fundraising. Religions for Peace has ample experience with determining the feasibility of collaborative efforts to raise the funds essential to engaging in the broad-scale work required.

Postscript on Mutuality and "Willed Isolation"

Recent studies on mutuality highlight the scandalous character of modern slavery and trafficking.[11] Mutuality, it can be argued, opens one to the other in transcendence and is the condition of possibility for language and human existence as we know it.

The world's religions have understood mutuality in their distinctive ways as extending to being in communion with one another in each person's openness to illimited sacred mystery. For example, Thomas Aquinas, emphasizing the relational character of personhood, understood love as *both* willing the well-being of the other *and* being in an appropriate union with them.[12] Modern slavery contradicts both dimensions of mutuality. While many focus on the wanton disregard for the well-being of the victims, it is equally valuable to focus on the distortion entailed in the refusal to be in union with them. This refusal to open oneself to and be in union with the radical depths of others is a form of "willed isolation," the opposite of the intentional union noted by Aquinas. Willed isolation simultaneously damages and distorts one's relations to oneself, others, and the sacred mystery attendant in all life. If sin is social, so, too, is grace.

Notes

1. International Labour Organization, *Global Estimates of Modern Slavery: Forced Labour and Forced Marriage* (Geneva: International Labour Organization, September 19, 2017), https://www.ilo.org/global/publications/books/WCMS_575479/lang--en/index.htm.

2. International Labour Organization, *Global Estimates of Modern Slavery*.
3. For a thoughtful review of the debates on the notion of the person implied in the Universal Declaration of Human Rights, see Jens David Ohlin, "Is the Concept of the Person Necessary for Human Rights?" *Cornell Law Faculty Publications*, paper 434 (2005), http://schloraship.law.cornell.edu/facpub/434.
4. Assets are only assets *for* something.
5. Most of these assets are *potential* assets insofar as they were not initially or intentionally built by religious communities for modern moral threats such as modern slavery and trafficking. Nevertheless, creatively engaged, they can be powerfully transformative.
6. Drivers are prioritized in terms of their significance and the relative strength of religious assets to impact them.
7. If these large assets are typically contributed freely by the religious communities, funding in the form of grants is typically required to mobilize, equip, and deploy them.
8. In is worth noting that rights protect, whereas virtues perfect, in the sense of actualizing human potential. Thus, rights and virtues are distinct but complement each another.
9. In addition, collaboration across religious traditions helps to build the trust, social trust, and solidarity essential for integral development.
10. Religions for Peace followed this approach in the HIV/AIDS pandemic by focusing on low-cost, high-impact, replicable interventions that can be operated on congregational platforms. This approach can be adapted for some of the specialized services needed, especially in areas underserved by specialized organizations.
11. "Mutuality" refers to person-to-person connection through "joint attention." For a collection of the burgeoning research on joint attention, see Naomi Elian, Christoph Hoerl, Teresa McCormack, and Johannes Roessler, eds., *Joint Attention: Communication and Other Minds: Issues in Philosophy and Psychology* (Oxford: Clarendon, 2005).
12. For a contemporary interpretation of Aquinas's virtue ethics in terms of mutuality, see Eleonore Stump, *Wandering in Darkness: Narrative and the Problem of Suffering* (Oxford: Clarendon, 2010).

CHAPTER 36

VIOLENCE AGAINST THE POOR AND ETHICS IN ACTION

SHARON COHN WU

Throughout the developing world, fear of violence is part of everyday life for the poor. A major study by the United Nations found that in developing and middle-income countries, poor people often name violence as their "greatest fear" or "main problem."[1] The scale of this "everyday" violence is massive. One in three women will be a victim of sexual violence.[2] Nearly two million children are exploited in the commercial sex industry.[3] In the developing world, impoverished children and families are uniquely vulnerable to violence because their justice systems do not protect them from violent people; they find that "police and official justice systems side with the rich, persecute poor people and make poor people more insecure, fearful and poorer."[4]

When the global antipoverty effort was set forth in the 2000 Millennium Development Goals, the issue of violence was not mentioned. Fifteen years later, the successors of the MDGs, the Sustainable Development Goals, contained a number of antiviolence goals and targets, demonstrating that the international community understands exploitation and abuse to be a crucial driver and sustainer of poverty.

The linkage between violence and poverty can be seen most starkly in modern slavery, in which tens of millions of people are exploited, degraded, and impoverished. In recognition of that reality, the SDGs include an antislavery goal, SDG Target 8.7: "Take immediate and effective measures to eradicate forced labour, end modern slavery and human trafficking and secure the prohibition and elimination of the worst forms of child labour, including recruitment and use of child soldiers, and by 2025 end child labour in all its forms."[5]

While a global consensus has emerged that slavery exists and must be extinguished, there is no consensus on a solution. Given the durability of the crime throughout history, it is reasonable to ask whether there is something inevitable about slavery that prevents its abolition. International Justice Mission (IJM) insists that there is nothing inevitable about slavery and that it can be abolished. IJM's convictions that slavery can be eradicated and that its victims and their communities can be restored are based on our nearly two decades of work in slavery-burdened countries. With our local government partners, we have developed innovative processes to rescue thousands of victims of labor slavery and commercial sexual exploitation and restore them to lives of freedom.

There is not a country in the world where slavery is legal; nonetheless, the crime is widespread, with numbers of victims in the tens of millions and profits upward of US$150 billion per year.[6] Simply put, unenforced laws are ineffective. Slavery, unlike other forms of violent crime (such as acts motivated by passion or ideology), is economically motivated. If there is little risk of criminal sanction to those engaging in slavery and trafficking *and* the profits are high, the continuation of the crime is virtually assured—and on a large scale.

Why do slavery-burdened countries fail to enforce their own antislavery laws? One significant factor is that development institutions such as the World Bank and most donor nations have not prioritized the development of effective enforcement of criminal laws. Police and security forces have a well-earned reputation for abuse of human rights. The international human rights agenda has focused for the past half-century on condemning and restraining such abuses by law enforcement agencies. Another challenge is that when donors engage with police forces that abuse authority, the donors themselves are implicated in the abuse. Such was the case for the United States, when human rights violations by U.S.-trained police forces in Southeast Asia and Latin America were so flagrant that in 1974, the U.S. Congress enacted a prohibition on assistance to police and prisons from the Agency for International Development.

The wariness of donors and development institutions is understandable, but the fact remains that for the poor and vulnerable, who are the most likely to be victimized by abusive police and security forces, there isn't an alternative source of protection. The well-functioning criminal justice system is the sole duty bearer and authority for the physical restraint of criminal behavior. The answer to poor policing is not *no* policing. In poor communities without professional policing and effective law enforcement, vulnerable people are left entirely at the mercy of criminals who can abuse and exploit them with impunity.

In its antislavery initiatives in Asia, Latin America, and Africa, IJM has seen that local police forces can improve greatly and successfully carry out their duties with respect for both victims and alleged perpetrators. It does not require a sea change in government for police forces, prosecutors, and courts to do their jobs, only a modicum of political will and well-designed technical assistance, including case-specific mentoring.

IJM has assisted two Southeast Asian countries, the Philippines and Cambodia, in confronting their considerable problem of the commercial sexual exploitation of children since the early 2000s. IJM provided mentoring and accompaniment of law enforcement officials (especially members of the two countries' police antitrafficking units) and represented child sex trafficking victims in court. The initiative in both countries greatly increased both the quality of care for survivors and the use of trauma-informed processes within the criminal justice system, ultimately resulting in the prosecution and conviction of hundreds of traffickers, pimps, brothel owners, and customers in both countries.

Perpetrators adjusted their activities quickly when Cambodian and Filipino police, prosecutors, and courts began to enforce laws. IJM conducted a baseline study of the prevalence of minor children in the commercial sex industry in the Philippines' second largest city of Cebu in 2006. Thereafter, IJM collaborated with local law enforcement to identify and remove victims from exploitation and apprehend and prosecute traffickers. Hundreds of pimps, brothel owners, and traffickers were apprehended and jailed pending trial, and their establishments were barred.

A subsequent prevalence study found that the availability of minor girls for sexual exploitation had been reduced by 79 percent. Similar studies in Manila, another IJM program site, found a 75 percent reduction in child availability in the sex industry. Encouraged by its success, the Philippines Government replicated the partnership in Pampanga and achieved a reduction of 86 percent over four years.

IJM observed a similar reduction in child sex trafficking in Cambodia. In the early 2000s, Phnom Penh became known for the availability of extremely young girls for sexual abuse. Although IJM did not conduct a baseline prevalence study, the Cambodian Government's own estimate was that children represented 15 to 30 percent of those in the commercial sex industry. IJM conducted a prevalence study in 2015 after a decade of collaboration with police and prosecutors to rescue children and prosecute traffickers, brothel owners, and customers. The study revealed that among those exploited in the commercial sex industry, 2.2 percent were minors. Young minors, under the age of 15 years, were almost impossible to find.

Trafficked labor must be investigated and prosecuted, no less than sex trafficking. The numbers of children, men, and women exploited in forced or bonded labor is staggering. A demographic study of eleven manual labor industries in the Indian state of Tamil Nadu revealed an average prevalence of bonded labor of 30 percent.[7] Some industries, including textiles, bricks, and quarried stone, had prevalence rates of bonded labor over 60 percent. But criminal investigation, prosecution, and conviction of perpetrators of labor slavery are rare, in part because forced laborers are difficult to identify, and access to law enforcement is limited for the most vulnerable.[8] The U.S. State Department's *Trafficking in Persons Report* contained global data on convictions for perpetrators of trafficking. Of 6,591 convictions for trafficking in 2017 (the most recent year for which data were available), only 298 were for labor trafficking cases. The remainder were sex trafficking cases.[9]

Notwithstanding the challenges, IJM has seen significant progress in India, Thailand, Ghana, and Cambodia, where we collaborate with local authorities to rescue and restore victims of bonded and trafficked labor and bring perpetrators to justice. When IJM began working on labor trafficking of Cambodians (both intranational and cross-border), the investigations and convictions of perpetrators were very few. But over the past several years, thirty perpetrators of labor trafficking have been convicted. Our experience demonstrates that technical assistance and mentoring on individual cases can successfully enable law enforcement to investigate and prosecute complex cases of cross-border labor trafficking.

Promising Approaches to Rescue and Slavery Deterrence

Case-Specific Mentoring and Accompaniment

IJM's approach to addressing gaps in government capacity to enforce antislavery and antitrafficking laws is to assist criminal justice officials on individual cases. IJM experts, including lawyers, investigators, and social workers, work alongside their counterparts in the police, prosecution service, and social welfare agencies on dozens of individual cases. The collaborative casework approach builds skill and reveals gaps and weaknesses in government capacity to investigate, apprehend, and prosecute suspected perpetrators, which can then be addressed, and builds capacity for trauma-informed care of survivors of slavery and trafficking.

Trauma-Focused Care for Survivors

In its collaborative casework model, IJM employs a "dual accompaniment" approach. In addition to equipping law enforcement officials, IJM lawyers and social workers also accompany survivors of sex and labor trafficking throughout the process of restoration and prosecution of abusers. Trauma-informed care equips professionals to understand the experiences and behaviors of survivors, to address challenges in a way that empowers recovery, and to identify when survivors need additional care. Ultimately, trauma-informed care at both the organizational and systemic levels minimizes the retraumatization of individual survivors.

Importantly, the appropriate treatment of trafficking and slavery survivors is also necessary for building a proper legal case against perpetrators. Making a legal case to prosecute traffickers and slave owners requires the testimony of victims. If they are treated poorly by the authorities or not provided with trauma-informed services, they can be traumatized by recounting abuses or will not participate in the legal process at all.

For survivors of crimes like forced labor slavery or sexual violence, the journey to restoration can be long. IJM's after-care teams and partners have served more than 6,800 survivors on their journey to restoration since 2013. Social workers provide services including immediate crisis care, ongoing counseling, legal process support, and job training and education, which help survivors to live in their communities with a low likelihood of being abused again.

Survivor Leadership

Survivors are critical stakeholders and subject matter experts on the crime of slavery. IJM has established an advisory council of survivor leaders to build a network of survivors who will be empowered to guide and publicly advocate for IJM's overall programming.

A vibrant grassroots organization of survivors of bonded labor slavery can be found in southern India, where survivors have formed a registered society to advocate for an end to bonded labor. The Released Bonded Labourers Associations (RBLAs) assist in the identification and rescue of people trapped in servitude, secure resources for their communities, and serve as local educators and watchdogs to end local exploitation. RBLAs have been registered in five districts in the state of Tamil Nadu (Tiruvallur, Kancheepuram, Vellore, Tiruvannamalai, and Villupuram) and include more than one thousand members.

RBLA members play a crucial role in identifying other laborers being held in illegal bondage and advocating for their release. One such case occurred in September 2019 when RBLA President Gopi in Kanchipuram District coordinated a successful operation with local authorities to free eight people from a rice mill, including four young children. In addition to bringing the families to safety, the RBLA and local authorities coordinated the resources the survivors needed to thrive: urgent medical care, warm meals, release certifications, rehabilitation funds, and safe transport back to their home village. Government officials ultimately filed charges against the rice mill owner and arrested him.

Slavery in Corporate Supply Chains

In recent years there has been increasing and welcome attention by the media and labor rights monitoring organizations of slavery in corporate supply chains.

Multinational corporations are regularly stigmatized when evidence emerges of forced or child labor in their supply chains.

No one should question the responsibility of importers and retailers to do their utmost to identify and eradicate their supply chains of slave-produced goods. But trafficking is a corrupt and usually hidden criminal activity that presents a huge challenge for corporations. Moreover, the nature of labor slavery in many industries is such that even the most scrupulous corporations will risk purveying slavery-tainted goods if the national governments of source countries do not take responsibility for the problem. It is governments, not corporations, who have the authority and the obligation to enforce national laws against forced labor, child labor, slavery, and trafficking. It is local and national police, prosecutors, and judges, not corporate executives, who can investigate, arrest, prosecute, and punish perpetrators.

The UN Guiding Principles on Business and Human Rights place the responsibility squarely on governments to protect individuals from slavery and exploitation. What, then, should be the responsibility of investors, importers, and retailers? Perhaps the first order of business is to be clear that labor slavery is quite different from other ethical issues, such as environmental or safety issues. Labor slavery—bonded labor, forced child labor, and trafficked labor—is a violent crime. Trafficked labor is not an unfortunate, accidental outcome; rather, it is the intentional action of perpetrators to profit from misery, without fear of apprehension. Protecting workers from harm and exploitation and deterring the crime requires that corporate stakeholders—investors, importers, shareholders, and retailers—know the landscape from which they are sourcing their supply. A crucial element of this landscape is the will and capacity of local and national authorities to enforce their laws against labor slavery.

Today, a range of actors are paying more attention and making greater efforts to address the issue of forced labor in the global supply chain, and each has a vital role to play. Local and international media bring issues of exploitation to light; nongovernmental organizations research specific industries, gather prevalence data, and identify victims; ethically inclined corporations scrutinize their supply chains, release information publicly, and improve procurement practices. But national governments, whose sovereign responsibility it is to protect the vulnerable and deter those who prey on them, are crucial players who all too often are simply missing from the conversation.

Governments of slavery-burdened countries need help. International donors should prioritize investment in functioning law enforcement, including antitrafficking police units, special prosecutors, and legal assistance for abused workers. Development agencies should help build government capacity to collect and analyze data on slavery. Workers, including migrants, should have access to a safe, anonymous mechanism for reporting exploitation and abuse. And corporations, investors, and shareholders should engage with their government contacts at all levels about labor trafficking and the necessity of the authorities to deal with it as the violent crime it is.

The inclusion of antislavery targets in the SDGs has the potential to animate a global effort to rid the world of the crime. To that end, it is necessary for national governments, donor nations, international organizations, and civil society to prioritize the enforcement of national laws that protect the vulnerable from exploitation and violence. Failure to do so will consign the global commitment to SDG Target 8.7 to irrelevance, to the detriment of the enslaved and the benefit of the enslavers.

Notes

1. Deepa Naraya, Raj Patel, Kai Schafft, Anne Rademacher, and Sarah Koch-Schulte, *Voices of the Poor: Can Anyone Hear Us?* (Oxford: Oxford University Press for the World Bank, 2000), http://documents.worldbank.org/curated/en/131441468779067441/Voices-of-the-poor-can-anyone-hear-us.
2. "Facts and Figures: Ending Violence Against Women," UN Women, updated March 2021, http://www.unwomen.org/en/what-we-do/ending-violence-against-women/facts-and-figures.
3. UNICEF, *State of the World's Children 2005: Childhood Under Threat* (New York: UNICEF, 2004).
4. Deepa Naraya, Robert Chambers, Meera K. Shah, and Patti Petesch, *Voices of the Poor: Crying Out for Change* (Oxford: Oxford University Press for the World Bank, 2000), 163.
5. "Sustainable Development Goals," United Nations, https://www.un.org/sustainabledevelopment/sustainable-development-goals/.
6. "ILO Says Forced Labour Generates Annual Profits of US$150 Billion," *ILO News*, May 20, 2014, https://www.ilo.org/global/about-the-ilo/newsroom/news/WCMS_243201/lang--en/index.htm.
7. International Justice Mission, *International Justice Mission Justice Review: A Journal on Protection and Justice for the Poor* (Washington, DC: IJM, 2018), https://ijmstoragelive.blob.core.windows.net/ijmna/documents/studies/IJM-Justice-Review.pdf?mtime=20210204220533&focal=none.
8. International Labour Office, *Hard to See, Harder to Count: Survey Guidelines to Estimate Forced Labour of Adults and Children*, 2nd ed. (Geneva: ILO, 2012), http://ilo.org/global/topics/forced-labour/publications/WCMS_182096/lang--en/index.htm.
9. U.S. Department of State, *2018 Trafficking in Persons Report: June 2018* (Washington, DC: U.S. Department of State, 2018), https://www.state.gov/j/tip/rls/tiprpt/2018/.

PART X

AN ETHICAL CONSENSUS ON SUSTAINABLE DEVELOPMENT

Indigenous Peoples

DECLARATION BY ETHICS IN ACTION ON THE SDGS AND THE MAGISTERIUM OF POPE FRANCIS FOR INDIGENOUS PEOPLES

Indigenous peoples, who have inhabited the world's lands before the arrival of later colonizers and settlers, are humanity's frontline stewards of sustainable development. Estimated to number around four hundred million people, they have been taking care of their environment since time immemorial, living in symbiosis with nature. Elders from Indigenous peoples have always taught their children that the earth does not belong to us. This point of view enhances their sense of responsibility toward the well-being of the environment—the community, the animals, and the flora. For them, the notion of interrelation is obvious. Thus, when the earth is suffering, her inhabitants are suffering; and when her inhabitants are suffering, She is suffering. In crucial ways, Indigenous peoples around the world are humanity's sentinels and pathfinders back to a path of integral and sustainable development.

Pope Francis, in his January 2018 meeting with Indigenous people of the Amazon, put it this way: "Those of us who do not live in these lands need your wisdom and knowledge to enable us to enter into, without destroying, the treasures that this region holds. And to hear an echo of the words that the Lord spoke to Moses: 'Remove the sandals from your feet, for the place on which you are standing is holy ground.'"

In a world beset by the greed of untrammeled corporate power, Indigenous peoples' experiences of the sacred root their philosophies in the concept of *buen vivir* (living well)—a form of shared well-being that includes the organic links among people, communities, and the earth. The holistic notion of *buen vivir* recalls the core truth that economies have a *telos*, a purpose, beyond profit.

That purpose is shared well-being, the flourishing of all in a common good that includes the earth.

This notion has analogues across Indigenous traditions. Examples include *Mino-bimadaziwin*, a cultural value and mandate of the Anishinaabe of North America to live a good, healthy life committed to the continuous rebirth of the gifts of creation; *Malama aina*, a Hawaiian cultural value that outlines humans' *kuleana* (responsibility) to take care of what nourishes us; and *Sumac Kawsay*, from the Quechua peoples of the Andes, a concept of living well in balance with the living earth and our human community.

Modern life, organized for individual self-interest and the pursuit of profit, has led humanity away from this profound notion of well-being, causing—in the words of Pope Francis—the "globalization of indifference." Indigenous peoples can help lead the world back to deeper purposes, in a holistic spirit of participation, mutual consent, dialogue, solidarity, and peace.

Indigenous peoples regard the modern world as overly influenced by a "masculine" value system—guided by such factors as speed, performance, logic, science, and the spirit of conquest. From this perspective, the imbalances within social and economic life reflect the imbalances within the human spirit. This idea also gels with Pope Francis's criticism in *Laudatos si'* of the "technocratic paradigm" and the "myths of a modernity grounded in a utilitarian mindset (individualism, unlimited progress, competition, consumerism, [and] the unregulated market)," alongside his call for restoring "the various levels of ecological equilibrium, establishing harmony within ourselves, with others, with nature and other living creatures, and with God."[1] Indigenous wisdom calls upon human beings to restore balance not only by supporting women and defending Mother Earth but by embracing more "feminine" values such as introspection, slowness, listening, meditation, creativity, collaboration, and patience. Nevertheless, Indigenous peoples—like all others—should reflect on the value of the equal dignity of women, proposed by modern feminism, against all forms of paternalism and machismo.

The profound moral paradox of Indigenous peoples is that, almost everywhere, they have been marginalized and repressed, enslaved and slaughtered, sometimes to the extent of genocide (or what we call today "cultural genocide") by colonizer and settler populations. Throughout the history of globalization of the past half-millennium, colonizers arrived with voracious appetites for the resources of the lands, rivers, and oceans—in particular gold and silver—viewing the Indigenous peoples they met as obstacles to their pursuit of wealth and power. More often than not, the colonizers brought powerful weapons, advanced technologies, and pathogens from abroad that enabled them to suppress Indigenous peoples. Those who were able to avoid enslavement and slaughter were forced onto reservations or pushed onto marginal lands where it became increasingly difficult to feed their families. There were a few important exceptions to this—most notably the famous Jesuit missions, which were an example of socioeconomic development for the common good and against forced labor.

The story of Indigenous peoples is also one of profound moral strength, of their determination to survive, thrive, and safeguard their cultures and languages on their own terms, despite hardships and oppression. Indigenous peoples teach us the virtue of resilience, hope, and faith in the face of overwhelming hostile forces. In all parts of the world, Indigenous peoples heroically strive to maintain their culture, languages, and connections with ancestors and ancestral lands, and to build a world of meaning for their own societies and for the world at large. In this way, the world's Indigenous peoples teach all humanity how to maintain hope and purpose in the face of great challenges.

The United Nations has recognized the unique role and vulnerability of Indigenous peoples, adopting the UN Declaration on the Rights of Indigenous Peoples in 2007. This declaration makes explicit many core human rights, both individual and collective, of Indigenous peoples, including the following:

- To be free from discrimination (Article 2)
- To self-determination (Article 3)
- To autonomy or self-government in matters relating to their internal and local affairs (Article 4)
- To maintain and strengthen their distinct political, legal, economic, social, and cultural institutions (Article 5)
- To not be subjected to forced assimilation or destruction of their culture (Article 8)
- To not be forcibly removed from their lands or territories (Article 10)
- To practice and revitalize their cultural traditions and customs (Article 11)
- To establish and control their educational systems and institutions providing education in their own languages (Article 14)
- To participate in decision-making in matters that would affect their rights (Article 18)
- To the improvement of economic and social conditions (Article 21)
- To the lands, territories, and resources that they have traditionally owned, occupied, or otherwise used or acquired (Article 26)
- To free, prior, and informed consent, and consideration in good faith, prior to the approval of any project affecting their lands or territories (Article 32)
- To maintain and develop contacts, relations, and cooperation, including activities for spiritual, cultural, political, economic, and social purposes, with their own members as well as other peoples across borders (Article 36)

The role of Indigenous peoples in the 2030 Agenda and the Sustainable Development Goals is also recognized at several points. The 2030 Agenda recognizes the vulnerabilities and needs of Indigenous peoples in line with the overriding objective "to leave no one behind." The unique needs and gifts of Indigenous peoples in education, agriculture, hunting, trapping, picking, and fishing are also noted.

At the same time, Pope Francis recognizes the importance of what he terms "ecological debt": "Inequity affects not only individuals but entire countries; it compels us to consider an ethics of international relations. A true "ecological debt" exists, particularly between the global north and south, connected to commercial imbalances with effects on the environment, and the disproportionate use of natural resources by certain countries over long periods of time."[2]

Yet the harsh reality is that Indigenous peoples have been not only on the frontlines of the struggle for sustainable development but also continue to face neocolonial power encroaching on their lands with ever greater and multinational force. Throughout the world, Indigenous peoples are facing land grabs and displacement, pollution caused by extractive industries, and expulsion from their lands because of decisions to lay pipelines, harvest minerals, log forests, and for other purposes—without their say, approval, or even participation.

Recently, for example, the Standing Rock Sioux Tribe of the Dakotas in the United States tried to resist the Keystone XL Pipeline but were ultimately overwhelmed by the power of the U.S. government in the service of private industry. Similarly, First Nations peoples in British Columbia are struggling against the Canadian federal government in resisting the Trans Mountain oil pipeline project that crosses many reserve lands. Many similar situations exist in the Amazon, Argentina, Chile, the Philippines, Polynesia, Africa, and other regions, where Indigenous peoples continue to struggle against the abuse of their rights and their lands.

Pope Francis calls on the world, especially the centers of power and wealth, to engage in dialogue with Indigenous peoples, recognizing that they have unique knowledge regarding pathways to sustainable development. Dialogue also advances human rights, since participation in determining one's future is itself a fundamental right, as well as the most powerful instrument to guarantee the fulfillment of other rights.

Above all, Ethics in Action calls for the recognition of Indigenous peoples, including their history and cultures, and for dialogue in solidarity and humility with the world's Indigenous peoples based on a commitment to mutual respect, mutual responsibility, and mutual learning. This dialogue must be genuine, not a one-way monologue of the powerful making demands on the powerless. It must be a true conversation that recognizes shared fates and responsibilities, acknowledging that Indigenous peoples possess not only rights but also unique kinds of knowledge that the entire world needs today.

Ethics in Action calls for a new framework of global sharing with Indigenous peoples in the spirit of the universal destination of goods and in line with

Laudato si'. In many places in the world, Indigenous communities are impoverished from centuries of displacement, discrimination, exploitation, attempted extermination, or assimilation. Restoring to Indigenous peoples the financial and material means for *buen vivir* is a not just a matter of solidarity and charity but also of justice and compensation for past misdeeds by colonizers and the iniquities of today's bullies.

Ethics in Action likewise calls for a new moral vision that recognizes the indispensable contributions of Indigenous peoples in the fulfillment of the SDGs. Their incomparable experience of the environments they live in, their traditional knowledge, and their unique understanding of biodiversity must inform and reorient our global conception of development and our own relationship with the environment and the natural resources it provides.

General Recommendations: Practical Actions for the Sustainable Development of Indigenous Peoples

(1) The promotion of the voices of Indigenous peoples in UN processes within the 2030 Agenda, the SDGs, and the Paris Agreement on climate change
(2) The recognition of a special ecological and human debt in favor of the Indigenous peoples by colonizers to compensate the former for the destruction of their habitats and forests, as well as to atone for the unjust extermination of many populations
(3) The encouragement and technical support of bold plans of action by Indigenous peoples to achieve the SDGs, adapted to the cultures and contexts of Indigenous peoples and in accordance with the concept of *buen vivir*
(4) The support of Indigenous communities to access new official and private development financing needed to achieve sustainable development
(5) The active and continuing partnership of Ethics in Action with the Interfaith Rainforest Initiative to protect and restore global forests, recognizing them not only as Indigenous peoples' ancestral domains and sacred spaces but also as vital habitats for diverse flora and fauna, as well as critical carbon sinks
(6) A call on the UN Commission on Human Rights, the UN Conference on Trade and Development, and the World Trade Organization to step up their efforts to stop land grabs and environmental damage to Indigenous lands
(7) A call upon UNESCO and governments to revise education guidelines, encouraging content relevant to local challenges and *planes de vida plena* (full life plans) and *buen vivir* for Indigenous students; to revise the content presented to both non-Indigenous and Indigenous students in ways that address existing misconceptions and prejudice against Indigenous peoples; and to promote Indigenous forms of education and ways of knowing, particularly Indigenous sciences such as agriculture, medicine, and navigation

(8) Support for the leadership and safety of Indigenous women, and indeed all initiatives helping women and promoting feminine values
(9) A call on countries with Indigenous peoples to legally recognize their rights and set aside territories for those peoples on a scale that allows them to practice traditional lifestyles and cultures
(10) Support for cooperation and exchange among Indigenous peoples on best practices to secure rights and promote development pathways that take into account the cultural and socioeconomic context of Indigenous peoples
(11) Recognition that Indigenous territories are an important complement to national parks and other protected areas, because such territories often have lower levels of deforestation and environmental degradation than those of national parks and other protected areas
(12) Support for programs allowing Indigenous peoples to reconnect with their identity, cultures, and languages so that they can participate more actively in the elaboration of their destiny and the destiny of the world
(13) Move Humanity's support for Indigenous communities around the world, including the mobilization of philanthropic contributions toward sustainable development for Indigenous peoples
(14) Support for the upcoming Synod of Bishops for the Pan-Amazon Region called by Pope Francis for October 2019. Move Humanity and its partners will work to establish a new fund for the sustainable development of the Amazon, calling upon Brazil's leaders, high-net-worth individuals, and key businesses to take leadership in providing necessary support

Notes

1. Francis, *Laudato si'* (Vatican City: Libreria Editrice Vaticana, 2015), sec. 210.
2. Francis, *Laudato si'*, sec. 51.

CHAPTER 37

CARE OF THE EARTH, CARE OF THE SOUL

Indigenous Communities and Inner Climate Change

T8AMINIK (DOMINIQUE) RANKIN, MARIE-JOSÉE TARDIF, AND DANIEL G. GROODY

The way we see the world shapes the way we treat it. If a mountain is a deity, not a pile of ore; if a river is one of the veins of the land, not potential irrigation water; if a forest is a sacred grove, not timber; if other species are biological kin, not resources; if the planet is our mother, not an opportunity—then we will treat one other with greater respect. This is the challenge: to look at the world from a different perspective.[1]

Creation is a sacred gift. It is given to every creature by a loving Creator. Because this Creator is the source of all life, all life is interconnected. Even our own inner life is intimately intertwined with all creation. These beliefs have been at the heart of Indigenous communities throughout the centuries.

Since all of creation shares an inextricable bond with both the Creator and other creatures, our ancestors have affirmed throughout the ages, "When the Earth suffers, human beings suffer. And when human beings suffer, the Earth suffers."[2] At the present moment, however, the earth—and the human community—are groaning in pain. The current challenges posed by climate change and other issues raise far-reaching questions and put even our very survival at risk.

Given these unprecedented challenges, what insight can Indigenous communities offer into our current crises? How can the teachings of the ancestors help us gain wisdom in the face of global disorders? And what ethical insight can native peoples offer as a resource for life-giving action? These are some of the questions we will explore in this chapter.

Here we will examine three teachings on creation from the Indigenous tradition that can help us in our troubled times. The first dimension is the interconnectedness of all creation. The second is the circle of life in all creation. And the third is the integration of the masculine and feminine in all creation. At first glance these notions may appear to be very basic. But like a river, their currents run deep into the heart and soul of the universe. At their core, they are foundational to our survival and ultimately flow toward a vision of healing and reconciliation among Creator, creature, and creation.

The Interconnectedness of All Creation

Among the many and diverse Indigenous communities around the world, our people are known as Algonquins. The word "Algonquin," however, does not belong to our language; it was first used by the French missionaries. We refer to ourselves rather as Anicinape (pronounced "anishinabe"), which means simply "human being." At the core of our teachings is the belief that the fundamental task of life is to learn how to become a genuine and authentic person. Ethics is the path to becoming human, which means becoming an integrated and whole person.

Though our teachings originate within our people, they are universal in scope. To be human means to bring into harmony the inner world with the outer world, or one's own nature with the natural world. If any human being lives according to these principles, they are Anicinape—regardless of their skin color, their religion, their bloodline.

The path of integrating the inner world with the outer world is the path of spirituality for Indigenous peoples. We often refer to the spiritual path as the red path: the path of the heart. Because ethics and action flow first from the heart, if the inner life of a person is not healthy, the outer world is also at risk. This means that while we are concerned about the effects of climate change in the outer world, we are also concerned about climate change within people.

If our actions in the outside world are directly related to our inner world, we could say that current climate change reflects our own overheated, or disordered, inner climate. That is why we think an "inner climate change" must accompany our efforts to save our planet: Without a reordering and healing of the heart, our world will never find peace and well-being.

Our inner climate ultimately shapes our choices and our actions. We believe that our choices are not simply the consequence of rational logic; our decisions are also shaped by our thoughts and our emotions. Without inner transformation, we act out of impulse and fear, stemming from unresolved and inner conflicts, which keep us from becoming whole. Unless values become virtues, we fall short of the task of becoming human, of being Anicinape.

To become Anicinape involves the long road of aligning one's humanity with the Creator and his designs in creation. This process highlights for us the need for prayer and contemplation. When we take a moment to pray every morning, surrounded by the extraordinary beauty of nature, and sincerely thank the Creator for the gift of a new day, our whole being aligns with the universe.

For us, the earth is not an object but a living being. And not just any living thing! The earth is our mother: generous, protective, beneficent, nourishing. She is a wonderful entity who compels us to admire, respect, and love her. All Indigenous peoples share this profound love and connection with Mother Earth. If the earth is our mother, how can we try to manipulate, own, and exploit her? Creation for us is not something to be conquered and subdued; it is not something that can be owned and acquired, bartered and sold; it is a gift given by a loving Creator, and we are called to revere and respect it, and above all to give thanks for its bounty and provision.

Our knowledge that we are possessed by the Creator, who has given us the gift of this creation, releases us from our obsessive need to possess. Our mothers and fathers have always taught to "give yourself everything," which means we should always nourish our life with the best food for our body, heart, and spirit. Only when we learn to respect ourselves in this way are we able to sincerely respect others.

The Circle of Life in All Creation

Another essential foundation in Indigenous philosophy is the circle. The interconnectedness of all creation reveals itself as a circle of life. We often speak of the circle of family, the circle of community, and the circle of existence. In contrast to the Western world, which often sees relationships through the lens of a pyramid—in which there are a few on the top and many more on the bottom—Anicinape see the world as a circle. The most important values are located around the center of the circle. All life emanates from the circle, and our world is organized and centered around this circle. In this great sacred circle, Anicinape are not superior to other terrestrial creatures; rather, they are shaped by a sense of egalitarianism and interbeing.

This cosmic vision is based on concentric circles. At the heart of these circles is a sacred center. Here we honor sacred objects that are a reminder of what brings cohesion to our lives. In the next circle are guardians of the sacred objects: male and female elders, medicine men and medicine women, designated to protect, share, and convey our philosophy and traditional medicine. Then we have a circle of children. Around them is a circle of mothers, supported by a circle of grandmothers. The last circle is that of men. They are the proud protectors of women, children, the medicine, and Mother Earth.

The circle puts us in touch with the reality of the universe; when we let the great movement of life flow freely, with its moments of mourning and the joys of renewal, human beings suffer much less and find the peace the Creator desires for all creatures. It speaks to the cyclical movement of seasons, as well as the cycles we experience in one day, one year, one life, one breath. Following the sacred cycles of life, the nomadic Anicinape people managed to survive in the forest for millennia.

In all our prayers, our legends, and our philosophical teachings, we learn how to observe nature, to respect it, to love it, and to be loved by it. In our language, we call the moon our grandmother, the sun our grandfather. The stones also bear the names *moshom* (grandfather) and *kokom* (grandmother). The sky and earth are father and mother to us as well. We know we share an intimate and inextricable bond with creation, which fills us with love.

In this community circle, everyone is both free and responsible. In many ways—especially in the capitalist system—people have promoted freedom without fostering at the same time responsibility. Our teachings find much in common with the Vietnamese Buddhist master Thich Nhat Hanh, who one said that the United States should erect a Statue of Responsibility next to the Statue of Liberty![3]

In our circles, everyone is tasked with looking after one another; each has a place in the community, and each takes pride in their respective responsibilities for the common good. Because each person is an integral part of the whole of Creation, they must also be a guardian of the whole. For us, every decision we make today should always be based on the well-being of the next seven generations, in the same way that the seven generations that preceded us have affected our lives.

The Integration of the Masculine and Feminine in All Creation

In the path toward becoming authentically free and responsible, our teachings also emphasize the need to develop and integrate the "masculine" and "feminine" energies within ourselves. This sacred male–female dynamic creates movement; it is fundamental in our vision of Creation. Respect for this relationship must be expressed in our actions.

This is why we see the need to foster the complementary relationship between men and women. Even as feminine and masculine energies may at first appear to oppose each other, integral human development involves bringing these poles together in such a way that we allow the sacred beauty of these two energies to come together and blossom within ourselves.

In the present world, we often see these energies at variance with each other. Not only are women often diminished, but the feminine aspects within us are also frequently neglected. It is as if the inner climate of our heart is polluted with an overly emphasized masculine value system that manifests itself through action, speed, intellect, rationality, science, conquest, competition, and exteriority, especially with the material world.

The current environmental challenges are a reflection of this great imbalance. Even if we must acknowledge the value of science in helping us master some of the forces of nature, we do great violence to ourselves when we neglect our other half. Alongside the masculine qualities, we must relearn feminine qualities such as rest, patience, reflection, intuition, creativity, gestation, collaboration, and interiority, especially with the invisible world.

For many Indigenous peoples, the great symbol of these feminine qualities is the turtle. Patient and connected to earth and water, the turtle knows how to withdraw into itself in order to find solutions. It has faith in silence and interiority. It surrenders and trusts, without trying to control things in the outer world. Maybe what the world most needs is a renewed sense of contemplation, or a rediscovery of the turtle within us. Those societies that most respect the feminine are the ones that treat the earth with the utmost respect. The kind of integral development we need requires not only informing the mind but also forming the heart. Only then can we truly transform the world.

Conclusion

Indigenous communities have much to share with the world, and thankfully we are now seeing greater interest in the wisdom that has been part of our communities across the ages. Even after years of conquest and domination, our prophecies have encouraged us to be patient; they have said that one day, "White people" will knock at our doors saying, "We have lost our balance. Can you help us?" It seems that the day has finally come to see the connection between our inner world and our outer world and our need to journey as one in this new era of crisis, change, and transformation.

The river is infinite. Let us paddle together.

Notes

1. David Suzuki, quoted in Bruce Parry, "Why Land Rights for Indigenous Peoples Could Be the Answer to Climate Change," *The Guardian*, November 29, 2016, https://www.theguardian.com/commentisfree/2016/nov/29/land-rights-indigenous-peoples-climate-change-deforestation-amazon.
2. Traditional Anicinape saying.
3. "Thich Nhat Hanh," *PBS: Religion and Ethics Newsweekly*, September 19, 2003, https://www.pbs.org/wnet/religionandethics/2003/09/19/september-19-2003-thich-nhat-hanh/1843/.

CHAPTER 38

PRACTICAL APPROACHES TO SUSTAINABLE DEVELOPMENT IN INDIGENOUS COMMUNITIES AND TRADITIONAL POPULATIONS OF THE AMAZON

VIRGILIO VIANA

In *Laudato si'*, Pope Francis calls for respect for the rights of Indigenous peoples. This respect includes the demarcation of Indigenous territories not yet under legal protection against invasion by land grabbers, illegal loggers, gold miners, and others. Such respect should also be given due consideration in the planning and implementation of large infrastructure projects such as roads and hydroelectric dams.

Indigenous peoples are responsible for the protection and management of more than 190 million hectares in Brazil, between legal territories and areas in the process of legal authorization (as of 2019). These ecosystems are vital to mitigating global climate change, conserving biodiversity, and meeting other global sustainable development goals. Despite this fact, Indigenous peoples face serious threats to their basic human rights, including violence and invasion of their ancestral territories.

Indigenous peoples across the globe play a key role in mitigating global climate change, conserving biodiversity, and protecting watersheds and other ecosystem services. Protecting the rights of Indigenous peoples is a moral and ethical obligation. In addition, supporting their challenges, including poverty and poor access to education and health, is essential to meeting the SDGs.

Indigenous peoples play a vital role in maintaining rich cultural diversity and ethnoecological knowledge. Indigenous people also provide a unique set of paradigms to guide societies from individualistic, consumption-driven happiness to collectivistic, community-oriented happiness. The conceptual frameworks of *buen vivir* (living well) and *planes de vida plena* (full life plans) provide an important refence for developing solutions to meet the global challenge of changing consumption patterns through reduced consumerism.

Indigenous peoples of the Amazon include 390 ethnicities, speak more than three hundred languages, and live in nine territories over an area of more than 170 million hectares of legalized territory.[1] These territories house 36 percent of all carbon stocks in the Amazon. The release of these greenhouse gases into the atmosphere would dramatically aggravate global climate change.

Rights of Indigenous Peoples

Recent reports by UN experts and human rights organizations document an alarming increase in violent attacks against and criminalization of Indigenous peoples defending their rights to their traditional lands and natural resources. Intensifying global competition over natural resources is increasingly making Indigenous communities, who act to protect their traditional lands and territories, targets of persecution by state and nonstate actors. Indigenous leaders and community members, who voice opposition to development or investment projects from extractive industries or large infrastructure projects like hydropower, are often subjected to criminalization, harassment, threats, violent attacks, and killings.

Over the last twenty years, Indigenous peoples' rights have been increasingly recognized through the adoption of international instruments and mechanisms, such as the UN Declaration on the Rights of Indigenous Peoples in 2007, the American Declaration on the Rights of Indigenous Peoples in 2016, twenty-three ratifications of the Indigenous and Tribal Peoples Convention of 1991, the establishment of the UN Permanent Forum on Indigenous Issues in 2000, the Expert Mechanism on the Rights of Indigenous Peoples in 2007, and the UN Special Rapporteur on the Rights of Indigenous Peoples in 2001.[2]

Indigenous Peoples and Global Sustainability

Pope Francis has called for greater respect for the rights of and increased support to meet the challenges currently faced by Indigenous peoples. In his January 2018 visit to Peru, he chose to begin his journey with a meeting with Indigenous people of the Amazon to highlight their importance for our common future. This followed an earlier decision to convene the Synod of Bishops for the Pan-Amazon Region in 2019.

Indigenous peoples across the world are facing the consequences of climate change. Indigenous peoples must, therefore, be heard and included in local, national, and global climate action. Indigenous peoples are particularly vulnerable to climate change and yet are the least responsible. Indigenous peoples have lifestyles and traditional knowledge from which others can learn in their efforts to address the climate crisis, and Indigenous peoples are highly motivated to drive solutions to overcome climate change. Many of the world's ecosystems and areas of biodiversity are protected and nurtured by Indigenous peoples. Their traditional knowledge and contributions to climate mitigation are increasingly being acknowledged and referenced in international agreements and declarations.

Indigenous peoples provide global benefits for global climate and nature conservation. Deforestation rates in Indigenous lands are lower than in other protected areas in many regions of the world, including the Amazon. The role of Indigenous peoples as stewards of nature conservation needs to be recognized, but this recognition needs to go beyond words to provide the basis for new, increased, effective, and efficient support mechanisms. This recognition also has an ethical and moral underpinning, as well as a science-based rationale for compensation for the ecosystem services provided by the territories protected by Indigenous peoples.

In 2015, the Paris Agreement was adopted as a global action plan to mitigate and prevent further climate change, and Indigenous peoples are mentioned in the agreement: "Parties should respect, promote and consider their respective obligations on . . . the rights of indigenous peoples." At COP21, the 2015 UN Climate Change Conference in Paris, it was decided to establish a knowledge-sharing platform on climate action for Indigenous peoples. In 2016, UN member states agreed on the 2030 Agenda for Sustainable Development, which had an ambition of "leaving no one behind." The International Work Group for Indigenous Affairs sees these international commitments as a window of opportunity for truly including the knowledge, experiences, and rights of Indigenous peoples in climate action.

Practical Action to Support Indigenous Peoples

Practical action to support Indigenous peoples should not only respect the concept of former, prior, and informed consent but also go beyond by including effective, bottom-up, participatory design, planning, implementation, monitoring, and evaluation. Such participation is essential to incorporating the cultural perspectives and priorities of Indigenous peoples to guide practical action.

Practical action should be broad in scope and follow a holistic approach. This approach allows for a priority-setting process not guided by projects with a thematic focus but one that is wide open so as to accommodate bottom-up processes.

The practical experiences of the Foundation for Amazon Sustainability are based on a participatory planning, implementation, monitoring, and evaluation strategy that has been refined over the last eleven years. The foundation's work includes more than two thousand projects and more than 1,200 participatory workshops to define investment priorities in 581 communities, both Indigenous and non-Indigenous. These priority investments were grouped into eight thematic areas that embrace the seventeen SDGs:

- Income generation
- Community infrastructure
- Environmental conservation
- Management and transparency
- Research, development, and innovation
- Health
- Education and citizenship
- Empowerment

The implementation of the *Floresta em Pé* Program has led to a reduction in poverty in 581 communities, with an average increase in income of 244 percent. This outcome is a result of a combination of (i) payment for ecosystem services through cash transfers (*Bolsa Floresta* Program); (ii) investment in income generation in sustainable production chains (*Floresta em Pé* Program); (iii) investment in entrepreneurship and professional education; and (iv) investment in the empowerment of grassroots associations. These investments created synergies and positive feedback loops that increased the success of income generation activities.

The implementation of the *Floresta em Pé* Program has led to a reduction in deforestation by 244 percent. This is a result of a combination of (i) making forests worth more standing than cut through investments in income generation for sustainable forestry and fisheries; (ii) commitments made by participants of the *Bolsa Floresta* Program for zero deforestation of primary forests and the adoption of fire prevention measures; (iii) environmental education at all levels (from early childhood to adult); and (iv) support for the social control of forest management practices.

Conclusion

Indigenous communities and traditional populations play a key role in reducing deforestation and forest fires in the Amazon. This role depends on practical actions to support sustainable development pathways that bring environmental and social goals together by making forests worth more standing than cut. These goals are reached through appropriate methodologies that combine participatory methods and holistic approaches.

The large-scale experience of the Foundation for Amazon Sustainability has shown that it is possible to combine social and environmental goals supported by sustainable income generation. Through this work deforestation has been reduced while incomes have increased.

Support for practical approaches to sustainable development in Indigenous communities and traditional populations of the Amazon should include the following:

- New mechanisms to provide large-scale funding for sustainable development in Indigenous territories and protected areas inhabited by traditional populations
- The concepts and values of Indigenous peoples and traditional populations, including the concept of *buen vivir*
- Strengthened mechanisms to secure the protection of the rights and territories of Indigenous peoples and traditional populations
- The empowerment of organizations that represent Indigenous peoples and traditional populations at subnational, national, and international levels

Notes

This chapter is based on a presentation given by the author at the Seminar on Science and Actions for Indigenous Peoples at the Pontifical Academy of Sciences, Vatican City, June 27 and 28, 2018.

1. "Coordinadora de las Organizaciones Indígenas de la Cuenca Amazónica," Coica, https://coica.org.ec/.
2. See "Indigenous Peoples," World Bank, https://www.worldbank.org/en/topic/indigenous peoples.

PART XI

AN ETHICAL CONSENSUS ON SUSTAINABLE DEVELOPMENT

Corruption

ETHICS IN ACTION ON CORRUPTION

Consensus Statement

Corruption is the abuse of power to pursue particular interests over the common good and constitutes a major impediment to integral and sustainable human development. The function of government is to enhance the common good, but this function is often distorted for corrupt purposes. It is for this reason that SDG 16 calls for countries to "promote peaceful and inclusive societies for sustainable development, provide access to justice for all and build effective, accountable and inclusive institutions at all levels," with targets pertaining to reducing illicit financial flows, corruption, and bribery and building effective, accountable, and transparent institutions at all levels. Yet corruption remains endemic in today's global economy.

Corruption encompasses bribery and theft of state assets, but it is broader than this—especially today, it includes the deliberate construction of tax havens and secrecy jurisdictions, which allow powerful elites and multilateral corporations to evade taxes, gain market advantage, and sometimes even launder the proceeds of crime as well as human and drug trafficking. As the Vatican noted in its 2018 document calling for an ethical economic and financial system,

> [Offshore sites] have removed decisive resources from the actual economy and contributed to the creation of economic systems founded on inequality. Furthermore, it is not possible to ignore the fact that those *offshore* sites, on more occasions, have become usual places of recycling dirty money, which is the fruit of illicit income (thefts, frauds, corruption, criminal associations, mafia, war booties, etc.).[1]

The whole business model of these havens serves to protect elite money and power from scrutiny, with no questions asked as to sources of income or the accountability of power. Estimates suggest that between US$20 and $30 trillion are now escaping taxation and accountability measures in these havens. And some estimates suggest that as much as half of all world trade passes through these havens. This ability to "escape" responsibility to the common good, both national and global, with impunity is a major—if not the most important—barrier to the implementation of the SDGs.

Corruption has a clear and devastating economic impact. By allowing the wealthy to avoid taxes, corruption not only deprives governments of revenue but also delegitimizes the entire tax system and system of government. On the spending side, instead of investing in health, education, and environmental conservation, corruption—through kickbacks and other means—induces politicians to favor special interests with little social value. By limiting the redistributive function of the government, corruption gravely increases poverty and inequality. Corruption also hurts private investment by generating uncertainty and encouraging rent-seeking rather than the creation of social value. Corruption is a particular problem among state-owned enterprises, given the close proximity between politicians and commercial decisions; the financial markets, which often put monetary value above all other considerations; and the extractive industries, which are often guided by a rapacious mentality that seeks financial gain at the expense of the poor and the planet.

The effects of corruption go beyond the economic sphere. Corruption undermines trust and social cohesion and can lead to political instability and conflict—the Arab Spring being a case in point. It has a debilitating impact on youth, who understand that connections matter more than capability—thereby reducing incentives to gains skills and education and weakening social ties. A high degree of corruption has been demonstrated to contribute to reduced levels of reported national well-being, even after accounting for national income. Indeed, human beings experience great levels of personal emotional duress when they live in societies marked by high levels of government corruption.

The scourge of corruption runs deeper than individual moral failure. People are often trapped in corrupt systems, in which they are coaxed to adapt to the prevailing corrupt culture to survive. In this sense, corruption embodies a "structure of sin," poisoning the organs of society and degrading the virtues, values, and ethics needed to uphold the common good. It constitutes an economy or *polis* with no values other than the "golden calf" and "money theism." In this sense, corruption weakens democracy and leads to plutocracy, a system in which the powerful seek only their own financial gain, oblivious to the common good. Corruption undermines participation and political representation: Oligarchic tendencies allow the wealthy to buy political power and, in the case of tax havens in particular, break the vital link between taxation and representation.

The challenge of corruption highlights the importance of embedding the economy in a proper ethical framework centered on the common good. As *Oeconomicae et pecuniariae quaestiones* states,

> Experience and evidence over the last decades has demonstrated, on the one hand, how naive is the belief in a presumed self-sufficiency of the markets, independent of any ethics, and on the other hand, the compelling necessity of an appropriate regulation that at the same time unites the freedom and protection of every person . . . In this sense, political and economic-financial powers must remain . . . directed, beyond all proximate harms, towards the realization of a good that is basically common, and not reserved only for a few privileged persons.[2]

The religious traditions represented by Ethics in Action affirmed that corruption has deleterious effects not only on the economy and society but also on what it means to be a human being. In this sense, corruption has not only a moral dimension but also an ontological one. It leads to "disintegrated being," a form of double life that threatens the integrity and wholeness of the human being by destroying the very relationships that define this wholeness: with God, with the rest of humanity, and with creation. Corruption reflects the temptation of power, greed, self-centered desires, narcissism, and self-interest. It reflects a disordered desire to acquire, possess, and accumulate in a violent matter, disregarding the other and nature. It offers a false and illusory path to fulfillment and flourishing. Pope Francis also affirms the gravity of the sin of corruption. Regarding "the difference between a sinner and a corrupt person," he writes that "those who lead a double life are corrupt" and "God does not forgive Christians who lead a double life." Hypocrisy and deception are two components of corruption, and where there is deception, according to Pope Francis, there is no Spirit of God. From his perspective, all individual sins are pardonable except for the social sin of corruption, for corruption disintegrates the very integrity of the social body.[3] In substance, Pope Francis considers corruption a sin against the Holy Spirit. According to Jesus Christ, this is an "eternal sin" that "will never be forgiven."[4]

Tackling corruption will require institutional change and conversion, predicated on transparency, accountability, and a commitment to the common good. Given the inability of even virtuous individuals to effect change in a corrupt system, top-down reform is essential. An effective reform strategy would prioritize legal codes, independent judiciaries, public participation in politics, parliamentary over presidential systems, term limits for officeholders, free speech, a free press, transparency, countervailing power, active civil societies, regulatory reform to stop rent-seeking, stronger financial sector oversight, and increased corporate accountability.

In addition to national solutions, tackling corruption will require global coordination, especially since "those deputed to exercise political power are often disoriented and rendered powerless by supranational agents and by the volatility of the capital they manage."[5] At the international level, anticorruption solutions include greater reliance on the UN Charter and the Universal Declaration of Human Rights, full implementation of the Paris Agreement on climate change to reduce the corruption associated with the fossil fuel industry, stronger Indigenous rights, globally adopted anticorruption laws and codes, increased transparency over international contracting, ending tax and secrecy havens, and stronger cross-border financial regulation.

Ending corruption will also require the development of social norms that deter fraud, bribery, theft, impunity, human trafficking, the drug trade, and tax evasion. It calls for the development of personal and social virtues that emphasize the common good over greed, self-interest, and a might-is-right mentality. All this in turn requires better role models, ethical education, and a strong role for religion.

General Recommendations: Policy Proposals for Consideration and Implementation by the United Nations, the OECD, and Other National and International Bodies

(1) An international plan agreed by the UN General Assembly to phase out all tax and secrecy havens by 2030, in line with and under the guidance of the SDGs
(2) The introduction of wealth taxes on accounts in tax and secrecy havens to overcome tax evasion, with the proceeds directed toward achieving the SDGs in low-income countries
(3) A call on the International Monetary Fund, the Organisation for Economic Co-operation and Development, and other multilateral bodies to assist developing countries in collecting the revenues they are justly due from multinational companies and to prevent the abuse by these companies of tax and secrecy havens
(4) A UN conference on removing money from politics in line with the dictates of SDG 16
(5) A call on international banks, especially those with private wealth departments, to cease and desist their operation in tax and secrecy havens
(6) The end of impunity made possible by "post-office box corporations" that disguise the true beneficial ownership of accounts and thereby undermine corporate accountability

Notes

1. Congregation for the Doctrine of the Faith and the Dicastery for Promoting Integral Human Development, "'*Oeconomicae et pecuniariae quaestiones.*' Considerations for an Ethical Discernment Regarding Some Aspects of the Present Economic-Financial System," Vatican website, May 17, 2018, sec. 30, https://press.vatican.va/content/salastampa/en/bollettino/pubblico/2018/05/17/180517a.html.
2. Congregation for the Doctrine of the Faith and the Dicastery for Promoting Integral Human Development, "'*Oeconomicae et pecuniariae quaestiones,*'" sec. 21.
3. Homily, *Sanctae Marthae* Chapel, 11.XI.2013.
4. Mark 3:29.
5. Congregation for the Doctrine of the Faith and the Dicastery for Promoting Integral Human Development, "'*Oeconomicae et pecuniariae quaestiones,*'" sec. 12.

CHAPTER 39

UNDERSTANDING AND COMBATTING CORRUPTION

SEAN HAGAN

Over the past several years, a number of high-profile corruption scandals have, justifiably, generated considerable moral outrage. However, as important as this moral dimension may be, this chapter will explore the broader social, economic, and political impact of systemic corruption. As we shall see, there is considerable evidence that corruption—at least when it becomes systemic—can exacerbate poverty and inequality. Moreover, by compromising governmental legitimacy and creating the conditions for radicalization, it can also undermine national security.

What Is Corruption?

While there are different definitions of "corruption," one that is generally accepted is "the abuse of public office for private gain." Two aspects of this definition are worthy of note. First, what types of abuse are covered? At a minimum, the definition includes those acts that are universally understand as criminal—in particular, bribery and embezzlement; that is, the theft of funds.[1] There is a view—albeit not a universal one—that corrupt behavior should also be considered to include acts that, although not illegal, result in public policy being unduly compromised by private interests. This is particularly relevant in the political sphere, including in the area of electoral finance. Second, the abuse must involve a public office. Accordingly, abuses that take place entirely within the private sector are not included.[2] Private actors are, of course, subject to criminal sanction to the extent that they offer bribes to public officials. As will be discussed later on, holding those who offer bribes to account is a critical element of any anticorruption strategy.

When Does It Become Systemic?

Isolated acts of corruption occur in all countries. The primary focus of this chapter is situations in which corruption is systemic; that is, where corrupt behavior is not the exception to the norm but, rather, has become the norm itself. Evidence suggests that there are a number of "enabling environments" in which corruption becomes systemic. One is an economy dominated by the exploitation of natural resources. As has been well documented, in these economies, bribery and embezzlement often become prevalent at each stage of the "value chain": from the awarding of the natural resource concession contract (which can often be tainted by bribes and kickbacks), through the extraction phase (in which portions of the resource may be stolen), to the investment of the proceeds of the sale (where the investment vehicles may be owned or controlled by public officials, giving rise to conflict-of-interest problems).[3] Moreover, in these economies, government accountability is undermined by the fact that public revenue is derived from ownership of a valuable resource. Since revenue is not dependent on taxes being levied on the private sector, there is no real incentive for the government to provide honest and effective public services.

Corruption also flourishes when the government is so weak that it has been effectively "captured" by private interests. This can occur when organized crime is entrenched: Through the bribery of public officials, criminal organizations are able not only to act with impunity but also to use the state to enforce their control over segments of the economy. Capture can also exist in post-conflict situations (in which governmental legitimacy and control are tenuous) and in countries that have recently gone through a transition from a centrally planned to a market economy (in which the owners of newly privatized industries effectively control regulatory institutions that have only recently been established).[4]

Assessing the Impact

There is considerable evidence that corruption can have a devastating economic impact by undermining the operation of key state functions.[5] One such function is fiscal; that is, the capacity of the state to raise money through taxation and to spend it effectively on critical services. With respect to the taxation function, where tax evasion is widespread—particularly among the affluent, professional classes—and where the tax authorities are complicit in this practice, the entire system can become delegitimized. This can pose threats to fiscal sustainability and give rise to debt crisis. Perhaps even more important, however, is the effect that corruption can have on the way the government actually spends its money. First, there is the problem of waste, which occurs when kickbacks are pervasive in the procurement process. Second, and more fundamentally, systemic corruption distorts spending decisions. Instead of investing in schools and public health, both of which are necessary to support inclusive growth, corrupt officials will prefer to build conference centers where kickbacks can be generated. Since the poor are the most dependent on social services that might be neglected, it is not surprising that systemic corruption is highly correlated with poverty and extreme inequality.

A second state function that is distorted by corruption is the government's role as a regulator. While views may differ on the appropriate design of regulations, it is generally recognized that any functioning market economy needs to be supported by an effective regulatory framework. However, when corruption is systemic, this framework can be abused in a number of ways. For example, government officials will demand bribes in exchange for licenses, permits, and other types of approval that are typically needed for investment. This not only makes investment more expensive but, perhaps more importantly, it also creates uncertainty, which discourages investment in the first place. When investors are considering making an initial capital investment in a country, they need to consider that, at every stage in the life of the project, they may be requested to pay another bribe to ensure the project's success. The uncertainty is, of course, exacerbated by the fact that corrupt agreements cannot be legally enforced. A second—and very different—way in which systemic corruption undermines the effective operation of a regulatory framework is through regulatory capture. In this situation, the pressure does not come from the public official seeking a bribe. Rather, it comes from the private actor offering a bribe to ensure that the regulator effectively suspends the application of regulatory requirements and "looks the other way." This can have catastrophic consequences on the economy. Indeed, one of the causes of the Asian financial crisis (1997–1998) was the excessive buildup of debt in the banking system. Regulators had effectively failed to exercise adequate supervisory oversight of these institutions because they were captured by powerful financial interests. Systemic corruption can also undermine public safety, with tragic humanitarian consequences. Think, for example, of bribes being paid by builders of factories to inspectors that result in the use of substandard materials and, eventually, the collapse of these factories and the deaths of workers.[6]

Finally, systemic corruption can lead to political instability and undermine national security. At a certain point, corruption can so enrage the population that it leads to civil disorder and armed conflict. It can also fuel radicalization and terrorism. When individuals are continually humiliated and harassed by government officials seeking payoffs, it is much easier for terrorist groups to recruit them. Moreover, as noted in a recent study, the financing of terrorist groups is often based on illicit transactions that rely on the bribery of public officials.[7]

Designing an Effective Anticorruption Strategy

When designing an anticorruption strategy, the instinctive reaction is to look for a "silver bullet." However, experience demonstrates that the most effective approaches are holistic and multifaceted. Moreover, it is important to avoid a "cookie cutter" approach. The societal forces that create systemic corruption in different countries will vary, and these differences will be of critical importance. As Einstein said, one should make solutions as simple as possible—but no simpler.

As has been noted by one commentator, corruption is a crime of calculation, not of passion.[8] Accordingly, it is critical to introduce measures that will change incentives. Of course, a key incentive for someone who is contemplating a corrupt act is the fear of going to jail. For this reason, a credible threat of prosecution is essential. However, the difficulty with the establishment of the rule of law is that it requires more than just the adoption of legislation. It requires the effective operation of institutions that have the capacity to implement the legislation. Where corruption is systemic, the police, the prosecution, and the courts are often corrupt, so enforcement is compromised. While it is critical to invest in efforts to strengthen the institutions charged with criminal enforcement, it is important to recognize that one cannot rely on criminalization alone. Not only will this take a considerable amount of time to become effective, but there is a risk that criminal enforcement will simply become politicized, with the incoming government using this strategy to prosecute—and weaken—members of the previous one. The reality is that, where corruption is systemic, the underlying problems are often a function of broader structural forces (as briefly described earlier) that require broader regulatory and administrative reform.

Indeed, experience demonstrates that the broader regulatory and administrative reforms needed to underpin a permanent and nonpoliticized anticorruption strategy are not, in fact, corruption specific. They have broader benefits. For example, in some countries excessively complex and nontransparent regulatory systems exist for the sole purpose of giving government officials the opportunity to extract bribes. Simplifying these systems is not only an effective anticorruption tool but also has broader efficiency benefits. Technology can also be very beneficial: Obtaining a license or permit online removes the human transaction that can often provide an opportunity for a bribe. Of course, there are limits to the degree to which regulation can be eliminated or streamlined.

As indicated earlier—and as we have learned through previous financial crises—an independent and effective regulatory framework in critical areas is essential. In cases in which regulators have been captured by private interests and no longer have the capacity to exercise their authority independently, the policy response is to strengthen—not cut back—this authority. This is why an understanding of the country-specific context is critical when designing an effective anticorruption strategy.

A second area of reform is increased transparency. Since corruption is normally considered a crime, it is, by definition, a hidden activity. Accordingly, increased transparency can play an important role in preventing it. For example, the theft of significant resources by senior public officials—often referred to as "grand corruption"—can be effectively addressed by ensuring greater transparency with respect to how governments manage their own finances. Requiring all governmental transactions to be part of a published, central budget subject to independent audits can play an important role.[9] Similarly, ensuring transparency and accountability with respect to state-owned enterprises—which is particularly important in countries that rely on natural resources—is essential.

A final—and critical—area is administrative reform. In the final analysis, the goal is to establish a professional civil service that is proud of its independence from private influence and public interference. Achieving this requires reforms in a number of areas. First, increases in salaries may need to be considered. In many countries, public sector salaries are so low that corruption is not an expression of greed but an act of survival. Second, it is important to put in place—and enforce—performance management systems that ensure that the hiring and firing of civil servants is conducted on the basis of merit. In Georgia, administrative reform played a central role in the elimination of corruption in the police: Although a large number were dismissed, those who were retained received significant salary increases.[10]

Implementation: The Political Dimension

As a result of experience, international best practices have been developed that provide considerable guidance on the design of the criminal, regulatory, and administrative reforms needed to combat systemic corruption. Indeed, in this area it can be said that the problem is no longer one of design but of implementation. And, when corruption is systemic, the problem of implementation often results from the absence of any political will to reform. While corruption creates many losers in society, it also generates a small number of winners, and these few are typically able to translate their economic power into political influence, influence used to defend a status quo that is the source of their wealth. In these circumstances, meaningful reform may require political realignment that often entails a change in government or a recognition among political leaders that their own political survival requires a change in approach. In many circumstances, democratic reform is critical. This was certainly the case in Indonesia: Significant reforms—including the establishment of the effective Corruption Eradication Commission—occurred only following the ouster of President Suharto and constitutional reforms that led to direct presidential elections. Democracy is no panacea, however. Experience has shown that it can happily coexist with corruption, as is the case in India. Indeed, in some cases, democracy can actually exacerbate corruption, including in those circumstances in which politicians actually buy votes by offering people official positions (as was the case in the United States during the nineteenth century, when the civil service was nothing more than a patronage system).[11]

Treating Corruption as a Global Problem

So far, the focus has been on strategies to be implemented by those countries where corruption is systemic. Over the years, however, it has been recognized that the problems in such countries are not entirely of their own making. Foreign investors from advanced countries—countries that may not be systemically corrupt themselves—are often part of the problem. Specifically, they are responsible for bribing foreign officials in order to gain access to the market in question. Moreover, once the bribe has been given, the public officials do not generally invest the proceeds of these crimes in their own countries. Rather, they transfer the funds overseas to financial institutions in advanced economies with the capacity to effectively conceal them.

Consequently, corruption needs to be seen as a global problem that requires a global solution. The good news is that many advanced economies are recognizing their responsibility in this regard and have made commitments to limit corruption by criminalizing the offering of bribes to foreign officials. The OECD Anti-Bribery Convention, signed by all OECD countries, requires countries to criminalize the bribery of foreign officials and is the leading legal instrument in this area. The Financial Action Task Force—consisting of thirty-eight members and including all major financial centers—addresses the problem of concealment, having put in place recommendations designed to prevent the laundering of funds of "politically exposed persons," which includes foreign public officials.

Unfortunately, the bad news is that the enforcement of these frameworks is very uneven. In the case of the OECD Anti-Bribery Convention, only a small handful of countries (led by the United States) actively prosecute the bribery of foreign officials.[12] With respect to the Financial Action Task Force recommendations, many countries have failed to effectively implement those measures generally needed to prevent money laundering and concealment. In particular, while the recommendations set forth standards for the disclosure of the beneficial owners of shell companies often used to launder funds and, in particular, the proceeds of corrupt transactions, a number of countries—including the United States—have not effectively implemented this requirement.[13]

Beyond Incentives: The Ethical Dimension

Increasing accountability and transparency and engaging in comprehensive regulatory, administrative, and criminal enforcement reform are essential to meaningful reform. However, it is not enough just to establish an incentive structure that prevents individuals from doing the wrong thing. It is also necessary to think about steps that will help people develop an ethical framework that enables them to do the right thing—even when no one is watching. How to do this? Inspired leadership can play a role in setting an example and establishing new norms of behavior. Increasingly, institutions of higher education are making ethics a required part of their curricula, with a view to shaping the attitudes of the next generation of leaders. As is demonstrated by the Ethics in Action initiative, spiritual leaders also play a central role in providing guidance on this issue. If the exclusive focus is on establishing a set of rules and penalties, there is a risk that a culture of compliance will dominate, when what is ultimately needed is the cultivation of a culture of values.

Notes

1. These crimes are recognized as "core" crimes of corruption under the UN Convention Against Corruption, a treaty that has been signed by 186 countries. The text of the convention can be found at https://www.unodc.org/documents/brussels/UN_Convention_Against_Corruption.pdf.
2. As is reflected in the UN Convention Against Corruption, many countries treat abuse that takes place exclusively within the private sector as a form of corruption; see Article 21 of the convention at https://www.unodc.org/documents/brussels/UN_Convention_Against_Corruption.pdf.
3. For a detailed analysis of the types of corrupt acts that can occur along the value chain involving natural resources, see OECD, *Corruption in the Extractive Value Chain: Typology of Risks, Mitigation Measures and Incentives* (Paris: OECD, 2016), http://www.oecd.org/dev/Corruption-in-the-extractive-value-chain.pdf.
4. For further analysis of corruption and post-conflict situations, see Susan Rose-Ackerman and Bonnie Palifka, *Corruption and Government: Causes, Consequences, and Reform*, 2nd ed. (Cambridge: Cambridge University Press, 2016), 295–306, 316–21.
5. The International Monetary Fund has conducted a study of the impact that systemic corruption can have on various state functions; see IMF, "Combatting Corruption:

Costs and Mitigating Strategies," *IMF Staff Discussion Note* (May 2016): 2–15, https://www.imf.org/en/Publications/Staff-Discussion-Notes/Issues/2016/12/31/Corruption-Costs-and-Mitigating-Strategies-43888.

6. See the discussion of the tragedy of the collapse of a garment factory in Bangladesh in 2013 in Rose-Ackerman and Palifka, *Corruption and Government*, 69–71.
7. See OECD, *Terrorism, Corruption and the Criminal Exploitation of Natural Resources* (Paris: OECD, October 2017), 1–4, https://www.oecd.org/investment/Terrorism-Corruption-Criminal-Exploitation-Natural-Resources-2017.pdf.
8. See Robert Klitgaard, "International Cooperation Against Corruption," *Finance and Development* (March 1998): 3–6, https://pdfs.semanticscholar.org/b6cf/ccb56a32cf-9124be07c07b3494b79e841f58.pdf.
9. The IMF's Fiscal Transparency Code sets forth best practices in this area; see IMF, *The Fiscal Transparency Code* (Washington, DC: IMF, 2019), https://www.imf.org/external/np/fad/trans/Code2019.pdf.
10. For an extensive analysis of the regulatory and administrative reforms introduced in Georgia to address systemic corruption, see World Bank, *Fighting Corruption in Public Services: Chronicling Georgia's Reforms* (Washington, DC: World Bank, 2012), http://documents.worldbank.org/curated/en/518301468256183463/pdf/664490PU-B0EPI0065774B09780821394755.pdf.
11. For a discussion of civil service reform in the United States during the nineteenth century, see Rose-Ackerman and Palifka, *Corruption and Government*, 423–25.
12. See Gillian Dell and Andrew McDevitt, *Exporting Corruption: Progress Report 2018—Assessing Enforcement of the OECD Anti-Bribery Convention* (Berlin: Transparency International, 2018), 5–8, https://images.transparencycdn.org/images/2018_Report_ExportingCorruption_English_200402_075046.pdf.
13. See Maíra Martini and Maggie Murphy, *G20 Leaders or Laggards? Reviewing G20 Promises on Ending Anonymous Companies* (Berlin: Transparency International, 2018), 9-14, https://images.transparencycdn.org/images/2018_G20_Leaders_or_Laggards_EN.pdf.

CHAPTER 40

THE ROLE OF INSTITUTIONS IN FIGHTING CORRUPTION

JERMYN BROOKS

This chapter presents the results of a number of surveys demonstrating the correlation between the existence of strong institutions and low levels of corruption in the same society. It also uses the example of state-owned enterprises, as institutions positioned between the public and private sectors, to illustrate how they can also contribute to curbing corruption.

The Millennium Development Goals, established in 2000 and to be achieved by 2015, did not include a specific reference to bribery and corruption. They focused on the broad challenges of reducing worldwide poverty, providing at least primary education to the poorest members of societies around the world, and ensuring that health care is available to all human beings, particularly with the aim of achieving fewer deaths at childbirth.[1] These goals appealed to all sectors of society, but for the business sector the UN Global Compact was created. The Global Compact encompassed ten principles covering business commitments to environmental, employment, human rights, and anticorruption standards.[2]

The internationally based discussions that led to the Sustainable Development Goals, established in 2015 and to be achieved by 2030, aimed to make the goals for governments, business, and civil society more specific and progress more easily monitored. The result was a total of seventeen goals, each with many subgoals called targets. SDG 16 calls for peace, justice, and strong institutions and has twelve targets, including "promote the rule of law at the national and international levels and ensure equal access to justice for all" (Target 16.3), "substantially reduce corruption and bribery in all their forms" (Target 16.5), and "develop effective, accountable and transparent institutions at all levels" (Target 16.6). The result is that while anticorruption activists welcome the first-time inclusion in an international document of such a commitment to reducing bribery and corruption, the focus on so many issues, particularly on environmental and human rights, means that anticorruption is just one of many concerns to be addressed to ensure progress in sustainable development in the world. Indeed, commentaries on SDG 16 scarcely mention Target 16.5 even though the UN's statistics show that the economic and human damage caused by bribery and corruption far outweigh the impact of most other areas referred to in the SDGs, let alone within SDG 16.[3]

An additional factor is that the UN Global Compact continues to be in force but is now overshadowed by the SDGs, so that less attention is being paid to the work of businesses in the Global Compact's working groups, and businesses now feel obliged to switch their focus to how they are responding to the challenges of the SDG Targets. Although integrity issues loom large in corporate analyses of material matters of concern to management and stakeholders, the focus of businesses on the SDGs rarely includes Target 16.5 (for example, refer to the materiality matrices published by many internationally active multinational corporations).

Including the importance of the rule of law and of strong institutions in the same SDG as the commitment to reduced bribery and corruption makes eminent sense, however.

In the experience of Transparency International, exhortations for moral behavior tend not to be very successful. Governments are more easily motivated to strengthen institutions with private sector support because, very frequently, stronger institutions lead to increased efficiencies, increased public revenues, and less corruption. These changes lead in turn to an improved environment for both citizens and business.

In a survey conducted by Transparency International,[4] a majority of interviewees (57 percent) reported not having faith in the efforts of their governments to fight corruption. The survey also found that worldwide, more than 30 percent of police, elected representatives, government officials, business executives, tax officials, judges, and magistrates were perceived as corrupt.

A justice system riddled with corruption undermines trust in the rule of law and opens many channels for corrupt acts aimed at sidelining the justice system. Further analysis by Transparency International has shown that countries with traditions of a strong rule of law are less prone to corruption.[5] This finding has also been demonstrated in many country-specific surveys. Institutional structures supporting the rule of law are the single most effective weapon in containing corruption. For this reason, Transparency International focuses on training for police and codes of conduct for judges and supports other organizations that train legal professionals. Institutionalized support for the rule of law is provided in some countries by independent anticorruption commissions, offices of ombudspersons, or auditors general.

If democracy can be defined as including strong and independent institutions, then it is interesting to note that in Transparency International's Corruption Perceptions Index (CPI), which ranks countries according to perceived levels of public sector corruption, no full democracy exists with a score below 50. A comparison between two countries of similar size, with some eighty million inhabitants, Germany and Turkey, illustrate these findings. Germany's CPI score is 80, whereas Turkey's 40. According to the European Research Centre for Anti-Corruption and State Building's Index of Public Integrity,[6] scores for independence of the judiciary on a 10-point scale are 8.54 for Germany and 3.96 for Turkey. Statistics from the OECD and Transparency International[7] show that whereas Germany ranks second after the United States in investigating and prosecuting cases of bribery and corruption, no such activities take place in Turkey. Germany is home to about four million people of Turkish origin. Criminal statistics from Germany over recent years indicate no significant variation in the incidence of crimes among people of different ethnic origins. It appears, then, that the existence of an effective rule of law in Germany has resulted in lower levels of crime than in Turkey, regardless of ethnic origin or practices that may continue to predominate in a country of origin.

Let us turn now to the institutions of an important sector but one that is focused on too little: state-owned enterprises (SOEs). SOEs have unique challenges because of their closeness to political decision-makers: They are vulnerable to misuse for political purposes and to situations of conflicts of interest.

In 2016, the OECD began research on SOEs with the aim of providing information on and guidance to governments regarding managing the practices of their SOEs. Some of this information is quite extraordinary, including that 81 percent of all bribes in analyses of bribery cases were promised, offered, or given to SOE officials. And the problems with SOEs are not limited to the Global South: Telia in Sweden, one of the least corrupt countries, paid large bribes to obtain mobile phone licences in Central Asia. In the United Kingdom, the National Health Service is estimated to lose £1.25 billion annually through fraud. And in 2017, U.S. authorities alleged that more than US$4.5 billion had been misappropriated in Malaysia by the now infamous sovereign wealth fund 1Malaysia Development Berhard.

Two cases emphasize some of the challenges and risks of SOEs. First is the case of Petrobras, the state oil and gas company partly privatized by the Brazilian state. Through the government's power to influence the activities of the company, the company was forced to pay 5 to 10 percent more than was necessary to suppliers, which was then passed back to the government. This laundering on the part of the Brazilian government, through an SOE, to its suppliers, and back to government became known as *Operação Lava Jato* (Operation Car Wash). In spite of high levels of corruption, the Brazilian justice system has managed to retain a degree of independence. It is proceeding against Petrobras and its major suppliers and has fined companies and so far imprisoned 107 executives and politicians involved in the scandal. Luiz Inácio Lula da Silva, a former president of Brazil, has been jailed for twelve years, and Grupo Odebrecht, a large construction conglomerate involved in the scandal, has been ordered to pay a fine of $3.5 billion. The investigations continue. This scandal is truly of a dimension that would have been thought impossible before it happened—and it would have been impossible without the public ownership and the infiltration of Petrobras by politicians and their appointees.[8]

The case of the Swedish telecommunications company Telia (formerly TeliaSonera) is less well known but instructive nonetheless. The company was partly owned by the Swedish and Finnish states and partly privatized. Its only opportunity to expand—because Scandinavian countries are rather small—was to move into Eastern European and Central Asian republics of the former Soviet Union. Unfortunately, in its eagerness to expand, particularly in Uzbekistan, Telia fell short on its due diligence and ignored many red flags. After much delay, the Swedish authorities have taken action against Telia, as have other countries. The whole of the company's senior management have been dismissed, and the company has completely revised both its anticorruption and human rights standards. Telia now represents what an ethically sound SOE should be doing in these areas.[9]

Some lessons: SOEs are extremely vulnerable to corruption and unethical practices owing to their proximity to government. Leadership and state-ownership oversight are often weak. In the case of Telia, directors and executives were complicit. They had false incentives to act corruptly. They deliberately ignored red flags and performed only superficial due diligence on third parties.

Why the focus on SOEs? Because they are incredibly important. The majority of SOEs are local. In Germany alone, there are sixteen thousand local SOEs active in many areas, such as utilities, waste collection, and waste disposal. They feature strongly in many countries' economies, with up to 50 percent of GDP often represented by SOE activities. They can also be global powerhouses: 10 percent of the world's GDP is reckoned to be in the hands of the world's SOEs. And China is home to colossal SOEs: 15 percent of the five hundred largest companies in the world are Chinese SOEs. SOEs are highly concentrated in the utilities sector, but also in finance, where public ownership is very common.

SOEs are huge employers. Seven out of the ten largest employers in the world are SOEs. They can therefore set the standards for employment and rights for the people they employ. Importantly, they have the potential to play a leading role in improving working conditions in the many challenging countries where they are major actors.

Most SOEs are averse to transparency. They often avoid public reporting. In repeated Transparency International reviews of public reporting on anticorruption policies and procedures,[10] SOEs score poorly when compared to the private sector, the latter driven by the reporting requirements of stock exchanges and corporate law in many countries. These same requirements tend not to apply to a large number of SOEs. If they are not required to report, they usually do not.

How did Transparency International react to these findings? It decided to complement the OECD guidance being prepared for government owners of SOEs by preparing an anticorruption code for the managements of SOEs.[11] To ensure it would have a positive impact, Transparency International used a multistakeholder approach to develop this code that included SOE representatives and members of civil society, governments, and international organizations such as the OECD. The guidance was published at the end of 2017.

The overriding feedback from the long discussions with SOEs and concerned stakeholders was that governance structures and the roles of boards and senior management must be clarified, above all in terms of the companies' relationships with government, political parties, and politicians. It became clear that to deal with corruption in SOEs, these companies need to be much more transparent about their structures and how they operate. Precisely because of their public ownership, SOEs have an opportunity to set best practices in how boards are managed, how ethical standards are set internally, and how supply chains can be positively influenced. An important purpose of Transparency International's ten anticorruption principles is to strengthen the resolve of SOE management in their often-difficult dealings with governments so they can say, "We are following the Transparency International best practices, and this is what we must do."

If governments wish to use their public ownership of commercial activities in this way, SOEs can become an important source of improved standards and reduced corruption in government and in business. For this reason, SOEs can become important institutions in the fight against corruption.

Notes

A version of this chapter was originally presented at the Ethics in Action meeting on corruption in Alpbach, Austria, in August 2018.

1. See https://www.un.org/millenniumgoals.
2. See https://www.globalcompact.org.
3. See https://sdg-tracker.org.
4. "Global Corruption Barometer: Citizens' Voices from Around the World," Transparency International, 2017, https://www.transparency.org/en/gcb/global/global-corruption-barometer-2017.

5. "Corruption Perceptions Index 2018," Transparency International, 2018, https://www.transparency.org/en/cpi/2018/index/dnk.
6. See https://www.hertie-school.org/en/ercas.
7. See "Exporting Corruption—Progress Report 2018: Assessing Enforcement of the OECD Anti-Bribery Convention," Transparency International, September 12, 2018, https://www.transparency.org/en/publications/exporting-corruption-2018.
8. For an account of *Operação Lava Jato*, see https://en.wikipedia.org/wiki/Operation_Car_Wash.
9. The Telia corruption scandal is well reported in https://en.wikipedia.org/wiki/Telecom_corruption_scandal.
10. "Transparency in Corporate Reporting: Assessing the World's Largest Companies (2014)," Transparency International, November 5, 2014, https://www.transparency.org/en/publications/transparency-in-corporate-reporting-assessing-worlds-largest-companies-2014.
11. "10 Anti-corruption Principles for State-Owned Enterprises: A Multi-stakeholder Initiative of Transparency International," Transparency International, November 28, 2017, https://www.transparency.org/en/publications/10-anti-corruption-principles-for-state-owned-enterprises.

PART XII

AN ETHICAL CONSENSUS ON SUSTAINABLE DEVELOPMENT

The Future of Work

ETHICS IN ACTION ON THE FUTURE OF WORK

Equipping individuals with skills and opportunities for decent work extends beyond merely meeting the economic need to "earn a living." Rather, the issue of decent work is a vital moral concern: Work is fundamental to becoming a full human being. Pope Francis states, "Work is indeed at the heart of the very vocation given by God to man, of prolonging his creative action and achieving, through his free initiative and judgment, dominion over other creatures, which translates not into despotic enslavement, but into harmony and respect." Pope John Paul II states in *Laborem exercens*, "Work is a good thing for man—a good thing for his humanity—because through work man not only transforms nature, adapting it to his own needs, but he also achieves fulfilment as a human being and indeed, in a sense, becomes 'more a human being.'"[1]

The notion of work as vocation, being central to an individual's fulfillment as a human being and integral to building up the common good, is shared among the world's major religious and ethical traditions. In Hinduism, work should comport with one's inner nature (*svabhava*), facilitate flourishing within one's community and context (*svadharma*), and contribute to universal flourishing (*lokasangraha*). In the Islamic tradition, work should be permissible (*halal*) in the sense that it promotes the common good and should engage the devotion of the worker, including faith in God. In the Chinese tradition, the "Stratagem of Da Yu" (in the *Book of History*) enjoins workers to "strengthen moral virtues, benefit others, and develop productions through peaceful means." The great philosophical traditions of virtue ethics, following Plato and Aristotle, which inspired monotheistic religions, envision work as part of the cultivation of virtues (*arete*), to become fully human and achieve happiness.

The issue of work also lies at the very center of the Church's social teachings. As Pope Leo XIII stated in *Rerum novarum* in 1891, in the midst of the Industrial Revolution, "At the time being, the condition of the working classes is the pressing question of the hour, and nothing can be of higher interest to all classes of the State than that it should be rightly and reasonably settled."[2]

Ninety years later, in 1981, Pope John Paul II presciently affirmed the continuing urgency of the issue of work in his magisterial encyclical *Laborem exercens*, in which he states that the issue of work is "always relevant and constantly demands renewed attention and decisive witness."[3] He continues,

> New developments in technological, economic and political conditions . . . will influence the world of work and production no less than the industrial revolution of the last century. There are many factors of a general nature: the widespread introduction of automation into many spheres of production, the increase in the cost of energy and raw materials, the growing realization that the heritage of nature is limited and that it is being intolerably polluted, and the emergence on the political scene of peoples who, after centuries of subjection, are demanding their rightful place among the nations and in international decision-making.[4]

These are indeed the greatest challenges of work today: automation accelerated by artificial intelligence, intolerable pollution, and marginalized populations excluded from full participation in society and decision-making.

The most important challenge of work is therefore not its specific type or sector—physical or mental, indoors or outdoors, good-producing or service—nor the earnings of a job, but whether work fulfills the deeper purposes of each person as a human being: the cultivation of virtues, the worker's dignity, and the fulfillment of our social roles as colleagues, friends, family members, and citizens. The type of work surely matters, as some work is dangerous, arduous, repetitive, and stultifying. The earnings of work also surely matter: Some work is so poorly remunerated that workers cannot meet their basic needs. Yet the yardstick for work should start from the moral perspective: How can decent work help each worker to become more fully human?

It is from this perspective that we should face the future of work. Consider the question of artificial intelligence. "Artificial intelligence" refers to a cluster of technologies that enable computers to perform tasks typically associated with human intelligence, such as pattern recognition, language translation, remote monitoring, feedback systems for automated machine processing, robotics, self-driving vehicles, and game playing at superhuman levels, among countless others. These technologies are advancing at an astonishing rate, and artificial intelligence systems are being rapidly introduced in applications across all sectors of the economy, including consumer devices. In combination with robotics and brain-computer interfaces, it may bring unique advances in medicine and care. By elucidating how we learn, it may also bring dramatic changes to education.

Artificial intelligence raises core issues about the future of work. The most widely discussed challenge brought by such technology is the question of machines replacing workers through artificial intelligence systems, robotics, expert systems, assembly-line automation, and the Internet of Things, as well as other modes by which artificial intelligence is being incorporated into economic, governance, and social systems. The integration of artificial intelligence into the economy has accelerated the long-term processes of automation catalyzed by the onset of the Industrial Revolution two centuries ago.

In principle, the rise in productivity from this technological advance can lead to many economic and social benefits. Increased output per worker affords the opportunity to raise living standards while also enjoying more time for family, friendship, voluntary work, education, cultural activities, health, sports, and leisure. We are reminded of the wise words of the Prophet Muhammad according to Ibn Abbas: "There are two blessings most people cheat themselves out of: health and leisure." Artificial intelligence can also provide vital benefits for health, education, and social inclusion. Artificial intelligence–enabled systems, for example, can provide tremendous and varied benefits for individuals with disabilities.

A vital concern, however, is that while artificial intelligence is likely to raise productivity and national income, it is also likely to impoverish parts of our communities. New technologies create new jobs and demand new skills but also end others, rendering many already-acquired professional skills obsolete and unmarketable. Since the beginning of the Industrial Revolution, technological advances have caused unemployment and falling wages for certain groups, especially for those jobs that are directly replaced by machines. Artificial intelligence tends to replace jobs that are repetitive, based on predictable physical labor, or characterized by massive data that can be used to "train" expert systems. Jobs based on emotions and personal contact—notably the care sector—will of course continue to provide vital jobs and meaningful work that cannot and should not be replaced by smart machines.

The phenomenon of technological unemployment raises a fundamental implication for artificial intelligence. The first round of consequences brought on by the new technologies will be mixed, with some workers benefiting from rising living standards and others suffering the loss of jobs and earnings. A *moral* response to technological change therefore will require effective social institutions to ensure that the gains from artificial intelligence are broadly shared and to facilitate the participation of underrepresented groups in new working modalities, including older people and individuals with disabilities. In this regard, Pope John Paul II wisely called for a "social order of work" to ensure that work broadly enhances well-being and all forms of health, rather than degrading it.[5]

To this end, for example, the social democracies of Scandinavia have created social institutions—including unions, universal access to health care and education, income transfers to vulnerable households, and worker retraining programs—that effectively ensure that the gains from technological advances such as artificial intelligence are broadly shared. As a result, opinion surveys find that workers in Scandinavia are far more optimistic about technological change than are their counterparts in countries lacking similar social democratic institutions. States should examine how the benefits deriving from artificial intelligence and robotics can be shared by all.

We must be clear that while artificial intelligence presents serious challenges, today's mass unemployment is actually driven more by policy choices than by technology. Moreover, the conversation about artificial intelligence and work is fundamentally impoverished unless we recognize how much uniquely human work there is to do—from care work for our young and old to the kind of political and intellectual leadership needed to address the world's urgent problems of poverty and climate change.

Artificial intelligence raises several crucial issues about work and well-being beyond the challenge of worker replacement. First, the actual functioning of the technology itself presents many possible complications that warrant further consideration. Will artificial intelligence systems be fair or biased, perhaps inadvertently and unknowingly? There are many reasons to worry that systems trained on social data may embed biases in those data. Another concern is that artificial intelligence systems and digital technologies more generally might trigger new psychological and social disorders. Many psychologists fear that "screen time" is becoming a widespread addiction among young people and that social media platforms such as Facebook are leading to an epidemic of anxiety and major depressive disorders as young people are bullied or feel that they do not live up to the glorified online images of their counterparts.

The Sustainable Development Goals direct the world's attention to several urgent concerns on the future of work. Decent work for all is a central theme of the 2030 Agenda, most notably in the targets outlined in SDG 8:

8.5 By 2030, achieve full and productive employment and decent work for all women and men, including for young people and persons with disabilities, and equal pay for work of equal value

8.7 Take immediate and effective measures to eradicate forced labour, end modern slavery and human trafficking and secure the prohibition and elimination of the worst forms of child labour, including recruitment and use of child soldiers, and by 2025 end child labour in all its forms

8.8 Protect labour rights and promote safe and secure working environments for all workers, including migrant workers, in particular women migrants, and those in precarious employment

8.b By 2020, develop and operationalize a global strategy for youth employment and implement the Global Jobs Pact of the International Labour Organization

SDG 5, on gender equality, calls on society to honor, respect, and indirectly compensate the household work of women:

5.4 Recognize and value unpaid care and domestic work through the provision of public services, infrastructure and social protection policies and the promotion of shared responsibility within the household and the family as nationally appropriate

Pope Francis has mobilized the world to adopt SDG Target 8.7 to end all forms of modern slavery, a crime against humanity still afflicting tens of millions of people around the world. Modern slavery devastates the afflicted individuals and their families by curtailing the essential freedoms to find meaning at work, friendship in society, and the abiding love of family. We call on the world to heed SDG Target 8.7 and to use every form of legal enforcement and technology to end slavery, bonded labor, human trafficking, and child labor for all national and global supply chains.

The Church has long taught that a proper social order of work will eliminate the conflict between labor and capital. The Church makes clear that labor must take priority over capital.

As Saint Pope John Paul II stated,

> A labour system can be right, in the sense of being in conformity with the very essence of the issue, and in the sense of being intrinsically true and also morally legitimate, if in its very basis *it overcomes the opposition between labour and capital* through an effort at being shaped in accordance with the principle . . . of the substantial and real priority of labour, of the subjectivity of human labour and its effective participation in the whole production process, independently of the nature of the services provided by the worker.[6]

There also needs to be a renewed focus on the fundamental teaching, going back to *Rerum novarum*, that the profits of scientific innovation must be fairly shared with the people who do the work through a fair and orderly process of collective bargaining between capital and labor. Pope Leo XIII envisioned an economy in which the benefits of wealth creation are widely distributed through shared ownership. In our own day, too, more widespread capital ownership could reduce the dependence of labor on capital, thereby empowering people to allocate their labor in ways not typically valued by the market; for example, care work, crafts, entrepreneurship, and education.

General Recommendations: Institutional Reforms to Ensure the Real Priority of Labor

(1) A massive expansion of union coverage in places like the United States where the labor movement has been under political assault
(2) Reforms to company law to ensure that workers and other stakeholders contribute to the management of the enterprise
(3) The amplification of the cooperative sector, wherein workers and consumers directly share in the management and direction of the enterprise

In implementing these findings, Ethics in Action recommends the following actions.

Ethics in Action Working Group Action Items

(1) A call for the UN General Assembly, in its implementation of the 2030 Agenda, to recognize and respect the moral meaning of decent work for all. The effort to develop intelligent machines must remain continuously directed to the common good, reducing the poverty gap, and addressing general needs for health, education, happiness, and sustainability.
(2) The use of artificial intelligence and related technologies to promote solutions for decent work, including support for basic social services (health care, education), environmental monitoring, the prevention of illicit activities (such as illegal fishing or illegal transfers to tax havens), and the accountability of supply chains to eliminate all forms of modern slavery.
(3) Social actions to ensure that the benefits of artificial intelligence are widely shared and the misuse of private data is avoided, including new skills training, education programs, social protection for groups hard hit by new technologies, and a public ownership stake in the automation process itself through ownership in either the companies benefitting from artificial intelligence or the intellectual property underlying the data.
(4) Stepped-up efforts to reform the enterprise sector through increased union membership, reforming company law, promoting multistakeholder governance, and strengthening the cooperative sector.

(5) The development of an ethical framework for the deployment of artificial intelligence so that its use supports the common good.

(6) A call on all economic decision-makers to adopt values focused less on the pursuit of profit and power and more on the prioritization of people and the planet. Such values would include nonviolence, reverence, fairness, justice, truthfulness, friendship, and solidarity. Organizations should also encourage and support moral and virtue education—including the great virtues of love and compassion—in their everyday operations.

Notes

1. John Paul II, *Laborem exercens* (Vatican City: Libreria Editrice Vaticana, 1981), sec. 9, http://www.vatican.va/content/john-paul-ii/en/encyclicals/documents/hf_jp-ii_enc_14091981_laborem-exercens.html.
2. Leo XIII, *Rerum novarum* (Vatican City: Libreria Editrice Vaticana, 1891), sec. 60, http://www.vatican.va/content/leo-xiii/en/encyclicals/documents/hf_l-xiii_enc_15051891_rerum-novarum.html.
3. John Paul II, *Laborem exercens*, sec. 1.
4. John Paul II, *Laborem exercens*, sec. 1.
5. John Paul II, *Laborem exercens*, sec. 9.
6. John Paul II, *Laborem exercens*, sec. 13.

CHAPTER 41

UNIONS AND THE FUTURE OF WORK

SHARAN BURROW

The global workforce is around three billion people. There is currently much discussion of the future of work, given technological advances in production and rapid digitalization. However, few acknowledge that if we don't heal the fractures of today's workforce that have been caused by the current model of corporate greed, we will see even greater inequality in years to come.

The dignity of decent work is being undermined by an economic model founded on labor arbitrage in the endless quest for profit. The undermining of fundamental rights, the absence of minimum living wages, and the decline in collective bargaining have resulted in a global slump in labor income share, which, along with the failure to ensure universal social protection, has led to historic levels of inequality.

Unprecedented Levels of Inequality

Tragically, the vested interests of the few have been afforded preference over the interests of the great majority. Naked self-interest has been dressed up in modern economic models to justify why wealth is not being shared, why natural resources are being exploited unsustainably, why corporations and the wealthy pay little or no tax, and why there is a severe lack of resources for social protection and vital public services, including health care and education.

"Trickle-down" economic theory and, more recently, austerity have failed. Global trade and investment rules have favored finance and capital in developed economies. Global supply chains channel wealth to a handful of global corporations, while workers in those supply chains experience low wages, job insecurity, and often unsafe work. A war against unions and freedom of association is being waged, and democracy is being corrupted by concentrated wealth. Indeed, too many corporations and their associations are buying or bullying legislators and executive branch officials to influence public policy to the detriment of working people. The concentration of ownership of the media, social and traditional, has also given the rich a huge influence over elections.

The result is a global workforce in serious trouble:

- Fewer than 60 percent of workers are employed in the formal economy, and more than 50 percent of these workers are in precarious or insecure work.
- More than 40 percent of workers struggle to survive in the informal economy—with no rights, no minimum wages, and no social protection. With the displacement of refugees and the emergence of platform businesses, the number of people employed in informal work is growing rapidly in every country.
- More than forty-five million people are engaged in modern slavery or forced labor.
- Three-quarters of the world's population have inadequate or no social protection.

Global GDP has tripled since 1980, yet labor income share has declined, and vital investments in infrastructure, care, and the green economy are inadequate.

In addition to these challenges and the consequent vulnerability of the global economy, the potential for a vastly more unequal world is emerging with waves of new technology. There are major deficits in the regulatory environment for ensuring decent work in a climate of internet-mediated platforms and algorithms, and there is scant investment in new jobs to mitigate the displacement of workers in the face of the future of production.

Insecurity on the Rise

Global polling by the International Trade Union Confederation shows that the majority of people are not worried about new technology, but they are deeply anxious about their jobs.

With increased digitalization, new business models are being founded on the expectation that workers will have to give up employment contracts and social security and forget the notion of a regular working schedule in which work, family, and leisure can be balanced. These businesses are informal, sometimes not registered, paying no or little tax in the country where profit is earned, and they take no responsibility for an employment relationship. Without a social license to operate, they are effectively above the law. They operate outside jurisdictions and thereby disrupt key sectors, including transport, health care, hospitality, financial services, education, and more.

Workers providing services for such companies are left on their own to pay for their social security, taxes, and training—all while having no control over pricing, working conditions, safety, or their personal data. They compete against one other for an irregular and unpredictable supply of work. Those who own the big platforms reap billions from this model, and traditional companies are also starting to outsource work through platforms, encouraging wage dumping and avoiding responsibilities.

And now the huge tech companies are buying up all manner of products and services, flouting labor rights, national constraints, and competition policy. Companies like Amazon also have huge amounts of capital, including workers' capital, underpinning their operations with little or no return on that investment to date.

Just Transition

Unions know that technology itself is not the issue—innovation will either be successful or not on its own merits. Societies will embrace the potential of scientific advances in health care and many other areas where it is safe. Unions have been involved in shaping technological change for decades, and, in many cases, collective bargaining has quickened the pace of adoption of new technology by addressing workers' concerns, ensuring skills training, and sharing productivity gains. Now, many unions are organizing freelance workers, building digital cooperatives, facilitating skills development, and even providing services for start-up enterprises. Efforts such as organizing and collective action by workers and setting floors for contract prices, including through cooperatives, are increasing workers' power in all spheres of work.

As well as the Fourth Industrial Revolution, workers also face the twin challenges posed by rampant and worsening climate change and the measures needed to tackle it and create a low- to no-carbon economy. We know that there are no jobs to be had on a dead planet: Rising sea levels and more frequent extreme weather events are just two changes that will destroy jobs as well as communities. Unions know we need to take action to combat climate change. But that mustn't be done at the expense of working people's living standards.

Addressing the challenges of digitalization and climate change both require a "just transition": ensuring that workers come first when the economy changes and that the benefits of change are fairly distributed—for example, through guaranteed pensions for older workers, income support, and access to retraining and redeployment support for others—so that no one is left behind. Lifelong learning is also central, as technology and decarbonization impact the tasks that workers do, creating a continual need for workers to be able to upgrade their skills.

Unions successfully campaigned and lobbied for a just transition to enter the lexicon of the Paris Agreement on climate change, most notably in Warsaw in December 2018, but we need to develop the collective bargaining agenda on both climate and technology.

A New Social Contract

As the International Labour Organization marks its centenary in 2019, it is well placed to address the challenges of inequality, insecurity, and change. For today's workplace injustice and tomorrow's challenges, the basis of decent work remains the same:

- Freedom of association
- Minimum living wages and collective bargaining rights
- Nondiscrimination and freedom from forced and child labor
- Safe work and universal social protection
- Social dialogue

To change the way we do business, we need these fundamentals to be reaffirmed as a universal labor guarantee for *all* workers as a floor in a new "social contract." That social contract must include, among other measures, a just transition for climate and technological shifts; the right to lifelong learning; the equal economic participation of women and an end to violence perpetrated against them; fair taxation; the provision of care and vital public services including childcare, care for older people, and universal health care and education; and a renewed commitment to full employment with investment in jobs, beginning with infrastructure and care services.

We also demand that all businesses have a social license to operate such that they are registered and pay tax in each country of operation, they have an employment relationship with those they depend on for work, and they contribute to social protection. Equally, we expect governments to mandate due diligence for all business to mitigate violations of human and labor rights.

Much of this is reflected in a report from the International Labour Organization's high-level Global Commission on the Future of Work, cochaired by the South African president, Cyril Ramaphosa, and the Swedish prime minister, Stefan Löfven. (Employers and unions are also represented in the commission.) This report sets out the case for a human-centered future of work with increased investment in people's capabilities, the institutions of work, and decent and sustainable work.[1]

Above all, the report proposes a renewal of the social contact with a universal labour guarantee—covering all working people, regardless of their employment arrangements—that would apply the International Labour Organization's core labor standards,[2] provide greater autonomy for workers regarding their working hours and their health and safety at work, and provide decent minimum living wages for every job. We will strive to build on this ambition and fight for workers' rights and due diligence to be enshrined in global trade rules, for due diligence to be applied to global supply chains, and for changing business models to ensure a fair competition floor as a basis for a just globalization.

We want the centennial International Labour Conference in June 2019 to adopt a formal declaration on the future of work that begins the work of turning these proposals into practical international instruments while simultaneously adopting a convention on gender-based violence. The universal labour guarantee, in particular, must become the key safety net for all workers.

Conclusion: The Tasks Ahead

The Sustainable Development Goals provide a roadmap for achieving a zero-poverty, zero-carbon world. But in a world where multilateralism is in crisis and the necessary political maturity and vital morality are missing, conflict and hate are growing, and historic numbers of people are being displaced. However, we have the people on our side, with 85 percent of the world's population wanting to see the rules of the global economy changed. The world of work must change to reflect this wish.

It will take all of us to create a sustainable future. Unions will fight to ensure that full employment and decent work remain the foundations of any economy. A new social contract is a vital start if we are to rebuild trust in a fractured world.

Notes

1. International Labour Organization, *Work for a Brighter Future: Global Commission on the Future of Work* (Geneva: ILO, 2019), https://www.ilo.org/wcmsp5/groups/public/---dgreports/---cabinet/documents/publication/wcms_662410.pdf.
2. These standards include freedom from child and forced labor, freedom from discrimination at work, freedom of association, and free collective bargaining, including the right to strike.

CHAPTER 42

THE COMING AI REVOLUTION

Is This Time Different?

CARL BENEDIKT FREY

The question I would like to address in this chapter is this: Should we be all that reassured if the future of automation mirrors the past? The simple answer is that it depends on which episode of the past we are speaking about.

Economists agree that technological progress is welfare improving. It makes the pie larger. We also agree that it is rarely Pareto improving when robots take people's jobs and new and better-paid jobs are not made available for all displaced workers at once. Thus, we acknowledge that some displaced workers might suffer hardships in the short run. But we are not very specific about what we mean by the "short run," and it makes a difference if the short run is twenty days or twenty years. Indeed, in some cases, we have even seen that wages have been stagnant or falling for many decades, so that the short run became a lifetime for some. Thus, the rational response for some people was to opt against technological progress. The Luddites, who destroyed machines during the Industrial Revolution, are often portrayed as irrational enemies of progress. But they weren't the ones who stood to benefit from mechanization. So, in many ways, their opposition to the mechanized factory makes sense. And the Luddite riots were only part of a long wave of machinery riots that swept across Europe, India, and China.

Even if we take the most optimistic calculations of real wages during the British Industrial Revolution, wages started to rise only around 1820, five decades after the mechanized factory had arrived. And by more realistic calculations, real wages began to rise only in the 1840s. The reason I think that figure is more realistic is because it mirrors what we know from data on consumption and biological indicators. If we look at a biological indicator such as people's heights, we find that people born in 1840 were shorter than those born in 1760.

Many factors clearly that play into this fact. The factory cities that people moved into were surely unhealthy environments. But vanishing incomes, as jobs were replaced by machines, also led to poorer nutrition, which stunted growth. What the mechanized factory did was to replace middle-income artisan jobs with new ones, but for a very different breed of worker. The first spinning machines were specifically designed to be tended by children. It is true that some new, better-paying jobs were created in the factories for administrators and supervisors, but those also required a very different set of skills.

Many middle-aged artisan men lost out to technological change during this period. New spinning jobs were created for children, who became a growing share of the workforce. The hard part for economists and economic historians lies in explaining why workers would have agreed to participate in the industrialization process if it reduced their utility. The simple answer is that they didn't. They petitioned to Parliament to stop the spread of machinery. And they rioted against the mechanized factory, but they had no chance against the British Army with whom they clashed on several occasions. Property ownership remained a requirement for voting, meaning that the Luddites were essentially politically disenfranchised. Mechanization was enforced on a large part of the population. Indeed, by hiring children, the pioneers of industry were able to sap resistance from the adult population.

We are now in the midst of what can be called a technological revolution—a revolution in artificial intelligence. Various studies have attempted to estimate how many jobs can be replaced as a consequence of artificial intelligence, but it is important to remember that any market outcome now depends on acceptance of the technology. Unlike the Luddites, average people today in the industrial West have a political voice. And if people don't see technology as working in their interests, the outcome is extremely hard to predict. They might try to bring technological progress to a halt. We might be in for a period of social and political upheaval, which will also shape the speed of technology adoption and our economic trajectories in the future.

We know that wages have fallen behind productivity across the industrial world in recent decades. However, there is also significant variation in the falling labor share of income across countries. The reason for these outcomes is that technology is not a soloist but part of an ensemble. It interacts with institutions and other factors. At the same time, the hollowing-out of middle-income jobs is a common features across countries. As automation has taken over middle-income factory jobs, it has also created entirely new computer-using occupations for the highly skilled. And what happens when a new tech job is created in a city is that the person with that job goes out and spends much of their money in the local service economy. They go to the hairdresser, take a taxi, go grocery shopping, and so on, which creates demand for low-income service jobs.

Indeed, computer industries haven't created the abundant job opportunities for semiskilled workers that the automobile and electric machinery industries once did. Instead, much of recent job creation has centered on the labor multiplier, with skilled tech jobs creating new demand for low-skilled in-person services, of which many remain hard to automate. This has been the motor of job growth in recent years. But it has also been driving much of the economic polarization we are now seeing.

It is true that if we look at the unemployment statistics, things look rather good, despite the rise of the robots: U.S. unemployment is down to about 4 percent. But nonemployment has risen steadily among men with no more than a high school diploma who would have flocked into the factories before the dawn of automation. What's more, there are huge differences between unemployment and nonemployment around the world. Put differently, if you were to put one hand in a freezer and the other on a heated stove, you could say that, on average, you feel warm. But the reality, of course, is that you are experiencing both extreme cold and extreme heat. So while we can say that, on average, people today are doing well, different groups of people are facing very different realities. This situation has resulted in part because new tech jobs have clustered in cities with skilled populations, creating demand for local services in those places. At the same time, where manufacturing jobs have been automated away, communities are in despair, because those jobs had supported other jobs in the local service economy. Thus, we have seen places like the Bay Area in San Francisco prosper as a result of automation, while places like the Rust Belt (a region comprising parts of the northeastern and midwestern United States) have suffered. And where jobs have dried up, we have seen increases in crime, worsening health outcomes, and faltering marriage rates. These outcomes will, in turn, have implications for the next generation. The children growing up in those places today have steadily worsening future prospects, which creates a vicious cycle.

It we want to understand phenomena like the election of Donald Trump, we need to understand what's happening in the labor market. Three key states that swung the election in his favor in 2016—Michigan, Wisconsin, and Pennsylvania—had all been firmly in Democratic hands for every presidential election since 1992. It is hard to believe that if real wages were rising and good jobs were available in abundance across the Rust Belt, this swing would have happened. If we look at a map of the locations of industrial robots and vote shares for populists, there seems to be a link between the two, both in the United States and Europe.

Obviously, much of the populist backlash witnessed in recent times has been centered on globalization rather than automation. But it's important to remember that the effects of the two have been very similar. No one would dispute that globalization has had huge benefits, but we long chose to overlook the social costs associated with it for some communities. That, I believe, cost the economics profession some of its credibility. We need to make sure to avoid making the same mistake with automation, because the future of automation depends on people's acceptance of it. I was quite struck by a Pew Research survey suggesting that 85 percent of Americans favor policy to restrict automation beyond hazardous jobs. If automation is resisted, we will all be worse off as a consequence. And that would be nothing new: As I argue in my book, *The Technology Trap*, resistance to technologies that threaten people's jobs has been the historical norm rather than the exception.[1] So we need to try to manage the short-run dynamics.

Note

1. Carl Benedikt Frey, *The Technology Trap: Capital, Labor, and Power in the Age of Automation* (Princeton, NJ: Princeton University Press, 2019).

CHAPTER 43

SLOW BUT SURE

Cooperatives and Integral Ecology

NATHAN SCHNEIDER

During the first year of his pontificate, in November 2013, Pope Francis told a story via video at the Festival of the Social Doctrine of the Church in Verona. After recalling a meeting he had held a month earlier with leaders of the global cooperative business movement, he conjured a more distant memory:

> I remember—I was a teenager—I was 18 years old: it was 1954, and I heard my father speak on Christian cooperativism and from that moment I developed an enthusiasm for it, I saw that it was the way. It is precisely the road to equality, not to homogeneity, but to equality in difference. Even economically it goes slowly. I remember that reflection my father gave: it goes forward slowly, but it is sure.[1]

Dame Pauline Green, then the president of the International Cooperative Alliance, recalled the earlier meeting with Francis in a blog post:

> It's true to say that we were all mightily impressed with the Pope's intimate knowledge and understanding of our movement. All alone with no staff or advisers and not a note or briefing paper to be seen, he spoke our cooperative language during a 45-minute informal discussion.[2]

Such words as these, and other statements on the subject the pope has uttered since,[3] lend heft that might not otherwise be noticed in what is probably the pope's most important document to date, the encyclical *Laudato si'*. There, he refers to cooperativism three times,[4] each as an example of what a positive "integral ecology" might look like for a world seeking to heal from a condition of social and environmental crisis. These might seem like merely casual examples if it were not for his memory of the discovery, at a formative age, "that it was the way."

The significance of cooperativism for Catholic social teaching, or of the movement's role in prefiguring a sustainable future, is not unique to Francis. The cooperative movement emerged alongside the rise of industrial capitalism and the modern church.[5] It presented an alternative to investor-centered capitalism by forming businesses in which core participants—workers, customers, or clients—own and control their shared capital democratically. Catholic social teaching has had an outsized influence on the global co-op movement, starting especially in the wake of Pope Leo XIII's 1891 encyclical, *Rerum novarum*, which called for strategies "to induce as many as possible of the people to become owners."[6] Catholic leaders and motivations explicitly undergirded such endeavors as the world-renowned Italian co-op sector; the advent of credit unions in North America, especially thanks to Alphonse Desjardins in Quebec, Canada; the Antigonish cooperatives of Canada's Maritime provinces; the writings of the economist Monsignor John Ryan, an architect of the New Deal; the largest worker cooperative system in the world, Mondragon, founded by the Basque priest José María Arizmendiarrieta; and countless cooperatives formed through missionary organizations to strengthen bargaining power in disadvantaged regions of the world. These achievements are the "Christian cooperativism" of which Francis's father spoke—no mere theory but an economic phenomenon by then very much underway.

Environmental sustainability was not a paramount consideration during the early years of the cooperative movement in the mid-nineteenth century, although that period did bear precedents. The founders of a famous early co-op—a customer-owned store in Rochdale, England—sought safer, less adulterated foodstuffs than the underregulated markets supplied. The current set of international cooperative principles, which were roughly derived from Rochdale, ends with the principle of "concern for community."[7] Increasingly, this principle is being understood as having to do with stewardship for the environment and an inclusive economy[8]—both hallmarks of Francis's call for an integral ecology.

The global cooperative movement, which claims to provide 10 percent of the world's employment,[9] has made significant commitments to sustainability in recent years, particularly by being the first global business coalition to endorse the Sustainable Development Goals. Leading up to the 2015 ratification of the SDGs' resolution, which twice refers to co-ops, this process was already underway. The theme of the 2014 International Day of Cooperatives was "cooperative enterprises achieve sustainable development for all"; for the following year, it was "paths to achieving the Sustainable Development Goals." A triumvirate of the International Cooperative Alliance, the International Labour Organization, and the UN Department of Economic and Social Affairs has since promoted the SDG/co-op intersection with annual conferences, frequent publications, and coordinated pledges by various cooperatives to make SDG-aligned changes in how they do business.[10] The UN secretary-general issued a report titled *Cooperatives in Social Development* in July 2017, which was followed by the adoption of a resolution with the same name by the General Assembly that December.[11]

When I attended the 2016 International Summit of Cooperatives in Quebec, the emphasis on the SDGs was palpable, even to the point of being heavy-handed. The summit passed a formal declaration in support of the goals, citing 403 existing projects and 345 proposals from co-ops to that end.[12] Cooperative leaders evidently find the SDGs a useful rallying cry for uniting disparate co-ops around the world, claiming substantial overlap between the SDGs and what the co-op movement has long stood for. UN leaders, in turn, see fit to acknowledge co-ops as allies in advancing the cause of the SDGs.

What are cooperatives actually doing to support sustainable development, or integral ecology? That is a more complex question.

For instance, in the United States, the widespread rural-electric cooperative utilities have been slower to adopt renewable energy sources than have investor-owned utilities, thanks in part to capital constraints and decades-old, federally mandated coal investments. But with the right incentives, energy cooperatives globally have shown a capacity to enable transitions to renewables more rapidly than have other kinds of firms.[13] Because they are, in theory, accountable to community members rather than outside investors, cooperatives should be able to balance social considerations with economic ones in ways other firms cannot. Yet, in practice, cooperatives can become highly conservative institutions, constrained by external financial obligations and unwilling to make risky bets on emerging technologies with their members' money. Bucking this inertia can require pressure; in the United States, an organization called We Own It has formed to support co-op member organizing, which appears to be bringing about a more active embrace of renewable energy and local economic development.[14]

This capacity for enabling communities to set their own economic and environmental priorities does seem to set cooperative firms apart. Papers shared at UN convenings on cooperatives and the SDGs have frequently stressed the potential for local empowerment in poorer regions of the world, which would otherwise depend on multinational investment decisions made in distant capitals.[15] The SDG-related pledges that cooperatives have made vary as widely as cooperatives themselves do, ranging from aspirations for increasing local financial inclusion by a certain percentage (Uganda Cooperative Savings and

Credit Union) and generating more fair-trade produce sales (the United Kingdom's Co-operative Group) to reducing carbon dioxide emissions by 40 percent (Japanese Consumers' Co-operative Union) and focusing on the uplift of youth and women (National Cooperative Union of India). These are self-defined, self-enforced commitments, and it remains to be seen to what extent they will significantly advance the SDG process as a whole.

Historically, the orchestrated use of cooperatives has been effective in producing outcomes not otherwise feasible through existing markets. The U.S. electric co-op system came about thanks to a tailored federal loan program, established after it became clear that investor-owned utilities would not adequately invest in rural economic development. Thanks to such support, U.S. rural cooperatives have been able to develop a formidable capital infrastructure of their own. The system of "social cooperatives" in Italy, largely facilitated by the Catholic association Confcooperative, enabled the privatization of previously public services through accountable community control. Coffee producers' cooperatives around the equatorial world, in partnership with religious organizations and fair-trade advocates in wealthier regions, have lessened the exploitation that has long characterized that supply chain. Favorable public policies in countries such as Denmark and Germany have enabled cooperatives to facilitate rapid expansions of wind-energy production. In each case, the systematic deployment of cooperatives seeded markets that otherwise might not have arisen on their own. This capacity is vital for achieving goals such as those set out in the SDGs, which does not seem feasible through political interventions or profit-seeking businesses alone. Cooperatives inhabit precisely the strategic intersection between the public and the private, between values and markets.

The 2017 General Assembly resolution on cooperatives in social development includes statements encouraging governments to actively support the cooperative sector. In economies oriented toward enabling highly capitalized, investor-owned firms, cooperatives face considerable obstacles when competing on their own. U.S. worker cooperatives, for instance, remain relatively rare, as they enjoy less policy support than do agricultural or utility cooperatives. More recently, in the emerging online economy, there has been growing interest in the prospect of "platform cooperatives," but, in lieu of a financing infrastructure, very few such start-ups have found the means to flourish.[16] In various times and places, cooperatives have been able to achieve rapid growth and transformative effects, but doing so requires ambitious imaginations and intentional deployments of infrastructure.

Achieving shifts as significant as those required for a genuinely integral ecology, or for meeting the SDG Targets, calls for similarly significant shifts in how business is done. This is a point Pope Francis—together with his predecessor, Benedict XVI—has reiterated many times, often with explicit reference to cooperatives and the legacy of Catholic cooperativism. The scope of Francis's vision was striking enough to Pauline Green that she made note of it in her recollection of their meeting on her blog: "The Pope argued that global leaders need to understand that co-ops are not just something for moments of crisis, but the way in which economic life should go in the future."[17]

Notes

1. Francis, "Video-Message of Pope Francis for the Third Festival of the Social Doctrine of the Church Held in Verona," Vatican website, November 2013, http://w2.vatican.va/content/francesco/en/messages/pont-messages/2013/documents/papa-francesco_20131121_videomessaggio-festival-dottrina-sociale.html.
2. Pauline Green, "Positive Coop Actions Crowned by a Unique Moment," *Dame Pauline Green*, October 23, 2013, https://damepaulinegreen.wordpress.com/2013/10/.
3. E.g., Francis, "Participation at the Second World Meeting of Popular Movements: Address of the Holy Father, Expo Feria Exhibition Centre, Santa Cruz de la Sierra (Bolivia)," Vatican website, July 9, 2015, http://w2.vatican.va/content/francesco/en/speeches/2015/july/documents/papa-francesco_20150709_bolivia-movimenti-popolari.html.
4. Francis, *Laudato si'* (Vatican City: Libreria Editrice Vaticana, 2015), sec. 112, 179, 180.
5. Ed Mayo, *A Short History of Co-operation and Mutuality* (Manchester: Co-operatives UK, 2017); Nathan Schneider, *Everything for Everyone: The Radical Tradition that Is Shaping the Next Economy* (New York: Nation, 2018).
6. Leo XIII, *Rerum novarum* (Vatican City: Libreria Editrice Vaticana, 1891), sec. 46, http://www.vatican.va/content/leo-xiii/en/encyclicals/documents/hf_l-xiii_enc_15051891_rerum-novarum.html; on Catholic cooperativism, see also Race Mathews, *Jobs of Our Own: Building a Stakeholder Society—Alternatives to the Market and the State* (Irving, TX: Distributist Review, 2009); Nathan Schneider, " 'Truly, Much Can Be Done': Cooperative Economics from the Book of Acts to Pope Francis," in *Care for the World: Laudato si' and Catholic Social Thought in an Era of Climate Crisis*, ed. Frank Pasquale (Cambridge: Cambridge University Press, 2019).
7. "Cooperative Identity, Values and Principles," International Cooperative Alliance, accessed April 30, 2019, https://ica.coop/en/whats-co-op/co-operative-identity-values-principles.
8. Principles Committee, *Guidance Notes to the Co-operative Principles* (Brussels: International Co-operative Alliance, 2015), https://www.ica.coop/sites/default/files/basic-page-attachments/guidance-notes-en-221700169.pdf.
9. "Facts and Figures," International Cooperative Alliance, accessed April 30, 2019, https://ica.coop/en/facts-and-figures.
10. The International Cooperative Alliance's portal on SDGs programs is available at https://coops4dev.coop/en/node/14779; the UN's cooperative portal is available at https://www.un.org/development/desa/cooperatives/. Representative statements include Frederick O. Wanyama, *Cooperatives and the Sustainable Development Goals: A Contribution to the Post-2015 Development Debate* (Geneva and Brussels: International Labour Organization and International Co-operative Alliance, 2014), https://www.ilo.org/wcmsp5/groups/public/---ed_emp/documents/publication/wcms_240640.pdf; Guy Ryder, "Cooperatives for Sustainable Development," Co-operatives of the Americas, International Labour Organization, July 5, 2014, https://www.aciamericas.coop/Message-by-ILO-Director-General; Ed Mayo, "Co-operatives Making a Difference Through . . . Sustainable

Development," *Coop News*, July 2018, https://www.thenews.coop/129403/sector/banking-and-insurance/co-operatives-making-difference-sustainable-development/.

11. UN Secretary-General, *Cooperatives in Social Development: Report of the Secretary-General* (New York: UN, July 2017); UN General Assembly, *Cooperatives in Social Development* (New York: UN, December 2017).
12. International Summit of Cooperatives, "Cooperatives: The Power to Act on the United Nations' Sustainable Development Goals," International Co-operative Alliance, October 2016.
13. Eric Viardot, "The Role of Cooperatives in Overcoming the Barriers to Adoption of Renewable Energy," *Energy Policy* 63 (December 2013): 756–64, https://doi.org/10.1016/j.enpol.2013.08.034; Marieke Oteman, Mark Wiering, and Jan-Kees Helderman, "The Institutional Space of Community Initiatives for Renewable Energy: A Comparative Case Study of the Netherlands, Germany and Denmark," *Energy, Sustainability and Society* 4, no. 1 (May 2014): 11, https://doi.org/10.1186/2192-0567-4-11; Nathan Schneider, "Economic Democracy and the Billion-Dollar Co-op," *The Nation*, May 8, 2017, https://www.thenation.com/article/archive/economic-democracy-and-the-billion-dollar-co-op/.
14. Disclosure: I am a member of the We Own It board.
15. For a list of these meetings, see https://www.un.org/development/desa/cooperatives/what-we-do/meetings-and-workshops.html.
16. Nathan Schneider, "An Internet of Ownership: Democratic Design for the Online Economy," *The Sociological Review* 66, no. 2 (March 2018); Schneider, *Everything for Everyone*. This is beginning to change—for instance, through a US$1 million grant to the Platform Cooperativism Consortium from Google.org, a small but significant step.
17. Green, "Positive Coop Actions."

CHAPTER 44

THE END OF WORK AS WE KNOW IT

A Muslim Perspective

HAMZA YUSUF

The Qur'an states, "Surely your worldly endeavors are sundry and diverse."[1]

The accelerating rate of change in our world today, catalyzed by technological advances, has radically splintered and specialized traditional notions of work, causing great harm to the spiritual nature of humanity.

When we consider the end of work, especially the threats faced by labor and white-collar jobs, we must account for the pervasive and all-consuming influence of technology, and more specifically the force of artificial intelligence (AI) and the so-called Fourth Industrial Revolution, a post-digital era in which technology deeply embeds in society and even in human beings. Work means something today that it did not mean yesterday, and tomorrow will bring even greater, perhaps unimaginable, change. We must also ask, What does the aim or purpose of work entail? This might be the most important question about work, and the answer has also begun to mean something today that it did not mean yesterday.

The increasing complexity of our societies has fragmented professions and antiquated traditional work. Where once we had a dentist treating a patient, we now have an entire dental industry that includes general dentists, oral surgeons, orthodontists, periodontists, implant specialists, lab technicians, dental assistants, hygienists, prosthetics artisans, administrative workers, and manufacturers of medical equipment, not to mention trade schools and colleges offering specialized degrees for all of them. Meanwhile, software apps and AI have begun to redefine—or replace—many professions, from truck and cab drivers to legal and surgical assistants. Disintermediation has destroyed countless jobs and their brick-and-mortar workplaces owing to online shopping and digital transactions.

Technological advances not only increase both efficiency and profits but also enable instantaneous and mindless consumption that numbs the soul. And the weight of modern complexity threatens to collapse in on itself owing to the fragility of the digital world. A massive solar pulse could devastate entire industries that rely on electricity to power extremely fragile systems of immense complexity.

Undoubtedly, this advanced, complex, and specialized world has also contributed to an unparalleled rise in standards of living for many. For example, infant mortality in Saudi Arabia in 1962 was 270 per one thousand births. Today it is less than thirty-five. Statistics show that in the past century, life expectancy has increased, time doing laundry has decreased, accidental deaths in the workplace have gone down, and the average IQ score has gone up. But these statistics mask the underlying malaise and the damage to the spiritual nature of human beings wrought by the very technological advancement that produced those advances. The trailblazing scientists and engineers encouraging the fusion of technologies that blurs the line between the physical and digital spheres appear unaware of the impact of their accomplishments on the personal and social fronts. The disruptions they spawn continue to accelerate conditions of angst, alienation, and uncertainty that invariably lead to higher rates of depression, anxiety, and even suicide.

***

Technology and profits, the driving forces of modernity, have smothered and suffocated the spiritual impulse of human beings. Work can enrich a person's life materially but diminish it spiritually; conversely, work can enrich a person's spiritual experience but provide a meager living. We must look to the past, if only to learn what can, and ought to be, restored and resurrected, and what can still nurture our spiritual nature.

Traditional work often involved learning a craft through apprenticeship and discipline at the hands of a master craftsperson. Even the art of cooking for women who spent their lives at home had its own tradition of painstaking learning, a transmission of the art from mother to daughter that often required many hours during a day to prepare nutritious and delicious homemade meals. Sewing, with embroidery work that museums now display as works of art, took time and apprenticeship to learn. Everything from the reed mats to the bowls that graced kitchens tended to reflect symbolic art, often with embedded cosmologies. The materials that adorned a home or even a workplace were products of intentional and meaningful work that nourished the soul of the worker.

Skills in arts and crafts were acquired through a rigorous apprenticeship system often accompanied by a spiritual practice in Buddhist, Hindu, Christian, and Muslim societies. The Jewish communities, as a result of their diaspora, worked within the structures of these major civilizations.

During the height of Muslim civilization, the arts were highly appreciated, and wealthy patrons provided endowments to sustain the work of artists. Handicrafts, such as mosaic works, beautified and bedecked homes, and calligraphy and book copying were respected professions with guilds and union-like organizations. With the transition from the premodern world to an industrial one, many crafts were lost or relegated to niche markets. The twentieth-century Swiss scholar Titus Burckhardt relates the story of Abd al-Aziz, a Moroccan craftsman who makes combs from the horns of an ox. As a traditional artisan, he now finds himself caught in the transition from the old world to the new and articulates what he sees being lost in the transition as he observes the spread of plastic combs:

> It is not only a pity that today, solely on account of price, poor-quality combs from a factory are being preferred to much more durable horn combs. . . . My work may seem crude to you; but it harbors a subtle meaning which cannot be explained in words. . . . This craft can be traced back from apprentice to master until one reaches our Lord Seth, the son of Adam. It was he who first taught it to men, and what a Prophet brings—for Seth was a Prophet—must clearly have a special purpose, both outwardly and inwardly. I gradually came to understand that there is nothing fortuitous about this craft, that each movement and each procedure is the bearer of an element of wisdom. . . . But even if one does not know this, it is still stupid and reprehensible to rob men of the inheritance of Prophets, and to put them in front of a machine where, day in and day out, they must perform a meaningless task.[2]

In the Islamic tradition, the devotional nature of work is highlighted in two fundamental areas: the nature of the work itself and the attitude of the worker toward the work.

The nature of the work must entail what the Qur'an terms *halal and tayyib* work: in other words, permissible and pure work. Permissible work refers to anything that promotes the common good, whereas prohibited work involves anything that is deemed harmful either to the person doing the work or to the society. Usury, for instance, is prohibited in the Qur'an in the harshest terms, as are any occupations that involve or promote interest-bearing loans. Also, occupations that involve high risk (*gharar*), such as selling mining rights in an area without certainty that it contains the resource or selling speculative goods, such as future commodities are prohibited. Scholars have penned many books on livelihoods and have emphasized that all permitted and productive work including labor bestows honor on those involved. The Qur'an encourages economic trading with the condition of mutual satisfaction among the traders stating, "Let your commerce be of fair and agreeable trade." Crafts and skilled labor that provide goods and services are also encouraged. The Qur'an mentions that the prophet David was a blacksmith who fashioned chain mail. The prophet Muhammad said, "The best food a person can eat is that gained from the labor of his own hands, and even David the Prophet of God used to eat from the labor of his own hands."

The second aspect of the devotional nature of work involves the attitude of the worker toward his or her work. The Islamic tradition highlights the stoical importance of trust in God for one's provision; the Qur'an declares that the provision for each of us was decreed and measured out before birth. The idea of striving is very important, and any lack of productivity without a valid excuse is seen as undesirable. The prophet Muhammad said, "That one of you should take rope, gather firewood with it, and then sell it in the marketplace such that it suffices for his livelihood fares far better than asking people for a handout, irrespective of whether they give or withhold."[3]

In one prophetic tradition, a man once complained to the Prophet of intense poverty, such that he feared his children would starve. The Prophet asks the man what he owns and is told he owns some blankets and bowls. The Prophet then sells the materials to his companions for two silver coins (*dirhams*). He tells the man to buy some food for his family with one *dirham* and buy an axe with the other. He then tells the man to go with his axe to a particular valley, collect firewood, and sell it in the market. Two weeks later, the man returns to thank the Prophet for improving his circumstances.[4] The lesson is clear: It is better to learn to fish than to be given fish. The traditional entrepreneur made an honest living through hard work that provided a service to others instead of simply begging for food or money. In many poor Muslim countries today, one will find people cleaning car windows at traffic lights for small change rather than begging.

***

Another important aspect of modern society that should be seriously addressed relates to the meaninglessness of many of the soul-deadening and alienating jobs modern forms of production entail. The robotic nature of so many service jobs that too many workers face day in and day out takes a tragic toll on those involved. Added to this, the increasing sophistication of advertising and its ability to manufacture a demand for unnecessary goods and services drive people to seek ever-increasing income in order to pay for it all. This consumptive thirst—seemingly unquenchable—often results in otherwise talented people working at vacuous, unfulfilling jobs that pay well but that can be as morally, spiritually, and psychologically devastating as unemployment. In an essay titled "Why Work?" the twentieth-century poet and essayist Dorothy Sayers, says, "A society in which consumption has to be artificially stimulated in order to keep production going is a society founded on trash and waste, and such a society is a house built upon sand." Gracing such trash and waste with the euphemism "high standard of living" doesn't alter its baser reality.

Speaking of trash, the amount of trash produced in industrial and postindustrial societies now threatens oceans and water tables around the globe. William Leach's "Land of Desire" documents the corporate marketing campaigns (a military term) between the late 1800s and early 1900s that transformed the United States from a thrifty society of Spartan citizens who lived within their means to a spendthrift consumer society brought low by the Great Depression. The subsequent period of the Second World War demanded thrift, reuse, and recycling only to be abandoned in the jubilant postwar years of massive consumption and the advent of a throwaway culture fueled by plastics and the cheap production of goods designed with "planned obsolescence" as a corporate profit-making maneuver.

The West has, for decades now, thrived with a "high standard of living," but its system of consumer capitalism has externalities bound to haunt us in the future. Global warming, polluted rivers and oceans, diminishing water tables, ravaged rainforests, and nuclear devastation are looming on the horizon. The myth of the Wheel of Fortune, described in book II of Boethius's *Consolation of Philosophy*, remains a useful one to understand the rise-and-fall phenomenon in which no one remains at the top of the wheel forever. At the twelve o'clock position of the wheel is the ruler, confident and full of joy, with *regno* ("I rule") as his motto. At three o'clock is *regnavi* ("I have ruled"), with fear was the corresponding emotion. At six o'clock is *Sum sine regno* ("I have no kingdom"), with grief as the corresponding emotion. And at nine o'clock is *regnabo* ("I will rule"), with hope as the corresponding virtue. The myth reminds us that no group can remain on top forever, and no way of life will always be sustainable. Hence, a twelve o'clock technological victory may well result in a number of six o'clock calamities such as global warming, mass starvation, or nuclear devastation. In the West, many live on top but do so at the cost of countless people around the world living at the bottom, too often seething with resentment at the imbalance. But "the wheel's still in spin," as a poet said, and "and there's no telling who it is naming."

***

C. S. Lewis pointed out that premodern people saw the world as finite and human desire as infinite. Their resolution to this quandary was to regulate the infinite human appetites to conform to a finite reality. That proved to be a largely successful approach for centuries and brought human nature into balance with the natural world. The modern zeitgeist of the West, on the other hand, arrogantly attempts to force finite reality to accommodate, and conform to, our infinite appetites. If all societies on our planet were to live at the standard of middle-class Americans or Europeans, three Earths would be needed to sustain us. The German statesman Willy Brandt warned of this dangerous situation in the early 1980s. Later that decade, the psychologist Robert Ornstein and the biologist Paul Ehrlich coauthored the book *New World, New Mind*, pointing out that humans have a dire inability to act on gradually creeping crises. Our brains are wired to respond well to immediate crises owing to the way they developed; hence, emergency medicine is very effective, while long-term care is too often simply disease management that sorely lacks the ability to promote health. The authors argue that we must think in new ways or else suffer the devastating consequences of a crisis that has been slow in building but will wreak devastating havoc on us when it reaches critical mass.

Sound religious traditions are often best equipped to promote corrective measures that can begin to wean people away from the seductions of modernity so they can better tend to their spiritual well-being. This corrective effort must start with a restoration of the prophetic principles, found in all the world's religious traditions, of "doing without," of a liberating detachment from the "stuff of the world" and of the virtues of a Spartan lifestyle. For instance, the simplicity movement of John Wanamaker in the late 1800s and early 1900s, inspired by his commitment to early Christian ideals, resonated with many Americans still steeped in the frugality of the Protestant tradition and work ethic. It is in the nature of human beings to seek quietude, to be in nature, to walk in the woods or by the ocean, and to feel the embrace of God's creation, what the Japanese refer to as "forest bathing." However, most people today are far too distracted by the digital world to reap the benefits of the ever-present natural one.

Our religious leaders must work to restore and revive leisure as an activity of human development, a spiritual pursuit that leads to human flourishing, and they must help people move away from the idea of leisure as a time for indulgence in mindless entertainment, intoxication, and other escapist strategies. Work ought to be seen as an avenue for contributing to the common good, a way to supplement our spiritual practice, not simply as a means for acquiring wealth to be spent on feeding our baser appetites.

The end of work, meaning the mass disruptions that automation, artificial intelligence, and other technological developments will continue to have on our societies, has not yet been addressed. It poses a grave threat to postindustrial societies and will perpetuate more suffering in the developing world. Religious traditions, with immense spiritual resources to address these pressing issues, must be enlisted as part of the solution to the struggles and seductions of this novel secular modernity that promises so much and often delivers so little.

Moreover, we must individually and collectively rise to the current challenges and present a vision of the end—the purpose—of work that aligns more appropriately with our spiritual natures and human need for meaningful production.

Notes

1. Qur'an 92:4.
2. Titus Burckhardt, *Fez, City of Islam* (Cambridge: Islamic Texts Society, 1992), 76–79.
3. *Musnad* of Imam Ahmad ibn Hanbal
4. *Sunan Abi Dawud* by Imam Abu Dawud al-Sijistani

CONCLUSION

Toward a Moral Economy

ANTHONY ANNETT AND JESSE THORSON

A Multireligious Consensus on a Moral Economy

Over the course of two years of meetings, the Ethics in Action Working Group drew attention to an overlapping ethical consensus centered on the Sustainable Development Goals shared by a diverse group of religious and secular ethical traditions. These traditions consistently—if imperfectly—summon their adherents and others to live their lives more deeply in accord with a vision of human dignity and the common good, marked by environmental sustainability, a just distribution of resources, and a holistic sense of well-being centered on the fullest development of capacities. What emerged most clearly in our program was a strong multireligious emphasis on the need for a truly moral economy in order to achieve the SDGs and integral and sustainable development.

A moral economy is one that explicitly recognizes universal human dignity and is oriented toward the common good: a vision of the good held in common that sufficiently recognizes our interdependence and shared well-being. It is fueled neither by relentless profit-seeking nor by a belief that technological advances can solve all problems. It seeks the well-being of all, with special attention paid to the poor and the least well-off. A moral economy is concerned not only with the strictly financial or transactional dimensions of society but also its social and political components, which also shape and guide how individuals and groups relate to and interact with one other. A moral economy not only guides us to act in the interest of the common good but also teaches us to truly desire the common good, a task in which religious institutions ought to be playing a large role.

Virtue Ethics

But how is it that one learns to desire the common good and to act accordingly? The Ethics in Action Working Group consistently emphasized the need for a new kind of virtue ethics that would attend to the reality of our social, economic, and ecological interdependence. A virtue ethics for the twenty-first century would have much in common with previous secular and religious virtue ethics traditions, including a focus on moral habits, rationality, and the *telos* of human beings, but it would also incorporate knowledge from modern science.

Contributors to this volume have sketched out some necessary (though not sufficient) components of a modern virtue ethics, including some of the hallmark virtues that should be located at its center. Justice is an important virtue for a modern ethic, for both individuals and corporate bodies, especially an attentiveness to distributive justice or economic fairness. (For more on this, refer to the Catholic doctrine of the universal destination of goods).

Moderation and temperance are necessary virtues for restraining our sometimes monstrous and insatiable desires. Consumerist capitalism has manufactured overwhelming desire and demand for products far beyond the need or even good of any well-intentioned individual. This challenge requires an active counter-formation, and we need to leverage the technical capacity and moral resources of both religious communities and secular institutions of education to teach and lead people into practices—and ultimately lives—of moderation.

Along with moderation comes prudence, which is a quality traditionally ascribed to those who govern their lives according to reason. A life of prudence is a life of practical wisdom. The prudent person is thoughtful, sometimes intentionally slow, even cautious. Prudence demands patience, anticipation of the consequences of one's actions, and humility when these consequences are uncertain. The precautionary principle, for instance, urges prudence in the face of environmental or technological change and innovation. And a prudent economy not only attends to waste and inefficiency but also to the distribution of wealth and resources. Though not utilitarian in the final analysis, prudence involves the weighing of costs and benefits, value judgments, and taking responsibility for the effects of one's actions as a consumer in the economy or as a social creature in the political body.

Solidarity and love are unifying virtues for a moral economy. Love demands that one value and work toward the well-being and good of the other, which is, in fact, shared well-being. Solidarity leads us to identify with and prioritize the needs of the poor, the marginalized, and the least well-off. From those at the top of the socioeconomic ladder, solidarity might require sacrifice and lifestyle change. Love, in a moral economy, is no mere symbol of sentimentalism.

Instead, love reflects a binding commitment to work toward the good of our neighbors and the world.

A modern virtue ethics must also be oriented toward the achievement of some end or goal (a *telos*) for an individual or society. It must center on the of notion of an agreed common good, the achievement of which requires democratic deliberation in both politics and economics. Our initiative demonstrates that there is a widely shared, unforced consensus on what this common good is, as outlined by the SDGs. True, there may be variations and substantial disagreements among traditions on several issues (with regard to gender and education, for instance), but there remains a wide array of opportunity for cooperation and collaboration on issues of common concern. Our group agreed on most of the changes and progress we would like to see realized in the world—and that is the important outcome of this program. Together, we see that there is a gap between business as usual or the status quo and the ethically ideal state of affairs and that there are plenty of opportunities for religious and secular communities to collaborate with one other on shared goals.

To reach this end, a modern virtue ethics would emphasize the development of good habits and practices in the individual and social dimensions—to help us act in ways that are aligned with the individual and common good. This suggests a role for moral education and training.

Modern science has much to bear on this. For example, ecological and climatological sciences would help inform the *telos* of moral action, because a world characterized by climate stability and environmental sustainability can be brought about only by a certain kind of person in the world. Indeed, acknowledging our dependence on the natural world (and others, even God) elicits a certain kind of restrained behavior, moderation, or humility, which leads us to think more carefully about our influence on the natural world and others. Similarly, the psychological and neuroscientific sciences tell us that true human nature has a strongly relational and purpose-driven component, pointing again to the kinds of virtues needed to live life in accord with human nature.

Education

We've noted that modern science will be of assistance in updating virtue ethics for our contemporary situation, in the sense of better informing us of the kinds of ends (and virtues) necessary for a sustainable future and what would best facilitate training in virtuous behavior and comportment. For both of these purposes, we need to more broadly develop and promulgate ethical education.

Ethical education plays a large role in determining our desires and shaping what we consider to be the *telos* or purpose of our lives. The liberal arts give us a goal to reach, an aim to strive for. We need to recognize and do what we can to steward the power of education while recognizing that education extends beyond schools and other institutions we would recognize more formally as educational. Human beings are educated and shaped by all the diverse ways we participate in social life. We are formed by our families, of course, but also by community organizations, local and national political bodies, the media, churches, synagogues, mosques, temples, and after-school clubs. And these groups all exert some ethical influence (though often unarticulated and subtle), in the sense that they train us to recognize and pursue particular visions of human flourishing, happiness, or well-being. Education is always guided by values, so there is no such thing as an ethically neutral education, curriculum, or pedagogical method. We must always be oriented toward some sort of good, as Aristotle, Augustine, and others have cogently demonstrated.

The good that education should aim us toward is that of sustainable and integral human development. Through classroom learning, religious training, and practical instruction, ethical education will encourage students to attend to the complex challenges of sustainable development, including the thematic topics discussed throughout this compendium. Ethical education will teach virtues to illuminate the path toward sustainable development, not only for individuals and religious groups but for corporations, communities, and other institutions.

In short, ethical education will develop students more fit for life and work in the moral economy. It will challenge conventions of market capitalism and its encouragement of egocentric motives and self-interest. A renovated virtue ethics will render students more attentive to the challenges of distributive justice and environmental degradation. Ultimately, an ethical education that orients students toward sustainable development will produce better business leaders, educators, politicians, and practitioners.

Ethics in Action: The Contours of a Global Common Good

The ten meetings of the Ethics in Action initiative gathered together a diverse group of leaders and thinkers to reflect on the ethical underpinnings of the most urgent challenges facing our world today. The group advocated for a moral economy based on the dignity of all people, in which each person is enabled to actualize their capabilities and live a full life across all dimensions. The ethical traditions represented by the initiative all agree that such a summons applies to both people and communities. It is therefore intrinsically linked to the common good and entails the development of both social and individual virtues. In the final analysis, no person may be excluded from this holistic sense of development. Likewise, this notion of integral and sustainable human development requires a commitment to care for and nurture the environment, our common home.

This represents the shared moral vision of Ethics in Action. And though the initiative was not intended to produce comprehensive and exhaustive solutions, our group understood the practical dimension to be central: Our goal was to outline some of the practical contours of the global common good. Against this backdrop, each meeting presented a list of concrete action items centered on the moral economy and sustainable and integral human development at the local, national, and global levels (see the consensus statements in part 3 of this volume). These actions can be summarized as follows:

- *Ethics in action to end poverty*. This meeting recognized that the juxtaposition of extreme poverty amid global plenty is a moral scandal and that the integral development of people is blocked when they are beset by poverty. Accordingly, the initiative called for financial transfers from rich to poor, both within and between nations. It called for ending the gross ethical misallocation of resources. And it called for social activism and the empowerment of the poor so that they can become active agents of their own development.
- *Ethics in action to promote peace*. This meeting acknowledged that peace is not merely the absence of war and conflict but also has a positive dimension. In this sense, peace is a necessary condition for people to be able to unfold their potential free from impediment. It is thus central to the global common good. The initiative proposed stepped-up advocacy and engagement to build a culture of peace, especially through the involvement of religious leaders and communities.
- *Ethics in action to support migrants and refugees*. The initiative acknowledged that, owing to prolonged conflict, climate change, and a lack of sustainable development, migrants and refugees are increasingly vulnerable populations. The group therefore called for ethical actions to aid these people, especially in terms of meeting their basic material needs, supporting their political participation, and fostering openness to their full integration into their adopted homelands. The initiative recognized that supporting migrants and refugees also means supporting peace and sustainable development in their lands of origin.
- *Ethics in action for businesses to support the common good*. The initiative recognized the positive role of business in promoting the common good through investing in sustainable development. There are, however, many areas in which business falls far short of what is required of a moral economy, focused instead on a thirst for maximum profit. Ethics in Action recommended that business adopt a duty to a wide array of stakeholders, not merely financial shareholders. Business must be responsible to workers, the environment, and society at large.
- *Ethics in action for education*. Another necessary component of any moral economy is comprehensive education. All people must have access to decent

education, at least to the secondary level, so they are enabled to unfold their capabilities and contribute to the common good. Accordingly, the initiative called for a global fund to invest in education in developing countries. It also stressed the importance of moral education, given the importance of virtue to integral human development.

- *Ethics in action for climate justice.* The initiative recognized that climate change is one of the most important and urgent moral challenges facing the world today. It called for the immediate implementation of the Paris Agreement on climate change and for providing assistance to the countries that suffer most from the effects of climate change. In terms of concrete strategies, the initiative recommended actions to phase out fossil fuel technologies, as well as fossil fuel divestment and lawsuits against offending companies.
- *Ethics in action on modern slavery, human trafficking, and access to justice for the poor and vulnerable.* The initiative affirmed that modern slavery and human trafficking are among the most egregious offenses in the global economy today and are the very antithesis of a moral economy. Its concrete recommendations included better supply-chain management; clamping down on modern slavery, prostitution, sex trafficking, and the sale of organs; and improved criminal justice for the poor.
- *Ethics in action to support Indigenous peoples.* The world's Indigenous peoples are some of the most vulnerable people in the world today, subject to systemic abuse, discrimination, and injustice. To rectify this, the initiative called for stepped-up financial support for Indigenous communities to allow them to invest in sustainable development in accord with their own values, greater legal protection for Indigenous rights and lands, and the promotion of Indigenous voices in international fora.
- *Ethics in action to combat corruption.* The group agreed that corruption corrodes any attempt to develop a moral economy. To combat corruption, the initiative called for the elimination of tax havens and secrecy jurisdictions and recognized an option to impose wealth taxes on accounts shielded behind these screens. It also called for the removal of money from politics.
- *Ethics in action to promote the dignity of work.* An ethical consensus emerged that decent work has the nature of a vocation and is therefore a critical avenue for humans to unfold their inherent dignity and creativity. It remains central to the moral economy. To support the dignity of work, the initiative called for greater support for unions and for the cooperative sector. It also called for the development of an ethical framework for the deployment of artificial intelligence so that it supports the common good rather than private gain.

The proceedings of the Ethics in Action initiative reflect a shared vision of the primary contours of a truly global common good. Following the initiative's recommendations would encourage the development of a moral economy and, ultimately, the realization of sustainable and integral human development.

CONTRIBUTORS

Emmanuel Adamakis, Metropolitan of France, Ecumenical Patriarchate, was born in Greece in 1958. He studied education, philosophy, and theology in Greece, France, and the United States. He serves as Metropolitan of France (Ecumenical Patriarchate—Orthodox Church) and represents the Ecumenical Patriarchate at the academic dialogues with Judaism and Islam. He is a comoderator of the World Conference of Religions for Peace and the chairman of the board of directors of the King Abdullah Bin Abdulaziz International Centre for Interreligious and Intercultural Dialogue (KAICIID).

Anthony Annett, Senior Advisor, UN Sustainable Development Solutions Network, is a Gabelli fellow at Fordham University and a senior advisor at the UN Sustainable Development Solutions Network. He has a PhD in economics from Columbia University and spent two decades at the International Monetary Fund, where he worked as a speechwriter for the managing director. He is also a member of the College of Fellows of the Dominican School of Philosophy and Theology.

R. Scott Appleby is a professor of history and the founding dean of the Keough School of Global Affairs at the University of Notre Dame. The author or editor of fifteen books on religion, violence, and peace building in the modern world, including *The Fundamentalism Project* (five volumes) and *The Ambivalence of the Sacred: Religion, Violence and Reconciliation*, he is the recipient of four honorary doctorates, a fellow of the American Academy of Arts and Sciences, and a fellow of the American Academy of Political and Social Sciences.

Jermyn Brooks, Member, International Council, Transparency International, has worked for the international anticorruption nongovernmental organization Transparency International (TI) since the year 2000 in a wide range of roles, including management, board membership, and leadership of the organization's business-facing anticorruption programs. He is currently a member of TI's International Council. His work for TI has involved developing anticorruption tools and guidelines for business via multistakeholder taskforces, overseeing multiyear projects for the Siemens Integrity Initiative, and membership of the Wolfsberg anti–money laundering group. Jermyn was a founding member of the World Economic Forum's Partnering Against Corruption Initiative, and he led the tenth principle working group of the UN Global Compact. He is a frequent conference lecturer on anticorruption issues and advises companies on implementing compliance systems after ethical breaches.

Christina Lee Brown, Cofounder, Christina Lee Brown Envirome Institute, University of Louisville, founded the Institute for Healthy Air, Water and Soil in 2012 to promote an interconnected vision of health incorporating citizen science and advocacy. That vision expanded broadly in 2018 through a partnership with the University of Louisville to form the Envirome Institute, which researches the effects of the environment on health, promotes holistic scholarship, and bridges academic research with community engagement, transforming the city of Louisville, Kentucky, into an urban laboratory. Guided by the Circle of Health and Harmony, a tool she developed, Brown's philanthropy promotes responsible decision making through the lens of health. The circle reveals interrelationships among environmental and human health and the need to bring all forms of social and individual health into balance. She currently serves on the boards of the Sustainable Food Trust, the Berry Center, the Center for Interfaith Relations, the Louisville Orchestra, and the National Trust for Historic Preservation.

Sharan Burrow is the general secretary of the International Trade Union Confederation, which represents 200 million workers in 163 countries and territories and has 332 national affiliates. The president of the Australian Council of Trade Unions from 2000 to 2010, Burrow is a passionate advocate and campaigner for social justice, women's rights, the environment, and labor law reforms and has led union negotiations on major economic reforms and labor rights campaigns in her home country of Australia and globally.

Jacqueline Corbelli is the founder of the US Coalition on Sustainability (USCS), a nonprofit organization established in collaboration with the United Nations with the single purpose of unifying and accelerating progress toward achieving the Sustainable Development Goals by 2030. To achieve this mission, she created SustainChain, a digital and machine learning technology platform that unites innovators, impact investors, purpose-driven brands, nongovernmental organizations, and alliances with a shared vision of rebuilding global supply.

She is also the founder and CEO of BrightLine, a technology company that is a cornerstone of advanced media for America's leading television broadcasters. Before BrightLine, Jacqueline was the president of Aston Associates and directed the organizational redesign of ten major corporations worldwide. She serves on the Leadership Council of the UN Sustainable Development Solutions Network and is a board member of the SDG Center for Africa. She is also a board governor of the New York Academy of Sciences.

Owen Flanagan is James B. Duke Distinguished University Professor of Philosophy at Duke University. He serves as the codirector of the Center for Comparative Philosophy and holds appointments in psychology and neuroscience, in addition to being a faculty fellow in cognitive neuroscience and a steering committee member of the Philosophy, Arts, and Literature program. Flanagan's work is in philosophy of mind and psychiatry, ethics, moral psychology, and cross-cultural philosophy. His latest book is *How to Do Things with Emotions: The Morality of Anger and Shame Across Cultures* (Princeton University Press, 2021).

Carl Benedikt Frey is Oxford Martin Citi Fellow at the University of Oxford, where he directs the program on the future of work at the Oxford Martin School. In 2012, Frey became an economics associate of Nuffield College and a senior fellow at the Institute for New Economic Thinking, both at the University of Oxford. In 2019, he joined the World Economic Forum's Global Future Council on the New Agenda for Economic Growth and Recovery, as well as the Bretton Woods Committee. In 2020, he became a member of the Global Partnership on Artificial Intelligence, a multistakeholder initiative to guide the responsible development and use of AI, hosted by the OECD. His most recent book, *The Technology Trap*, was selected as a *Financial Times* Best Book of the Year in 2019.

Daniel G. Groody is an associate professor of theology and global affairs and the vice-president and associate provost for undergraduate affairs at the University of Notre Dame. He is an internationally recognized expert on migration and refugee issues, and his books include *Globalization, Spirituality, and Justice: Navigating the Path to Peace* and *Border of Death, Valley of Life: An Immigrant Journey of Heart and Spirit*, as well as other edited and coedited volumes on poverty, justice, and migration. Father Groody has worked with the U.S. Congress, the U.S. Conference of Catholic Bishops, the World Council of Churches, the Vatican, and the United Nations on issues of theology, globalization, migration, and refugees.

Jennifer Gross cofounded the Blue Chip Foundation in 2015 to address the root causes of poverty through innovative solutions and partnerships using the framework of the United Nations Sustainable Development Goals. Ms. Gross serves on the Leadership Council of the UN Sustainable Development Solutions Network and as a director of the William, Jeff, and Jennifer Gross Family Foundation, Association of Sustainable Development Solutions Network,

Duke Global Health Institute, and VII Foundation. She has been a founding member of several initiatives at the Pontifical Academy of Sciences at the Vatican including Ethics in Action; Faith and Science for Happiness and Sustainable Development; Pan-American Judges Summit on Social Rights, the Environment and Social Justice; and African Female Judges and Prosecutors on Human Trafficking. Ms. Gross is involved with the production of the World Happiness Report and partners with the Global Happiness Council on the Global Happiness Policy Report. She has fifteen years of philanthropic and business experience and earned her undergraduate degree from Duke University.

Sean Hagan, Visiting Professor of Law, Georgetown University Law Center, was general counsel and director of the Legal Department of the International Monetary Fund from 2004 until his retirement in September 2018. Mr. Hagan has published extensively on both the law of the IMF and a broad range of legal issues relating to the prevention and resolution of financial crisis, with a particular emphasis on insolvency and the restructuring of debt, including sovereign debt. Previously, Mr. Hagan was in private practice, first in New York and then in Tokyo.

Radhika Iyengar is the director of education and a research scholar at the Center for Sustainable Development of Columbia University's Earth Institute. She leads the Education for Sustainable Development initiatives and promotes international development for education as a practitioner, researcher, teacher, and manager. In addition to directing education initiatives at the Center and conducting fieldwork in more than ten countries, she contributes to the scientific community, focusing on international educational development. She is the chair of the Environmental and Sustainability Special Interest Group at the Comparative International Education Society. In 2020, she received the Distinguished Early Career Award from Teachers College, Columbia University, for her service to the field of international education development. Her latest publication is "Education as the Path to a Sustainable Recovery from COVID-19" in the journal *Prospects*. She received a distinction from Teachers College, Columbia University, for her PhD dissertation, "Social Capital as a Determinant of Schooling in Rural India: A Mixed Methods Study." She received her master's degree in economics from the Delhi School of Economics, India.

Lise Johnson leads the Columbia Center on Sustainable Investment's work on investment law and policy. Her work centers on analyzing the contractual, legislative, and international legal frameworks governing international investment and shaping the impacts that those investments have on sustainable development objectives. She focuses on analyzing international investment treaties and the investor-state arbitrations that arise under them, examining the implications of those treaties and cases for host countries' domestic policies and development strategies. She has a BA from Yale University, a JD from the University of Arizona, and an LLM from Columbia Law School and is admitted to the bar in California.

Kerry Kennedy is the president of Robert F. Kennedy Human Rights. Since 1981, she has worked on diverse human rights issues including child labor, disappearances, Indigenous land rights, judicial independence, freedom of expression, ethnic violence, impunity, women's rights, and the environment. Kennedy is the author of the *New York Times* best-seller *Being Catholic Now*; *Robert F. Kennedy: Ripples of Hope*; and *Speak Truth to Power: Human Rights Defenders Who Are Changing Our World.*

Klaus M. Leisinger, Founder and President, Foundation Global Values Alliance, was a professor of sociology at the University of Basel specializing in business ethics, corporate responsibility, and sustainable development (1990–2020); president and CEO of the Novartis Foundation for Sustainable Development (1996–2013); special advisor on the Global Compact for UN Secretary-General Kofi Annan (2004–2005); CEO of Ciba-Geigy Pharmaceuticals in East and Central Africa with headquarters in Nairobi, Kenya (1978–1982); and a member of the Consultative Commission for International Cooperation of the Swiss government (1982–1995). He has also worked with management consultancies since 2013.

Erin Lothes is a theologian at Saint Elizabeth University in Morristown, New Jersey, a researcher in the field of energy ethics, and a scholar of the faith-based environmental movement. Dr. Lothes served as an Earth Institute Fellow at Columbia University researching environmental advocacy in diverse American congregations. She is the author of *Inspired Sustainability: Planting Seeds for Action* (Orbis, 2016) and *The Paradox of Christian Sacrifice: The Loss of Self, the Gift of Self* (Herder and Herder, 2007). She is the lead author of the coauthored article "Catholic Moral Traditions and Energy Ethics for the Twenty-First Century" in the *Journal of Moral Theology* and the author of essays on energy ethics and articles on faith-based environmentalism. Since 2003, Dr. Lothes has participated in the activism of the interfaith environmental and divestment movement through collaborations with groups such as GreenFaith, the Yale Forum on Religion and Ecology, the Catholic Climate Covenant, and the Global Catholic Climate Movement. She holds a PhD in systematic theology from Fordham University, an MA in theology from Boston College, and BA in English from Princeton University.

Ella Merrill works with the director of the Columbia Center on Sustainable Investment, Lisa Sachs, supporting the center's research and programmatic work in investment policy. She studied environmental policy and human rights at Barnard College. As an undergraduate, she worked as a research assistant to a leading environmental journalist and conducted her own research on the previously undetected persistence of herbicides in a Florida lagoon. She also worked on Barnard College's successful fossil fuel divestment campaign. Before joining CCSI, she worked in energy and sustainability management.

Veerabhadran Ramanathan, Edward A. Frieman Endowed Presidential Chair in Climate Sustainability, Scripps Institution of Oceanography, University of

California, San Diego, is an international leader in the science of climate change and in developing solutions for slowing global warming. In 1975, he discovered that the greenhouse effects of non–carbon dioxide pollutant gases like chlorofluorocarbons could warm the planet in significant ways. With Madden, was the first to predict that global warming by carbon dioxide would be detected above the weather noise by the year 2000. He led the Indian Ocean experiment that helped bring to light the presence of widespread atmospheric brown clouds. Based on his research, the UN created the Climate and Clean Air Coalition to mitigate non–carbon dioxide super-pollutants. He served as the science advisor for Pope Francis's Holy See delegation to the historic 2015 Paris climate summit and advised the former governor of California Jerry Brown on climate mitigation. In 2013, the UN environment program named him Champion of Earth, and, in 2018, he (along with James Hansen) was honored as the Tang Laureate for Sustainability Science. He is the coeditor and coauthor of many books, including *Bending the Curve: Climate Change Solutions* (2019) and *Health of People, Health of Planet and Our Responsibility* (2020).

Anantanand Rambachan is a professor of religion at St. Olaf College. His books include *The Advaita Worldview: God, World and Humanity*, *A Hindu Theology of Liberation*, and *Essays in Hindu Theology*. Professor Rambachan serves as a copresident of Religions for Peace.

T8aminik (Dominique) Rankin, Elder, Anicinape Tradition, Canada. After a successful career in politics as the grand chief of the Anicinape (Algonquin) First Nation, Grandfather Dominique Rankin focused on the role of spiritual leader that his elders had intended for him since his childhood. He serves as a cochair of the World Council of Religions for Peace and is a member of the Order of Canada and a Knight of the National Order of Quebec. He is a founder of the Dominique Rankin Foundation and the nonprofit organization Kina8at-Together, two organizations dedicated to the preservation and transmission of Indigenous traditions around the world.

David Rosen, International Director of Interreligious Affairs, American Jewish Committee, is the former chief rabbi of Ireland and the international director of Interreligious Affairs for the AJC. He is a member of the Chief Rabbinate of Israel's Committee for Interreligious Dialogue and serves on the Council of the Religious Institutions of the Holy Land. He is an international president of Religions for Peace, an honorary president of the International Council of Christians and Jews, and the only Jewish member of the board of directors of the King Abdullah International Center for Interreligious and Intercultural Dialogue. He is a past chair of the International Jewish Committee for Interreligious Consultations, which represents world Jewry to other world religious bodies. In 2005, Pope Benedict XVI bestowed upon Rabbi Rosen a papal knighthood in recognition of his contribution to promoting Catholic-Jewish reconciliation, and in 2010 he was awarded a CBE (Commander of the British Empire) by Queen Elizabeth II.

Jeffrey D. Sachs is a university professor and the director of the Center for Sustainable Development at Columbia University, where he directed the Earth Institute from 2002 until 2016. He is also the president of the UN Sustainable Development Solutions Network and a commissioner of the UN Broadband Commission for Development. He has been an advisor to three United Nations Secretaries-General and currently serves as an SDG advocate under Secretary-General António Guterres. His most recent book is *The Ages of Globalization: Geography, Technology, and Institutions* (2020).

Lisa Sachs is the director of the Columbia Center on Sustainable Investment. She teaches a master's seminar at Columbia Law School and Columbia's School of International and Public Affairs on extractive industries and sustainable development and lectures at Externado University in Colombia on international investment law. She has served on World Economic Forum Global Future Councils on International Governance and on Mining and Metals and is a cochair of the UN Sustainable Development Solutions Network's thematic group on the good governance of extractive and land resources. She was a 2020–2021 senior fellow of NAFSA, the Association of International Educators, and sits on several advisory boards, including of the Investor Alliance for Human Rights and SDG Academy. Before joining CCSI, she worked at the Interfaith Center on Corporate Responsibility and at Amnesty USA on shareholder engagement. She received a BA in economics from Harvard University and earned her JD and an MA in international affairs from Columbia University, where she was a James Kent Scholar and recipient of the Parker School Certificate in International and Comparative Law.

Marcelo Sánchez Sorondo, Chancellor, Pontifical Academy of Sciences and Pontifical Academy of Social Sciences, was born in Buenos Aires and ordained a priest in 1968. He has served as a lecturer and full professor in the history of philosophy at the Lateran University in Rome, where he also served as dean of the faculty of philosophy. He also served as full professor of the history of philosophy at the Libera Università Maria SS. Assunta in Rome. In 1998, he was appointed chancellor of the Pontifical Academies of Sciences and Social Sciences by Pope Saint John Paul II, who then consecrated him titular bishop of Vescovìo. He has been awarded the Italian *Cavaliere di Gran Croce* and the French *Légion d'honneur*, among other honors.

Nathan Schneider is an assistant professor of media studies at the University of Colorado Boulder, where he leads the Media Enterprise Design Lab. His most recent book is *Everything for Everyone: The Radical Tradition that Is Shaping the Next Economy*.

Ted Smith is an associate professor of environmental medicine at the University of Louisville School of Medicine and serves as the director of the Center for Healthy Air, Water and Soil at the Christina Lee Brown Envirome Institute. Dr. Smith is focused on creating a new vision of health that considers the scientific basis for reintegrating humans with our natural ecology. He was the cofounder of the AIR Louisville Project and is a coinvestigator on the ambitious Green Heart Louisville Project, a large clinical trial to demonstrate the effect of greenery in cities on cardiovascular outcomes.

Kyoichi Sugino is an ordained dharma teacher in Rissho Kosei-kai, a major Japanese Buddhist denomination, at the New York Center for Engaged Buddhism and the deputy secretary general of Religions for Peace. Rev. Sugino has been directly engaged in multireligious diplomacy and track II negotiations in Iraq, Syria, Sri Lanka, Myanmar, and other conflict zones. He has worked with senior leaders across various streams of Buddhism and coauthored the historic *Yogyakarta Statement on Shared Values* to overcome violent extremism and advance common action for peace. Earlier in his career, he served as a policy research officer at the Office of the United Nations High Commissioner for Refugees in Geneva (1996–1998) and published a well-cited article on a historical perspective of the role of humanitarian agencies in complex emergencies.

Anna Sun is an associate professor of religious studies at Duke University, a scholar of Confucianism in particular and of contemporary Chinese religious life in general. Her first book, *Confucianism as a World Religion: Contested Histories and Contemporary Realities* (Princeton University Press, 2013), received the Distinguished Book Award in the sociology of religion from the American Sociological Association and the Best First Book in the History of Religion Award from the American Academy of Religion. A coedited volume on the sociology of spirituality, *Situating Spirituality: Context, Practice, and Power* (with Brian Steensland and Jaime Kucinskas), was published by Oxford University Press in 2021.

Marie-Josée Tardif, Elder, Anicinape Tradition, Canada, spent the first fifteen years of her professional career as a journalist. In 2007, the elders of the Anicinape (Algonquin) First Nation extended an invitation for her to become a sacred pipe carrier. She accepted and then devoted her life to the study of the traditional medicine and culture of this ancient people. Marie-Josée Tardif is a founder of the Dominique Rankin Foundation and the nonprofit organization Kina8at-Together, two organizations dedicated to the preservation and transmission of Indigenous traditions around the world.

Jesse Thorson is a special project coordinator at the Center for Sustainable Development at Columbia University's Earth Institute. He received his BA in sustainable development from Columbia University, where he researched electricity access in Nicaragua for his thesis and was awarded departmental honors. Upon graduation, he was awarded the Henry Evans Travel Fellowship to meet and interview the poet, theologian, and revolutionary leader Ernesto Cardenal.

William F. Vendley is the secretary general emeritus of Religions for Peace International and a vice-president and senior advisor for religion at the Fetzer Institute. He advised President Obama through his service on the Multi-religious Cooperation and International Affairs Task Force of the White House Faith-Based Council, was appointed by U.S. Secretary of State Hillary Clinton and reappointed by Secretary of State John Kerry as one of ten members of the U.S. State Department's Advisory Committee on Strategic Partnership with Civil Society, and is a cochair of the State Department's Religion and Foreign Policy Working Group. He serves on the Leadership Council of the UN Sustainable Development Solutions Network. Dr. Vendley is a theologian and has served as a professor and dean in Roman Catholic graduate schools of theology. He earned his BA from Purdue University (1971) and received its Distinguished Alumni for Science Award in 2005. He has a PhD in systematic theology from Fordham University (1984).

Virgilio Viana, Director General, Amazonas Sustainability Foundation, is one of Brazil's leading experts on forestry, the environment, and sustainable development. He served as Brazil's secretary of state for environment and sustainable development from 2003 until 2008 and is currently the director general of the Amazonas Sustainability Foundation, an organization charged with the challenge of implementing the *Bolsa Floresta* Program, as well as providing the institutional framework to market the environmental services of Amazonas's forests. He is a professor of forest sciences at the Luiz de Queiroz College of Agriculture at the University of São Paulo and has served as a consultant to institutions such as the World Bank, the German Technical Cooperation (GTZ), the International Institute for Environment and Development, the Center for International Forestry Research, the World Wildlife Fund, and Greenpeace.

Maryanne Wolf is the director of the newly created Center for Dyslexia, Diverse Learners, and Social Justice at the UCLA Graduate School of Education and Information Studies. Previously she was the John DiBiaggio Professor of Citizenship and Public Service and the director of the Center for Reading and Language Research in the Eliot-Pearson Department of Child Study and Human Development at Tufts University. She is the author of *Proust and the Squid: The Story and Science of the Reading Brain* (HarperCollins, 2007), *Tales of Literacy for the 21st Century* (Oxford University Press, 2016), and *Reader, Come Home: The Reading Brain in a Digital World* (HarperCollins, 2018) and the editor of *Dyslexia, Fluency, and the Brain* (York, 2001).

Sharon Cohn Wu is the principal advisor on violence against women and children for International Justice Mission (IJM), leading IJM's center of excellence in addressing sexual violence against children and intimate partner violence and developing globally applicable best practices from IJM's extensive programmatic experience worldwide. Prior to her current role, Sharon was responsible for designing IJM's Justice System Transformation Model and directing IJM's global operations. She has testified before the U.S. Congress on the issue of trafficking and has given interviews for numerous news outlets. Sharon received her JD from Harvard Law School and her BA from the University of Virginia. Prior to joining IJM, she clerked for Hon. Richard L. Williams and was a corporate litigator for Arnold & Porter.

Hamza Yusuf currently serves as the president of Zaytuna College in Berkeley, California, the first accredited Muslim liberal arts college in the United States, with both bachelor's and master's degree programs. He was ranked by *The Muslim 500* as the twenty-third most influential Muslim worldwide. A proponent of the traditional liberal arts and great books education in both the Western and Muslim traditions, he has translated, authored, and coauthored numerous publications, including scholarly books, articles, and papers on major current areas of ethical concern. Hanson holds traditional advanced degrees (*ijazat*) in the Islamic sciences, as well as a BA in religious studies from San José State University and a PhD in African history from the Graduate Theological Union in Berkeley, California.

Stefano Zamagni is a professor of economics and a former dean of the economics faculty at the University of Bologna. He is the president of the Pontifical Academy of Social Sciences and a member of both the Scientific Committee of the Pontifical Council for Culture and the Scientific Committee of the Italian Social Impact Agenda. His work spans welfare economics, theory of consumer behavior, social choice theory, economic epistemology, ethics, and the history of economic thought and civil economy. His latest book is *Laudata Economia* (2021).

John D. Zizioulas, Metropolitan of Pergamon, Ecumenical Patriarchate was a professor of systematic theology at the Universities of Glasgow and Thessalonika. He has also taught at the Universities of Edinburgh and London (King's College), the Gregorian University (Rome), and the University of Geneva. Since 1993, he has been an ordinary fellow of the Athens Academy of Arts and Sciences. He holds honorary doctorates from several universities, including those of Munich and Münster, the *Institut Catholique de Paris*, and others. He has been a major contributor to modern ecumenical discussion and has served as the cochair of the official theological dialogue between the Roman Catholic and Orthodox Churches and as the chair of the Ecumenical Patriarchate's international symposia on religion, science, and the environment. His publications, translated into many languages, include *Being as Communion* (St. Vladimir's Seminary Press, 1985), *Communion and Otherness* (T&T Clark/Continuum, 2006), and *The Eucharistic Communion and the World* (T&T Clark/Continuum, 2011).

INDEX

Printed and bound by CPI Group (UK) Ltd, Croydon, CR0 4YY
16/04/2025
14658571-0001